AF531444

BIOLOGY
OF
HELMINTHES

BIOLOGY OF HELMINTHES

By

Dr. D.R. Khanna
Reader
Reader in Zoology
Gurukul Kangri University
Haridwar (Uttaranchal)

&

Dr. P.R Yadav
Reader
Department of Zoology
D.A.V. College
Muzaffarnagar (U.P.)

DISCOVERY PUBLISHING HOUSE
NEW DELHI-110002

First Published-2004

ISBN 81-7141-909-7

Published by

DISCOVERY PUBLISHING HOUSE
4831/24, Ansari Road, Prahlad Street,
Darya Ganj, New Delhi-110002 (India)
Phone: 23279245 • Fax: 91-11-23253475
E-mail:dphtemp@indiatimes.com

Printed at:
Tarun Offset Printers,
Delhi-110 053

PREFACE

The present title "Biology of Helminthes" has been carefully organized and clearly written to meet the requirements of the undergraduate, postgraduate and those involved in the various competitive examinations. The text has been designed to approach the classification, morphology, anatomy, physiology and development of all required types in a simple, lucid and straightforward manner.

The text is not only encyclopaedic in scope, but also of introductory coverage, commensurate, concise, comprehensive yet exhaustive presentation we sincerely feel that this publication will adequately meet the long felt need of the students for a simple and upto date book and hope that the method of presentations will be liked and found beneficial.

The approach to the discussion is very simple so as to impact to the students a clear and vivid understanding. According to the scheme of treatment all important animal types of the Phylum have been dealt in a easy understandable manner. Efforts have also been made to present the text with complete, authentic and uptodate account. Further separate chapters on topics of significance and general interest pertaining to the Phylum have also been added to make the treatment more elaborate. Profusely illustrated the work has become very informative.

It is difficult to acknowledge adequately the assistance of all who have contributed to the preparation of this project. Authors wish to thank all friends and colleagues whose continuous inspiration have initiated them in bringing out the present title.

There can be no claim of originality except in the manner of treatment and much of the information has been extracted from the books and scientific journals available in different libraries.

Though every care has been taken to bring out foolproof book, it is quite likely that some errors might have crept in which may be of very insignificant nature. However, suggestions to improve the book and all sorts of positive and creative criticism will gratefully be appreciated.

The authors express their gratitute to Mr. Wasan and staff of M/s Discovery Publishing House for their whole hearted co-operation in the publication of this book.

Authors

CONTENTS

1

ORIGIN OF METAZOA

Since the first Metazoa were almost certainly radial animals, the Bilateria must have sprung from a radial ancestor, and there must have been an alteration from radial to bilateral symmetry. This change constitutes a most difficult gap for phylogeneticists to bridge, and various highly speculative conjectures have been made. It is generally believed that bilateral symmetry originated in consequence of the assumption by a radial ancestor of creeping mode of life. We have already traced this ancestor to a planuloid form resembling the planula larva of the Cnidaria. This planuloid ancestor was an elongated, radially symmetrical organism, without mouth or archenteron, and consisted of a ciliated or flagellated epidermis, probably composed of epithelio-muscular cells, and a solid interior mass of digestive cells. There was probably a nerve net under the epidermis with an accumulation of neurosensory cells at the anterior pole, and indifferent mesenchyme cells, capable of differentiating into sex cells and other types of cells, were scattered through the interior mass.

Food catching and digestion remained wholly of the protozoan type so that the organism had no need of a mouth or archenteron. The creature was polarized with definite anterior or aboral and posterior or oral ends but lacked dorsoventral differentiation. This planuloid organism may be supposed to have given rise to the coelenterate line by formation of a mouth at the site of the closed blastopore, hollowing out of the interior to form a primitive gut, and sprouting of tentacles, thus becoming a simple medusa similar to an actinula. It is her assumed also to have given rise to the Bilateria by adopting a creeping mode of life, then flattening out and developing bilateral symmetry and dorsoventral differentiation. There are three chief theories of the steps from the radiate of to the bilateral condition: the *ctenophore-polyclad*, the *ctenophore-trochophore*, and the *planuloid-acoeloid* theories.

The ctenophore-polyclad theory was first suggested by Kowalevsky in 1880 as a result of his finding of the peculiar ctenophore *Coeloplana* but has been supported and elaborated chiefly by Arnold Lang, the eminent student of the polyclad flatworms, who based his conclusions on his studies of the anatomy and embryology of various turbellarians, and on the peculiar features of the aberrant ctenophores *Ctenoplana* and *Coeloplana*. These ctenophores resemble polyclads in their oval flattened shape with dorsoventral differentiation; in the presence of two dorsal tentacles; in creeping upon the entire ventral surface; in the centrally located ventral mouth and branched, blindly ending digestive canals; and in the radiat-ing, anastomosing nervous system. There is a certain amount of resemblance in the embryology, as both ctenophores and polyclads have determinate cleavage with the formation of micromeres and macromeres, and in both there is a large stomodaeal invagination which in both contributes to the ventral surface.

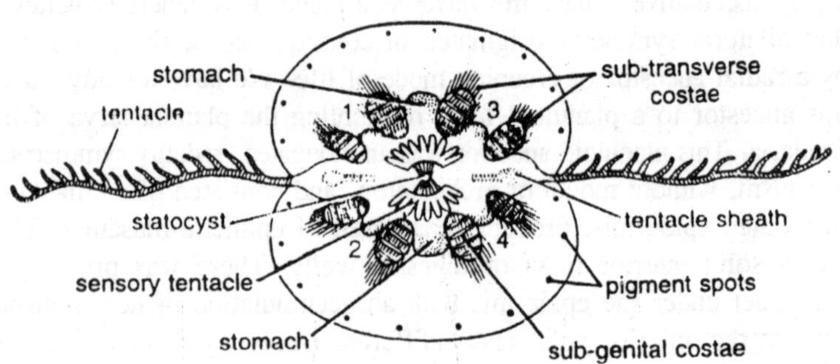

Fig. 1.1. Ctenoplana (Dorsal view).

Many polyclads have a swimming larval stage with eight ciliated lobes which seem comparable to the eight comb rows of ctenophores. However, as Lang fully realized, *Coeloplana* and *Ctenoplana* have no trace of real bilateral symmetry, being in fact biradially symmetrical, and their nervous center is in the middle of the dorsal surface, whereas it is characteristic of the Bilateria that the nervous center is anteriorly located. To bridge these differences, Lang postulated a series of stages from the platyctene ctenophores to the polyclads by a forward shift of the nerve center and tentacles and extensive branching of the digestive canals.

Lang's theory of the ctenophores as intermediate forms between the radiates and the polyclad flatworms had a wide and extended acceptance but is to be rejected on the following grounds. First, it is agreed by all students of ctenophores that the Platyctenea are simply highly aberrant ctenophores without phylogenetic significance. Second, the embryology of the two groups is in fact very different and the cleavage patterns follow very diverse plans, that of the ctenophores being biradial, and that of the polyclads spiral. Third, in the theory of Lang, the polyclads are necessarily regarded as the most primitive existing Bilateria; this is a mistaken idea, since it is now clear that the order Acoela occupies that position and the polyclads stem from acoel ancestors.

The ctenophore-trochophore theory is a variant of the trochophore theory, discussed below, and is based on alleged resemblances between ctenophores and the trochophore larva. These resemblances are best stated after the anatomy of the trochophore larva has been described. Here it may be pointed out that this theory proceeds from the ctenophores to the annelids and other higher Bilateria and fails to account for the origin of the lower Bilateria. Supporters of the theory have realized this difficulty and take one or the other of the two possible explanations. Some regard the larvae of the acoelomate Bilateria (*i.e.*, the flatworms and the nemertines) as earlier stages of the trochophore larva. But then such stages would necessarily fall between the ctenophore and the trochophore and should more nearly resemble the ctenophore than does the trochophore larva. This, however,

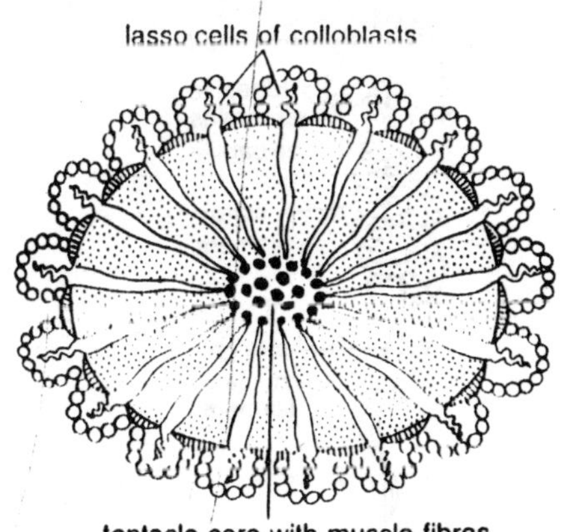

Fig. 1.2. Cross section of branch or pinna.

is not the case for in fact the ctenophore similarities are manifested only in the fully developed typical trochophore larva. The second explanation is to the effect that the acoelomate Bilateria are degenerated from the higher Bilateria. This cannot possibly be admitted.

It may be regarded as firmly established that the free-living flatworms, especially the order Acoela, are of exceedingly primitive structure, positively not degenerated from or modified from a higher anatomical type. The ctenophore-trochophore theory therefore appears to the author unacceptable, although it has been widely supported.

The planuloid-acoeloid theory is based upon the remarkable features of the flatworm order Acoela, whose primitive structure was first recognized in 1882 by Ludwig von Graff, the distinguished pioneer in the study of the free-living flatworms. The Acoela were nevertheless long regarded as degenerate forms and hence of no phylogenetic importance until recent years when their primitive nature has again been recognised by all students of the Turbellaria. The Acoela are group of very small flatworms with the following primitive characteristics. The ciliated epidermis is often syncytial, a in many coelenterates, and in several members has basal muscle fibers, as also in coelenterates. There is no distinct basement membrane beneath the epidermis. The interior is a solid mass of cells as in the planula larva without a digestive cavity, and comparison of the embryonic development with that of other flatworms indicates that this interior mass represents entoderm and ecto- and entomesoderm. There is a mouth in the center of the ventral surface, and this leads into the

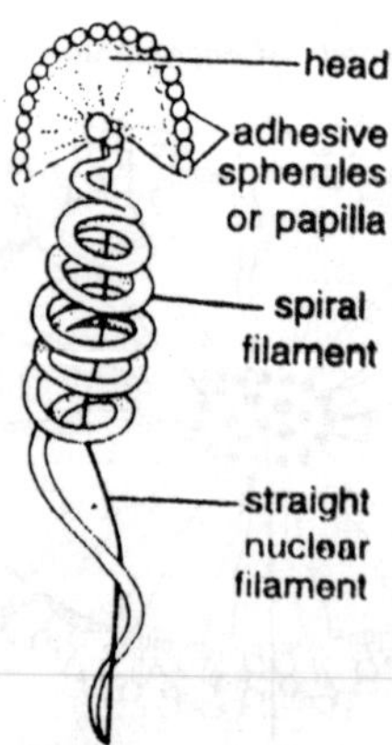

Fig. 1.3. A colloblast.

interior mass either directly or by way of a simple tubular pharynx. These worms feed by ingesting small animals through the mouth into the interior mass where intracellular digestion occurs; thus it is directly proved that an archenteron is not a necessity for the early bilateral holozoic animal, as supposed by the Haeckelian gastrula theory.

The nervous system is of primitive construction, consisting of a plexus beneath the epidermis with several more or less distinct longitudinal strands radiating from an anterior brain mass; but in some acoels the nerve plexus is situated in the basal part of the epidermis and the brain is a mere thickening also epidermally located. Most of the Acoela have a statocyst near or embedded in the brain; this structurally is more like the statocyst of medusa than of a ctenophore. There are no distinct gonads, but the sex cells differentiate out of the cells of the interior mass. Female ducts are wanting, and the eggs discharge either through the mouth or by rupture of the body surface. There are usually indistinct male ducts which lead to a copulatory

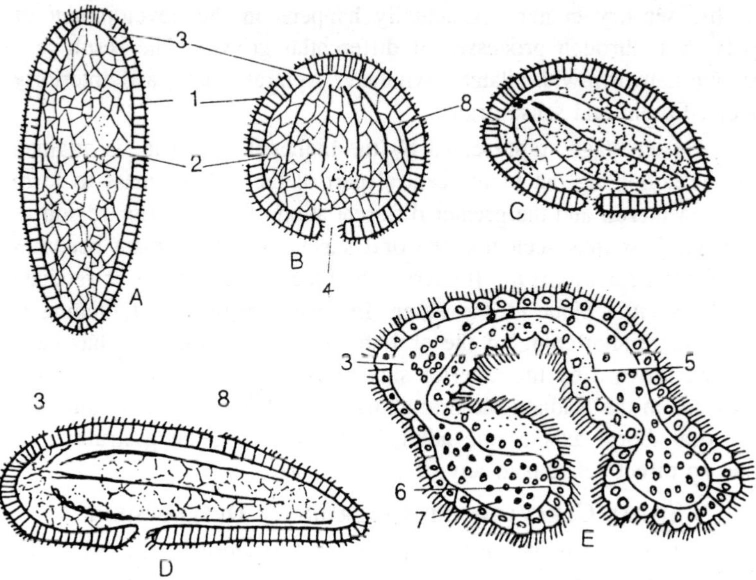

Fig. 1.4. Diagram illustrating the planuloid-acoeloid theory of the origin of the Bilateria. A—Planula larva, mouthless with apical nerve center. B—Mouth formed, oral-aboral axis shortening. C—Body elongating in a sagittal plane, nervous center shifting forward. D Acoeloid stage, a bilateral creeping worm with anterior nerve center. E—Later stage, archenteron formed; actual embryonic stage of a polyclad. 1-ectoderm; 2-entodermal mass; 3-nervous center; 4-mouth; 5-archenteron; 6-stomodaeum; 7-mesoderm; 8-nerve cord.

organ. Although the animals are hermaphroditic, cross-fertilization is the rule and is internal. The Acoela lack an excretory system. It is evident that there is no difficulty in passing from a planuloid type of ancestor to an acoeloid form. Epidermis, muscular system, interior mass, nervous system, and sex cells are, even in present Acoela, distinctly at a coelenterate stage or but slightly advanced from this stage. Consequently we accept the theory of the origin of the Bilateria from a planuloid ancestor by way of an acoeloid form.

The main changes necessary are the alteration from radial to bilateral symmetry and the forward shifting of the nervous center. As no intermediate forms are known, the steps by which these changes occurred may be inferred from the embryology of the lower Turbellaria. As the mouth of the Acoela occupies the site of the blastopore, it may be taken as a fixed point. It is then evident that, as postulated by Lang, the planuloid form must have flattened down in the oral-aboral axis and at the same time there must have occurred a forward migration of the sensory center, as actually happens in the development of polyclads, through processes of differential growth. The result is a simple flatworm with bilateral symmetry, ventral mouth, and anteriorly located brain and sense organs.

A later stage, represented by the higher Turbellaria, is attained by the hollowing out of the central part of the interior mass to form an archenteron and the greater differentiation of the anterior region as a head. It is thus seen that the oral surface of the Radiata becomes the ventral surface of the Bilateria while their sagittal plane is retained as the sagittal plane of the latter. But the dorsoventral flattening is not a mere shortening of the oral-aboral axis, for this axis has been curved forward by the anterior shifting of the nervous center. Thus, whereas in the Radiata the oral-aboral axis of the larva is retained as the oral-aboral axis of the adult, in the Bilateria the gastrular axis has no direct relation to any of the adult axes.

Spiral Cleavage and Determinate Development

Determinate or *mosaic* development, characteristic of most of the Protostomia, is usually associated with the spiral type of cleavage, found in polyclad flatworms, nemertines, annelids, and mollusks. Among these groups the cleavage pattern and the fate of the various blastomeres are so nearly identical that a common descent is scarcely to be doubted. In spiral cleavage, the spindle axes are oriented obliquely with respect to the polar axis of the egg. As a result successive tiers of blastomeres alternate, the cells of one tier resting in the angles between the cells below them. This displacement may be either clockwise (dextrotropic) or counterclockwise (levotropic).

Spiral cleavage is commonly holoblastic but unequal, and this inequality is often evidenced even at the first cleavage. In spiral cleavage, the fate of each blastomere can be determined, a type of study known as *cell lineage* and pursued with brilliant results by American embryologists. For clarity in describing the course of events, each blastomere is named. The four blastomeres formed by the first two meridional cleavages are labeled *A, B, C, D*. The third division is transverse and latitudinal and results in eight cells, four small animal ones called the *first quartet of micromeres* and four large cells below, called *macromeres*. The micromeres are named 1*a*, 1*b*, 1*c*, 1*d*, and the macromeres 1*A*, 1*B*, 1*C*, 1*D*. At the next cleavage, the macromeres give off above a *second quartet* of micromeres, called 2*a*, 2*b*, 2*c*, 2*d*, and are themselves then designated as 2*A*, 2*B*, 2*C*, 2*D*; and at the next cleavage the latter again divide, giving off a *third quartet*, 3*a*, etc., themselves becoming 3*A*, etc. Meantime the first and second quartets have cleaved, and their offspring are also numbered by a definite system. Thus the two daughter cells of 1*a* become $1a^1$ and $1a^2$; the exponent 1 is applied to the cell nearer the animal pole. When these divide again, exponents are added; thus the daughter cells of $1a^1$ are numbered $1a^{11}$ and $1a^{12}$, and of $1a^2$, $1a^{21}$ and $1a^{22}$, respectively.

This system obviously is capable of indefinite expansion and serves to designate exactly the origin of any particular cell. All the cleavages are oblique and are alternately dextrotropic and levotropic. All the blastomeres bearing one letter, as *a*, are offspring of one of the original four blastomeres, in this case *A*, and hence more or less occupy one quadrant of the embryo. When the third quartet of micromeres has been given off, the germ layers are already fixed (usually 32-cell stage, as the first and second quartets have cleaved). The three quartets and their offsprings form the whole of the ectoderm (and also the larval mesoderm, i.e., the ectomesoderm) and hence are often called *ectomeres*. The four macromeres, 3*A*, 3*B*, 3*C*, 3*D*, form the whole of the entoderm and the true or entomesoderm.

At their next division, they give off at the vegetal pole a fourth quartet of micromeres, 4*a*—4*d*, three of which are also purely entodermal. The fourth cell, 4*d*, is larger than the corresponding macromere, 4*D*, is called the *mesentoblast* cell, or *M*, and is the entire source of the entomesoderm. It was originally supposed that all

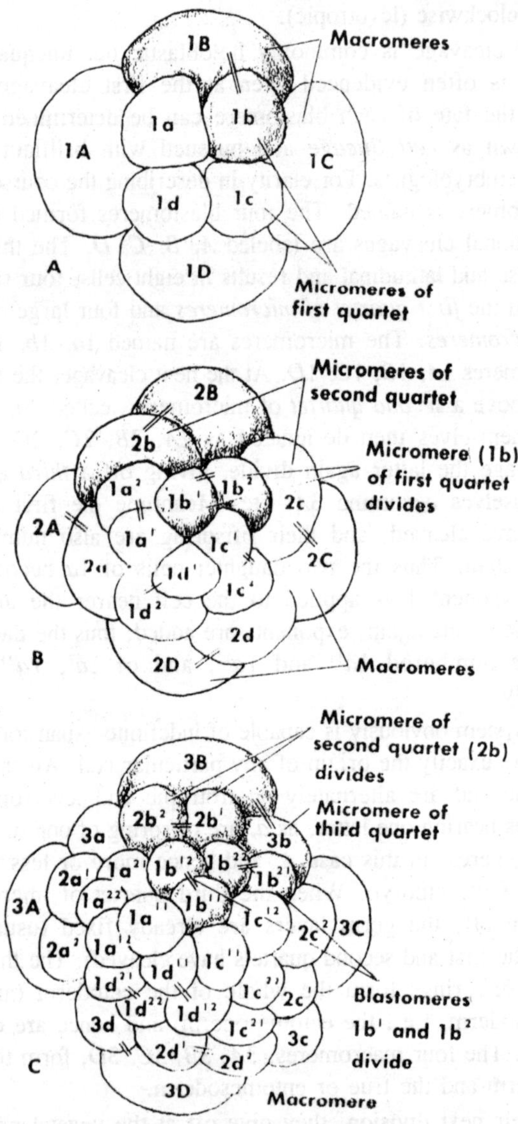

Fig. 1.5. Cell lineage, polar view.

its offsprings are mesodermal, but later researches have shown that a few of these remain entodermal and contribute to the gut wall. The main offspring of *4d*, however, in annelids and mollusks, are two cells known as the *teloblasts*, *pole cells*, or *primordial mesoderm cells*.

These take up a bilateral position just in front of the future anus and each proliferates a band of cells, the *mesoderm band*, the sole source of the entomesoderm.

The embryo becomes an inequal coeloblastula or, in the case of very yolky eggs, a stereoblastula. The ectomeres spread and grow down over the macromeres, which also invaginate. A typical invaginate gastrula thus arises partly by epiboly and partly by emboly. The mouth forms at the site of the blastopore by way of stomodaeal invagination. In forms with an anus, the blastopore elongates and closes together medially so as to leave its anterior end as mouth and its posterior end as anus. The anal end usually closes temporarily but later reopens as the permanent anus. A stomodaeum is universal throughout the Protostomia. The mouth is usually shifted forward by growth processes elsewhere in the embryo.

The embryo develops an apical tuft of cilia and an equatorial girdle of cilia and swims about as a free-living larva. The fate of the various cells of the cleavage pattern may be stated briefly. The micromeres of the first quartet form the ectoderm of the aboral part of the larva. The $1a^{111}$–$1d^{111}$ series occupy the animal pole where they make a four-rayed pattern termed the rosette which becomes the aboral sensory plate with its tuft of cilia. Their sister cells, the $1a^{112}$–$1d^{112}$ series, in annelids alternate with them as a cross-shaped figure called the *annelid cross*. A similar *molluscan cross* occurs in molluscan development but there is formed of the cells which lie between the arms of the annelid cross, namely, the $1a^{12}$–$1d^{12}$ cells, with some assistance from second quartet cells. Thus the arms of the cross are interradial in annelids, radial in mollusks. The cross, very evident in early stages, eventually becomes obliterated and plays no special role in development as such.

It is often regarded as a ctenophore reminiscence, and the ctenophore resemblance is enhanced in some cases by the bifurcation of the tips of the arms to form an eight-rayed figure. The $1a^2$–$1d^2$ descendants of the first quartet are termed the *primary trochoblasts*; they form four groups of four cells each, one group in each quadrant of the embryo near the equator. Each cell develops a transverse tuft of cilia reminiscent of a ctenophore comb. These four groups of ciliary tufts constitute the primitive ciliated girdle, or *prototroch*, of the larva. Later the prototroch is completed by the formation of cilia on the cells between the primary trochoblasts, termed *secondary trochoblasts*, which come from the first or second quartet or both. The cells of the

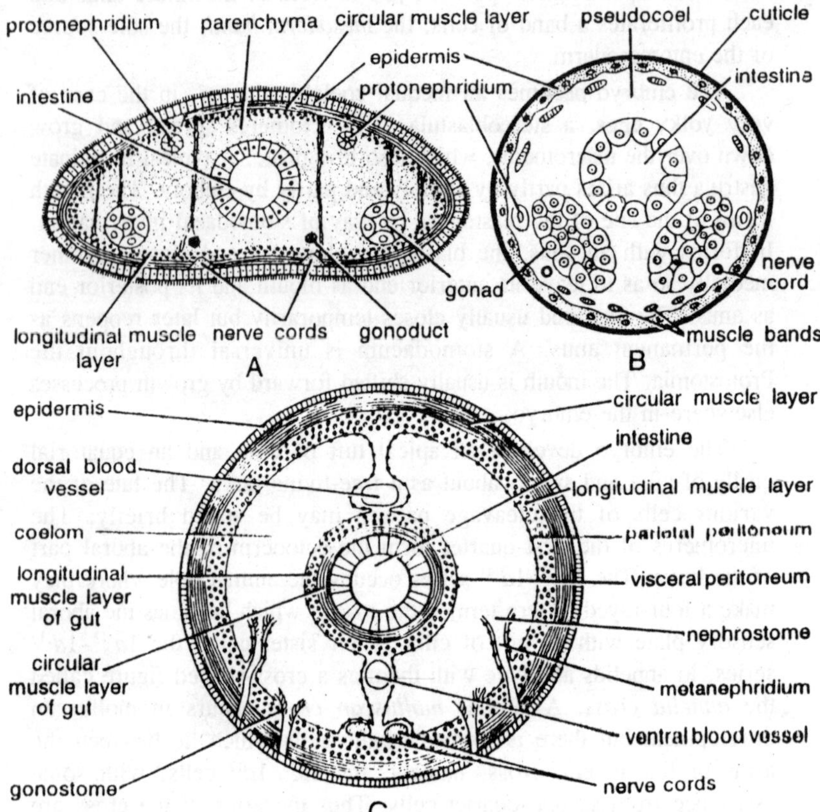

Fig. 1.6. Diagrammatic cross sections of grades of structure; A—Acoelomate grade; B—Pseudocoelomate grade; C—Eucoelomate grade.

second and third quartets contribute to the surface ectoderm, and certain ones, termed *stomatoblasts*, invaginate as the stomodaeum.

In annelids, the cell 2*d*, known as the first *somatoblast*, gives rise to practically the entire ectoderm of the adult trunk. Cells are also given off from the second and third quartets into the blastocoel as mesenchyme which differentiates into larval muscles. This mesenchyme constitutes the larval mesoderm or ectomesoderm and is undoubtedly a reminiscence of the original ectomesoderm of the ancestral stem form of the Bilateria. Descendants of $3c^2$ and $3d^2$ give rise in annelids to the larval excretory organs, the *archinephridia*. In the most primitive case, that of the annelid *Polygordius*, the archinephridium comes from two cells, one forming the tubule, and the other the solenocyte. The entoderm originates almost wholly from seven cells, namely, *4a, 4b, 4c, 4A, 4B, 4C,* and *4D*, which invaginate and produce the entodermal wall of the stomach and intestine of the larva.

As already noted, a few cells descended from *4d* become incorporated into the intestinal wall. The intestine (in the higher Protostomia) connects to the exterior by an anus formed at the site of the rear end of the blastopore. In producing the anus, the ectoderm invaginates slightly, and this ectodermal termination of the gut is called the *proctodaeum*. The two teloblasts, offspring of *4d*, lie in the blastocoel just in front of the anus. By repeated divisions each gives rise to a band of cells which extends forward into the blastocoel. These bilateral *mesoderm bands* are the whole of the entomesoderm and give rise to muscle, connective tissues, gonads, excretory system at least in part, blood vessels, etc.

The formation of typical mesoderm bands from teloblasts is confined to the phyla Annelida and Mollusca, but the bands are well evidenced in the Arthropoda. In the annelids and mollusks, spiral cleavage often results in a characteristic larva, known as the *trochophore*, and somewhat similar larvae occur in other protostomous phyla. As great phylogenetic significance has been attributed to the trochophore larva, it becomes necessary to describe its characters before proceeding further.

The Trochophore Larva, The Trochophore Theory, And The General Phylogenetic Significance of Larval Types

The trochophore larva is a somewhat biconical creature with a protruding equator. Its external surface consists of a one-layered epithelium (ectoderm) thickened at the apical pole into a sensory plate which bears a tuft of cilia. Around the equator there is a girdle of cilia termed the *prototroch* which passes above the mouth; and there may be present a second equatorial girdle, the *metatroch*, passing below the mouth, and sometimes also a ciliated circlet around the anus, the *paratroch*. A complete digestive tube is present extending in an L-shape from the mouth at the equator to the anus at the lower pole; it consists of a ciliated stomodaeum leading into an expanded rounded stomach from which the short narrowed intestine proceeds to the anus. Between the digestive tube and the ectoderm there is a spacious cavity which is the blastular cavity or blastocoel. This is more or less occupied by mesenchyme cells and well-developed muscle cells, all derived from the ectomesoderm.

The muscular system is often quite complicated consisting not only of bands acting on the digestive tube and ciliary girdles but also of bands under the ectoderm accompanying the nerves. The nervous system according to the best description may attain an astonishing degree of complexity, consisting of a ganglionic mass under the apical plate, a variable number of longitudinal nerves radiating from this mass, and one to several nerve rings connecting the radial nerves. The main nerve ring underlies the prototroch. Various sense organs, including eyes and statocysts, may be present. To either side of the intestine lies a mesoderm band derived from a teloblast, and near this there is on each side a nephridium consisting of a tubule whose inner end is closed by one or more solenocytes. A solenocyte is a long tubular cells having a flagellum playing in the tube.

Solenocytes occur in several invertebrate groups and also in *Amphioxus*, and nephridia with solenocytes are undoubtedly only a variant of nephridia with flame bulbs; the latter are general throughout the acoelomate and pseudocoelomate Bilateria. The *trochophore theory*, elaborately developed by Hatschek (1857), is to the effect that the trochophore is the larva of an ancestral form, the *trochozoon*, which was the common ancestor of most, if not all, the bilateral phyla, and which, of living forms, most nearly resembled a rotifer. It is to be noted that the trochophore is regarded as recapitula-ting the *larva* of the ancestor, not as being itself a replica of the ancestor.

The rotifer resemblances usually cited are the general pseudocoelomate grade of structure, the ciliary girdles, the structure and location of the brain and the nature of its attached sense organs, the muscle bands, the nephridia, the form of the digestive tract, and the presence of an anus. The rotifer-like nature of the hypothetical trochozoon was considered strongly supported by the discovery by Semper in 1859 in the rice fields of Philippines of a rotifer which he named *Trochosphaera* and which through its inflated spherical form and equatorial ciliary girdle bears considerable superficial resemblance to a trochophore larva. Time has detracted from the rotifer part of the trochophore theory, for those rotifers whose ciliary arrangements most suggest the girdles of the trochophore are now regarded as specialized types and the primitive ciliary apparatus of the rotifers has not much likeness to the trochophore girdles.

Further the rotifer *Trochosphaera*, which has been refound and restudied in various localities, is now known to be merely an aberrant type of no especial evolutionary significance. Nevertheless, while discarding the rotifer part of the trochophore theory, one must admit the general soundness of Hatschek's view that the trochophore is about at the structural level of a rotifer, *i.e.*, is of the pseudocoelo-mate

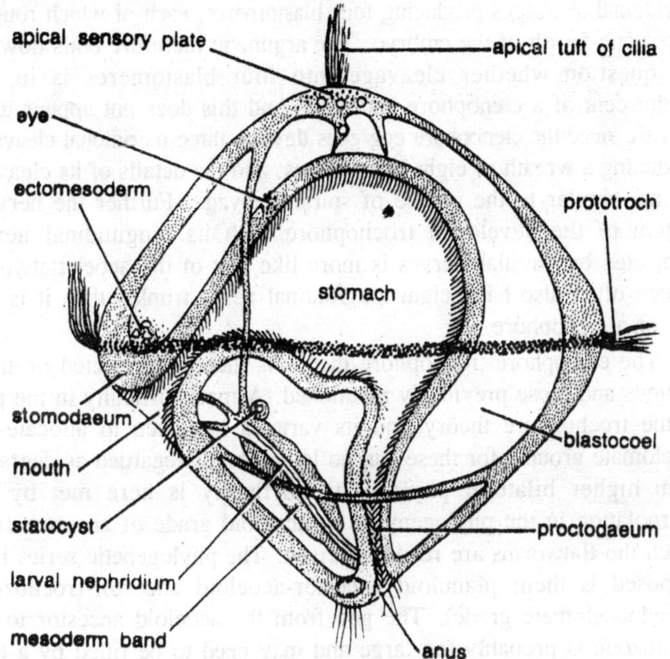

Fig. 1.7. A typical trochophore larva.

grade of structure. The fact further that nearly identical trochophores occur in two phyla in which the adults are anatomically very different, *i.e.*, the annelids and the mollusks, suggests that the trochophore is indeed a reminiscence of the common ancestor of the eucoelomate Protostomia and perhaps also of the pseudocoelomate groups. However, the origin of the acoelomate Bilateria remains unaccounted for on this theory. The ctenophore-trochophore theory, mentioned above, is a more recent and widely accepted version of the trochopore theory. It proposes to pass directly from a ctenophore-like ancestor to the trochozoon and so to bridge the gap between the radiate and bilateral animals. The chief ctenophore resemblances of the trochophore are: the apical nerve center with its attached sense organs, the radiating subectodermal nerves, often (but not always) eight in number, the occurrence of the "cross" in the cleavage pattern, and the origin of the prototroch from four groups of ciliated cells.

The apical nerve plate and sense organs are, however, also a feature of the planuloid ancestor, and the other resemblances are a necessary consequence of the quadripartite pattern characteristic of spiral cleavage. This quadripartite pattern is the result of the two meridional cleavages producing four blastomeres, each of which roughly forms one-fourth of the embryo. The argument therefore boils down to the question whether cleavage into four blastomeres is in fact reminiscent of a ctenophore ancestry. And this does not appear to be the case since the ctenophore egg goes through three meridional cleavages producing a wreath of eight blastomeres, and the details of its cleavage are not similar to the course of spiral cleavage. Further the nervous system of the developed trochophore with its longitudinal nerves connected by circular nerves is more like that of the acoel flatworms (which often also have eight longitudinal nerve trunks) than it is like that of a ctenophore.

The ctenophore-trochophore theory is therefore rejected on these grounds and those previously mentioned. A main difficulty in the path of the trochophore theory and its variants has been to allocate the acoelomate groups, for these can no longer be disregarded as degraded from higher bilateral phyla. This difficulty is here met by the interpolation in the phylogeny of an acoeloid grade of ancestor from which the flatworms are readily derived. The phylogenetic series here proposed is then: planuloid ancestor-acoeloid ancestor-trochozoon (pseudocoelomate grade). The gap from the acoeloid ancestor to the trochozoon is probably too large and may need to be filled by a type with a digestive tract but no anus, resembling stages seen in the development of polyclads. An attitude of conservatism should always be maintained in regard to the evaluation of larval types.

A larva is a developmental stage that leads a free existence. Larvae are practically limited to marine animals. The main reasons given for this are: that larvae are too delicate to withstand the much wider range of environmental factors in fresh water than in the sea; that they would be swept away by currents; and that fresh water is poor in the salts needed by many larvae. It results that animals, to reproduce successfully in fresh water, must produce larger and fewer eggs, better provided with food and salts than their marine relatives; the eggs must be protected with shells, jelly, etc.; must be fastened to objects or develop inside the mother; and the young animals must hatch in a fully developed condition. Consequently, the development of fresh-water animals is generally highly modified and furnishes poor material for phylogenetic speculations. It is therefore difficult to work out the relationships of groups of animals which are predominantly fresh-water.

Even among related marine groups, some may have free larvae and others not, without apparent reason. In such cases the question arises which type of development is the more primitive. Again as larvae must fend for themselves and find food, they are very apt to display adaptive modifications which are without phylogenetic significance. Pelagic larvae tend to present common characters adapted to pelagic life, as transparency and lightness of body, swimming apparatus mostly in the form of ciliated bands, lobes and projections for buoyancy, and a food-catching mechanism, mostly of the mucous-ciliary type. Evidently great caution must be exercised in the interpretation of larvae, and most weight should be placed on characters not common to pelagic animals. In the case of the trochophore larva, the most salient features are the form of the nervous system, the presence of an anus, and the occurrence of primitive nephridia. These suggest an ancestral form higher than a flatworm and lower than an annelid. The foregoing analysis carries the line of evolution from the radiates to an through the protostomous Bilateria. The relation of the Deuterostomia to this line of ascent remains very dubious, and this situation casts much doubt on the generally accepted ideas of the original modes of origin of mesoderm and coelom.

Modes or Origin of the Mesoderm

The term mesoderm is rather loosely applied to all cells, cell layers, or cell masses which occur in the embryo between the ectoderm and the entoderm. The mesoderm is also spoken of as the third germ layer. As already noted, there are two kinds of mesoderm which seem to be phylogeneti-cally distinct, the ectomesoderm and the entomesoderm. The ectomesoderm exists in both the Radiata and the Protostomia and is undoubtedly the oldest form of mesoderm. It is always mesenchymal and typically consists of inwandered ectodermal cells, in spiral cleavage of cells descended from the second and third quartets of micromeres. Ectomesoderm appears to be absent in the Deuterostomia.

The Protostomia presumably inherited the ectomesoderm from a radiate ancestor, and their entomesoderm is a later development originating in the evolution of the Protostomia from a radiate ancestor. The entomesoderm, the "true" or definite mesoderm, includes all mesoderm which arises from or with the entoderm. It may form as mesenchyme or as bands, plates, or sacs, more or less epithelial in character. The attempt of the Hertwigs (1882) to draw a sharp distinction between entomesoderm of mesenchymal origin and that of epithelial origin (mesothelium) was mistaken. The entomesoderm may arise by either method, and mesoderm which begins as epithelium may later become mesenchymal or vice versa. The time of origin of the entomesoderm is also variable. The following modes of formation of the entomesoderm occur among the Bilateria.

2

Platyhelminthes

Term *Platyhelminthes* was proposed by Gegenbaur in 1859, means flatworms which refers to their characteristic contour of flattened body. The term is derived from two Greek words-*platys* = flat; *helminthes* = worms. The animals show low organisation as they are without anus, skeletal, respiratory and circulatory system. The body is filled with mesenchymal cells which are mesodermal in origin.

Phylum: Plathelminthes (Flatworms)

This phylum groups dorso-ventrally flattened worms; their body is covered with a dermomuscular sac which consists of a single epithelial layer overlaying a layer of muscles (longitudinal, circular, and oblique). The plathelminths have no special organs of locomotion. Their mobility is effected by contractions of the dermomuscular sac. The body is oval or ribbonlike, strongly elongated. The loose parenchyma inside the dermomuscular sac surrounds internal organs.

The digestive system is primitive; it is totally absent in cestodes. The excretory system consists of protonephridia. The nervous system consists of paired cerebral ganglia and longitudinal nerve cords. The plathelminths are hermaphroditic. Their reproductive system is very well-developed and fills almost the entire body of a worm. Numerous fish parasites can be found in the following classes: the *Monogenea*, *Cestodaria*, *Cestoda*, and *Trematoda*.

Class: Monogenea (Monogenetic Trematodes)

Monogeneans are mostly small parasites, ranging from about 1 to several mm in length. A few marine species are longer than 20 mm. The body is flattened dorso-ventrally (seldom cylindrical), elongated, oval, or resembling a circular disc, slightly concave ventrally and convex dorsally. The monogeneans have characteristic adhesive organs used to attach the parasite to the host's body and to aid in translocation on the host. The attachment organs on the head occur as suckers and processes. Additionally, the anterior part of the body contains glands producing a viscous secretion ensuring adhesion to the host when it is feeding or moving. The posterior end of the parasite's body is provided with a complex adhesive organ called the opisthaptor, differing widely from species to species: it may contain suckers, hooks, clamps, all of them forming an attachment disc fastening the parasite to its host.

The monogenean body is very contractile: it may double in length or shorten by half compared to a resting animal. When moving from

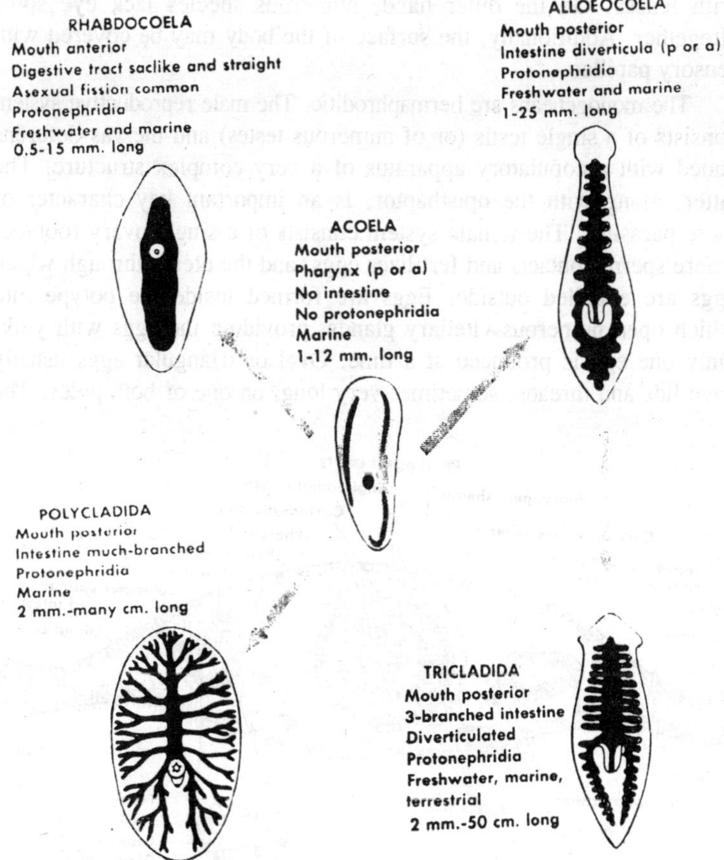

Fig. 2.1. Classic orders of class Turbellaria showing chif diagnostic characters (p–present; a–absent).

place to place, the parasite fastens its anterior end to the fish body, tears the opisthaptor off and transfers it forwards, then releases the anterior end and attaches it to a different spot on the fish body. The movements resemble those of a leech or a measuring worm. The digestive tract begins with the mouth, placed ventrally near the anterior end of the worm body. The mouth leads to the muscular pharynx acting as a sucking pump for food. The pharynx passes into a short oesophagus and further into the intestine, usually in the form of two bluntly ended caeca. Occasionally, the intestine is single, sac-like, or the two caeca fuse in the posterior part of the body. In some species, the two intestinal caecas have many branches. Sense organs occur in many species as two or more pairs of pigmented eye spots provided with lenses. On the other hand, numerous species lack eye spots altogether. Additionally, the surface of the body may be covered with sensory papillae.

The monogeneans are hermaphroditic. The male reproductive system consists of a single testis (or of numerous testes) and the vas deferens ended with a copulatory apparatus of a very complex structure. The latter, along with the opisthaptor, is an important key character of these parasites. The female system consists of a single ovary (ootype) where sperm contacts and fertilizes eggs, and the uterus through which eggs are expelled outside. Eggs are formed inside the ootype into which open numerous vitellary glands, providing the eggs with yolk. Only one egg is produced at a time. Oval or triangular eggs usually have lids and threads, sometimes very long, on one or both poles. The

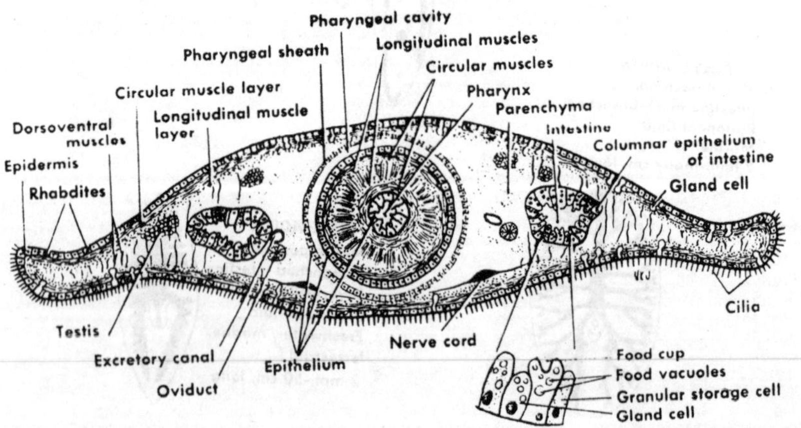

Fig. 2.2. Cross section through phryngeal region of planarian.

thread facilitates egg's adhesion to the fish mucus and aids the egg in floating in the water.

The *Monogenea* are mostly oviparous; some gyrodactylids only (e.g. *Gyrodactylus elegans*, *Macrogyrodactylus polypteri*) are viviparous. Development is direct, without metamorphosis. On hatching, the egg releases a ciliated larva, the oncomiracidium, its cilia arranged in zones. The larva possesses eye spots, pharynx, head, glands, and a developed, but inactive, opisthaptor with emerging hooks. Generally, the larval structure resembles that of an adult. The larva swims freely for some time, viable for 6—8 hours; on encounter with an appropriate fish host it adheres to the fish skin, loses its cilia and, crawling on the skin, gets to certain organs (the gills, mouth, cloaca) to settle there and develop, without metamorphosing, into an adult. In spite of

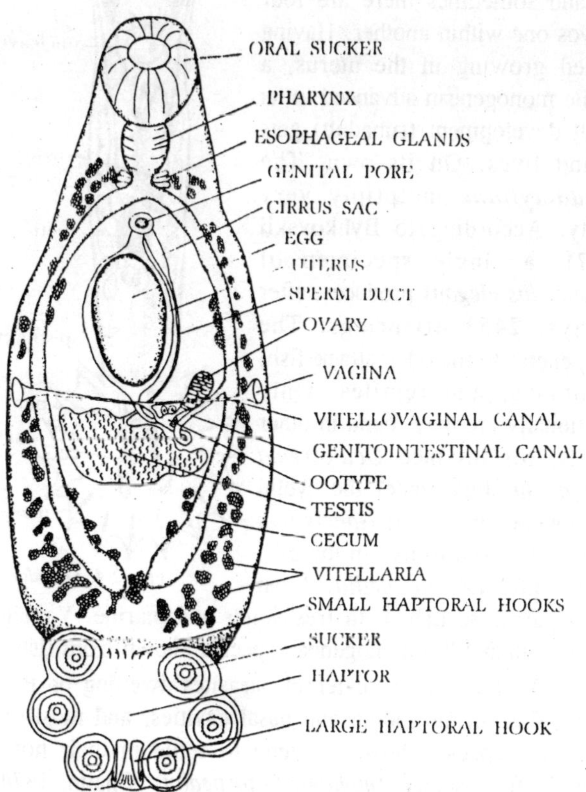

Fig. 2.3. Polystomoidella oblongum, a monogenetic trematode from the urinary bladder of turtles.

producing a single egg at time, the monogenetic trematodes multiply very rapidly. *Dactylogyrus vastator*, a common carp parasite, was observed to live for 10–12 days, from egg to death. During this time a single specimen produced 16 new ones, which in turn gave rise, before they died, to 310 offsprings.

In the viviparous species, the egg in the maternal uterus develops into an embryo; another one grows within it, the third embryo inside that, and sometimes there are four embryos one within another. Having finished growing in the uterus, a juvenile monogenean advanced in its overall development (subadult) gets out and lives. On its own. The *Gyrodactylidae* mulptiply very rapidly. According to Byhkovskii (1957), a single specimen of *Gyrodactylus elegans* produces, after 30 days, 2453 offsprings. The monogenetic trematodes attack fish, amphibians, and reptiles. Only exceptionally are they found in other animals, for instance *Oculotrema hippopotami*, fund under the eyelid of a hippopotamus, and *Isancistrum loliginis* observed to live in the gills of the squid *Loligo media*. Most species parasitise fish, both freshwater and marine. Yamaguti (1963) lists more than 1300 monogenean species known from fish.

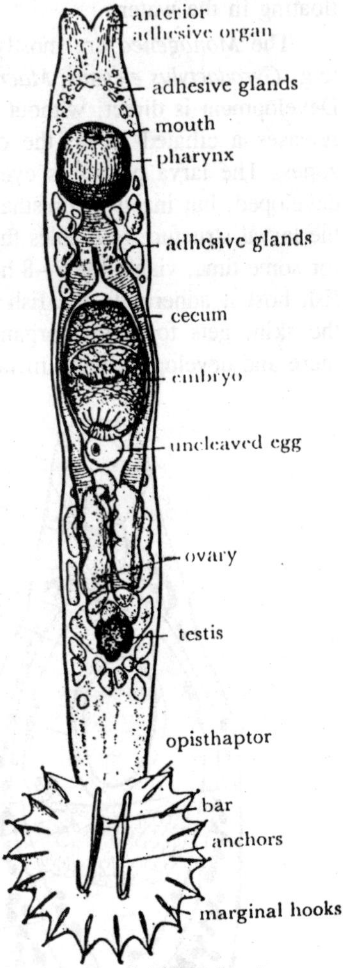

Fig. 2.4. Gyrodactylus.

The *Monogenea* are external parasites dwelling on the fish body surface, fins, in the mouth and nasal cavities, and most often on the gills. Some species, however, tend to settle in their host's internal organs, as for instance *Amphibdella torpedinis* (Chatin, 1874), a worm usually dwelling in the gill cavity of the electric rays, *Torpedo ocellata* and *T. marmorata*. Ruszkowski (1931) found the parasite, both adult individuals and eggs, in the *Torpedo* heart, *T. ocellata* invasion incidence reaching 35% and the eggs being found in more than 50% of the electric ray's specimens examined. Members of the genus *Acolpenteron* live in the urinary duct, while numerous species of *Calicotyle* settle in the cloaca of rays. Paperna (1963) described a new monogenean species, *Enterogyrus cichlidarum*, dwelling in the anterior part of the intestine of *Tilapia zilli* and *T. nilotica* from Israeli rivers.

These monogeneans are actually internal parasites and even show some adaptations to this mode of life. the body of *Enterogyrus cichlidarum* is covered with a thick cuticle; the cephalic adhesive organ is vestigial and the opisthaptor poorly developed. Such typically monogenean characters as the eye spots and posterior hooks are retained. On the other hand, no morphological adaptations to an altered environment were found in *Amphibdella torpedinis*. The examples described demonstrate one of the possible pathways to endoparasitism.

The monogeneans feed on fish muchus, peeled epithelium, and blood. The group contains numerous pathogenic species. The parasites attached to the host's body injure it with their hooks and produce ecchymoses; they damage gill and skin epithelium, the latter often peeling off in lobes as a strong invasion. Juvenile fish are particularly vulnerable. Frequently the irritation caused by the parasites results in epithelial outgrowths and tissue swelling, as was often observed in fish gills affected by *Dactylogyrus vastator*. Serious epizootics and mortality are known to occur in pond cultures. The monogeneans can be dangerous for aquaria-kept fish and also in maricultures.

It is exceptionally rare, however, for a monogenean invasion to cause a disease outbreak and mortality in fish under natural conditions. On the other hand, a high fish population density, undernourishment, water deoxygenation, altered pH, and pollution act in favour of parasites' proliferation: the parasites attack those fishes weakened by detrimental conditions of the environment and then produce substantial losses. Pond cultures in Poland often fall victim to invasions of gill-attacking monogeneans *Dactylogyrus vastator* producing such pathological conditions as dactylogyriasis, particularly dangerous for juvenile carp (the so-called July fry), and gyrodactyliasis, frequent in juvenile trout. Those interested in these diseases and in other species parasitising freshwater fish should consult fish diseases textbooks.

The largest monogenean specres and the most dangerous ones for fish belong to the family *Capsalidae*. The capsalids possess a large opisthaptor, frequently divided by septa into a number of radically arranged grooves which increase the parasite's force of adhesion. Moreover, the opisthaptor contains 14 small marginal hooks and 2—6 central ones. The opisthaptor clings to the host's body so strongly that it is sometimes easier to tear the parasite into pieces than to remove it from the fish body. *Capsala martinieri* (Bosc., 1811) belongs to the largest capsalid species. It is a flat oval monogenean, up to 30 mm long and 25 mm wide. It occurs on the gills of *Mola mola* and *Diodon* sp.

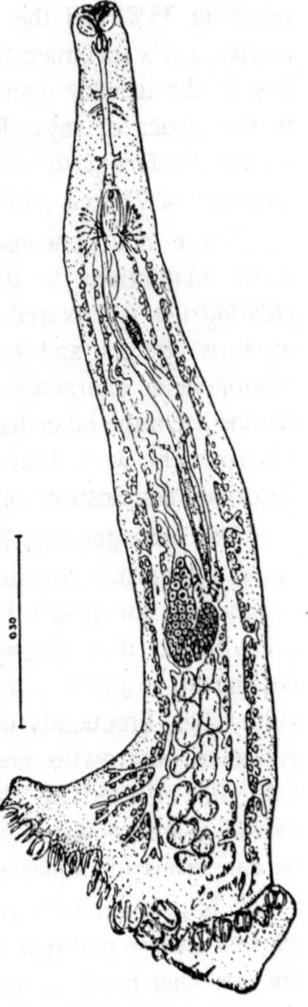

Fig. 2.5. Axinoides raphidoma, a monogenetic trematode.

Tristoma coccineum (Cuvier, 1817) frequently found on the gills of *Xiphias gladius*, is a large parasite with a flat, disc-like circular body more than 1 cm in diameter. The animal has three suckers near its mouth (hence the name *Tristoma*) and a large sucker-like opisthaptor divided by septa into additional adhesive areas. Frequently, an imprint of the parasite's opisthaptor, or even of its whole body, is seen on fish gills. The species of *Benedenia* show two large bow-like suckers on the head. *B. sciaenae* (Van Beneden, 1852) often occurs on the skin of meagre, *Sciaena aquila*. (Jahn and Kuhn 1932) described a strong invasion of *B. melleni* (MacCallum, 1927) in the serranids and lutjanids kept in New York marine aquaria, up to 2000 monogeneans being found on one fish specimen. The parasites attacked fish eyes, nasal cavity, and gills. Numerous fishes died as a result of the invasion. The parasites were introduced to the aquaria with newly delivered Atlantic fishes.

B. derzavini (Layman, 1930) a small (about 5 mm long) parasite is common in the gill cavity of *Sebastodes schlegeli* in the Sea of Japan. Japanese maricultures have often been affected by a mass occurrence of *B. seriolae* (Yamaguti, 1934) on the species of *Seriola* serious pathological changes and fish mortality being the outcome. The family *Capsalidae* also includes *Nitzschia sturionis* (Abildgaard,

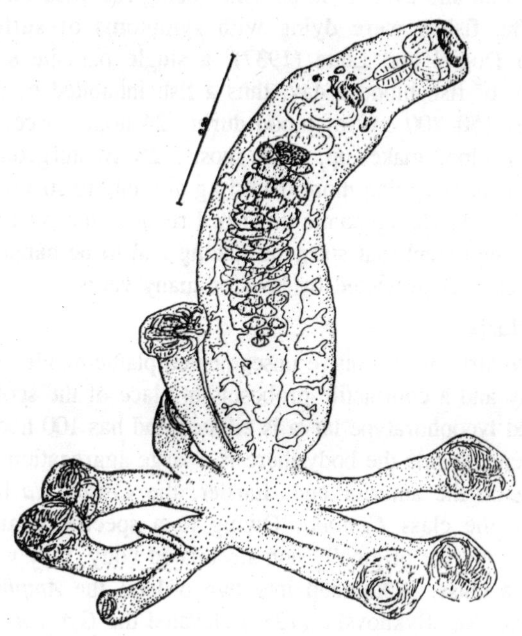

Fig. 2.6. Choricotyle louisianensis.

1794) a parasite of sturgeons in the Ponto-Caspian basin (the Caspian, Black, and Aral Seas). The parasite occurs on the gills, lips, palate and tongue, and even in the anterior part of the intestine. This large (1-2.5 cm long) monogenean has two attachment cylinders on the head and a sucker-like opisthaptor; the opisthaptor, in contrast to other capsalids, lacks septa. It contains 14 marginal hooks and 3 pairs of central ones. An extensive invasion of *N. sturionis* brings about emaciation and even death of the host. The parasite mechanically damages gill lobes, which results in hypertrophy of gill epithelium and causes necrosis in the damaged spot.

N. sturionis is common in Atlantic, Mediterranean, Caspian, and Black Sea sturgeons and usually produces no disease symptoms. However, a strong epizootic caused by the species was recorded in 1936; the disease was brought about by transplanting *Acipenser stellatus*, together with the parasite, from the Caspian to Aral Sea. The Aral Sea had been inhabited by an indigenous sturgeon species, *A. nudiventris*, *Nitzschia*-free and therefore lacking immunity to the parasite. *Nitzschia* attacked the defenseless Aral sturgeon and developed aboundantly on the gills, 100-600 and even 1000 parasites being recorded on a single specimen. The fishes were dying with symptoms of suffocation. According to Dogiel and Lutta (1937), a single parasite sucks out about 0.5 cm^3 of fish blood a day; thus a fish inhabited by 300–400 parasites loses 150–200 cm^3 of blood during 24 hours. According to Packov (1954), blood makes up, at the mos, 5.2% of sturgeon weight, so those sturgeons weighing more than 5 kg are able to survive such a large loss of blood. The epizootic described reduced the Aral sturgeon population to the extent that sturgeon fishing had to be banned in the area and the species protected by law for many years.

Class: Cestodaria

The *Cestodaria* are primitive unsegmented plathelminths, having a flattened body and a contractile proboscis in place of the scolex. The newly hatched lycophoratype larva is ciliated and has 100 hooklets on its posterior end. Inside the body there is a large aggregation of gland cells opening on the anterior end. Earlier, the *Cestodaria* had been placed within the class *Cestoda*. Owing to a specific character of their structure and the originally non-strobilated body, they were later separated as a class, subdivided into two orders: the *Amphilinoidea* and *Gyrocotyloidea*. Bykhovskii (1957) elevated the *Gyrocotyloidea* to the level of a separate class and demonstrated that they were more closely related to monogeneans than to cestodes.

In 1974, Dubinina rejected the name *Cestodaria* as redundant and formed the class *Amphilinoidea*, in her opinion phylogenetically related to monogeneans as well. The classification system adopted here is based on that developed by Baer and Joyeux (1961); the class *Cestodaria* is subdivided into the orders *Amphilinoidea* and *Gyrocotyloidea*. The attempts to alter the plathelminth classification system have only been briefly mentioned here.

Order: Amphilinoidea

This small parasitic taxon contains 7 genera with 10 species. The three species of the genus *Amphilina* are coelomic parasites of Holarctic acipenserids: *Amphilina foliacea* and *A. japonica* Goto and Ishii, 1936 parasitise European and Asian sturgeons, while the North American sturgeons are attacked by *A. bipunctata* Riser, 1948. The remaining species live in freshwater and marine tropical fish *e.g. Gephyrolina paragonopora* in catfish from India, *Gigantolina magna* in the Indopacific marine perciforms. *Amphilina foliacea* is the best-known member of the order. The parasite is common in sturgeons in the basins of the

Black and Caspian Seas, Lake Bailkal, and Siberian rivers (the Ob, yenisey, Irtish, Angara). The principal host is *Acipenser ruthenus*. Juvenile sturgeons show the strongest invasion: according to Dubinina (1974), invasion incidence in fish not older than 6 years may reach 70-89.9%. Other acipenserids: *Huso huso*, *A. stellatus*, *A. gueldenstaedti*, *A. nudiventris*, *A. baeri*, *A. schrencki*, *A. sturio*, and *A. medirostris* are less affected.

Amphilina foliacea is oval, dorsoventrally flattened, 60-65 mm long and 25-30 mm wide. The body is slightly convex dorsally and flat ventrally, milky white or yellowish in colour. The anterior end of the body is provided with a contractile proboscis, on the top of which open numerous unicellular glands located centrally in the body, between the gonads. The uterus opens next to the proboscis and releases eggs outside. The rear end of an adult body has 10 hooklets partly burrowed in the dermomuscular sac. The parasite is hermaphroditic. The male gonadal opening and the vaginal pore are placed close to each other on the posterior end of the body. Numerous testes are scattered between the uterine loops. The single ovary with bubble-like vitellaries on its sides lies posteriorly in the body. The intestine is absent.

Eggs are oval, lacking lids, each provided with a short process on one pole. Embryonic development proceeds in the uterus so that each egg, when ejected, contains a well-developed larva, the so-called lycophora. It is ciliated, with 10 hooks on its rear end. Once in the intestine of the intermediate host, an amphipod crustacean, the lycophora leaves the egg. Amphipods such as *Dikerogammarus haemobaphes*, *Pandorites platycheir*, *Corophium curvispinum*, *Carcinogammarus roeseli*, and *Rivulogammarus pulex* can serve as intermediate hosts.

The development of *A. foliacea* was studied by the Polish biologist Konstanty Janicki during his stay in Saratov on the river Volga in 1927-1928. Janicki put forward an interesting hypothesis that mesozoic reptiles were the original definitive hosts for *A. foliacea* and housed its adults, while the second larval stage, the plerocercoid, developed in sturgeons. The adult forms were extinct along with the reptiles and the plerocercoid acquired the faculty of neothenic reproduction, becoming thus the sexually mature *Amphilina* of today. Recently, Dubinina (1974) supplied a number of new pieces of information on development and structure of *A. foliacea*.

The parasite's eggs get into the water via the fish abdominal pores and sink down to the bottom where they are consumed by amphipods. The egg shell breaks under the pressure of gammarid maxillae and the motile lycophora, measuring 165–230 um × 70–190 μm, is liberated. The lycophora pushes through the gammarid intestine into its coelom and grows there into a 2–4 mm long larva, able to invade a sturgeon, which happens after 30 days at 27–30°C. This stage is a morphological equivalent of the adult *Amphilina*.

The fish become infected by consuming crustaceans, the basic item in their diet. Once in the fish intestine, the parasite makes its way to the body cavity where it undergoes further develpment, matures, and produces eggs passed to the water through the fish abdominal pores. The eggs develop in water. By mechanically injuring fish gonads, *A. foliacea* can bring about a partial or even complete castration, which has undoubtedly a bearing on the sturgeon population biology. The *Amphilinoidea* are phylogenetically very old animals, as are their hosts, both the definitive (acipenserids) and the intermediate (relict gammarids) ones.

Order: Gyrocotyloidea

This is a very small taxon consisting of 3 genera with 10 species only: the *Gyrocotyle*, *Amphiptyches*, and *Gyrocotyloides*. The *Gyrocotyloidea* parasitise exclusively chimaerids (*Holocephali*). The body consists of a single flattened segment, its sides strongly crenate. The parasites are rather large, up to a few cm long. The anterior end of the body is provided with the anterior sucker of an unknown function, while the posterior end has a muscular funnel-like attachment organ of more or less wrinkled margins. The funnel fastens the parasite to the chimaerid intestinal spiral valve. Some parasites (*Gyrocotyle*) show transverse striation on the surface, which then resembles segmentation. The intestine is absent, the parasite absorbing nourishment with the whole surface of the body.

The *Gyrocotyloidea* are hermaphroditic. Their development is known only incompletely; observations made by the Polish parasitologist Ruszkowski (1932) have greatly contributed to the present knowledge. Ruszkowski described the development of *Amphiptyches urna*, a common parasite of *Chimaera monstrosa* in the Norwegian Sea. The invasion incidence in that area was very high, up to 100%. *A. urna* is a rather large parasite. The living animal is translucent, pinkish in colour, very active and contractile. The *Amphiptyches* eggs, measuring 78–93 μm × 60–67 μm, have lids. Embryos develop within 25–30 days (under experimental conditions) after the eggs have been placed in water. The egg hatches into a lycophora-type larva, completely ciliated and with 10 hooks on the rear end, similar to the *Amphilina foliacea* larva.

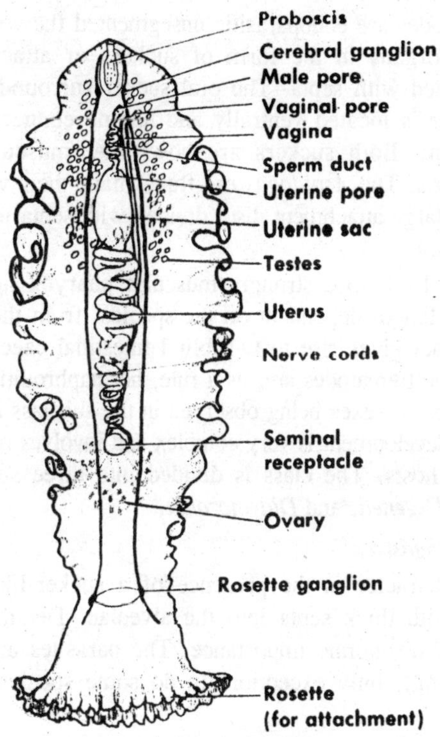

Fig. 2.7. Gyrocotyle.

However, the 120 μm long and 30 um wide *Amphiptyches* larva, as opposed to that of *Amphilina*, leaves the egg on its own by opening the lids with strong movements. The newly released larva swims rapidly, its movements resembling those of trematode miracidia. The presence of the lycophora in the *Gyrocotyloidea* life cycle is the only character they have in common with the *Amphilinoidea*.

The further larval development is unknown. The smallest larvae Ruszkowski found in the chimaerid intestines were 3-5 mm long and showed gonad formation in progress. Information on the *Gyrocotyloidea* pathogenicity is lacking. The parasites are of no importance for the fishing industry as the chimaderids are not caught for consumption, their flesh being unpalatable. The group has been discussed here as an interesting example of marine fish parasites, and for the sake of a complete presentation of parasitic plathelminths. The present state of knowledge of their biology is largely a result of contributions of Polish parasitologists.

Class: Trematoda (Flukes)

The trematodes are endoparasitic unsegmented flatworms provided with adhesive organs in the form of suckers or attachment discs, frequently divided with septa. The oral sucker surrounds the mouth; the other sucker is located ventrally and has no connection with the digestive system. Both suckers are bowl-like, muscular, strongly contractile organs. The *Aspidogastrea* flukes have their ventral sucker developed as a large attachment disc, divided with septa into secondary suckers (alveolae).

The mouth leads to a strongly muscular pharynx opening to the oesophagus, of length dependent on the species; from the oesophagus the digestive tract gives rise to two blind intestinal caeca which may be branched. The trematodes are, as a rule, hermaphroditic, a tendency toward separation of sexes being observed in the subclass *Didymozoida*. The trematode development is very complex and involves metamorphosis and change of hosts. The class is divided into three subclasses: the *Aspidogastrea*, *Digenea*, and *Didymozoida*.

Subclass: Aspidogastrea

The key character is the presence of a sucker-like attachment disc, divided with thick septa into the alveolae. Details of the disc structure are of taxonomic importance. The parasites are small (less than 10 mm long); only exceptionally do some species (*Macraspis*,

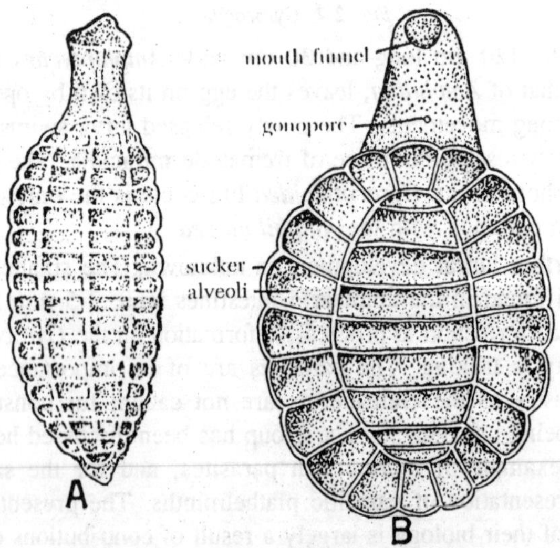

Fig. 2.8. Types of Aspidogastrea. A—Aspidogaster conchicola. B—Cotylaspis.

Stichocotyle) grow to attain a length of several centimetres. Most *Aspidogastrea* are parasites of molluscs, a few species only being found in marine fish: *Macraspis* in the intestine of *Chimaera monstrosa*: *Stichocotyle* in bile ducts of *Raja*; and *Cotylogaster* in the intestine of certain marine teleosts. Many cyprinids in the Black, Caspian, and Aral Seas are parasitized by *Aspidogaster limacoides* Diesing, 1835. Development of these trematodes is not known in detail. Presumably the development is direct, but juvenile stages of *Stichocotyle nephropis* (Cunningham, 1887) were found encysted externally on the intestine of lobsters *Nephrops norvegicus* and *Homarus americanus*. The adult trematodes dwell in bile ducts of *Raja clavata* and *R. laevis*. It is not, however, known how the fish becomes infected. The subclass is of negligible importance for the fishing industry.

Subclass: Digenea (Digenetic trematodes)

The subclass groups plathelminths which have two circular suckers with strong muscular walls: the oral sucker, placed on the anterior end of the body close to the mouth, and the ventral sucker. More than 2500 species dwelling in fishes have been described so far. The body is covered with the dermomuscular sac having a thin, elastic cuticle on the surface. Besides the suckers, the cuticle of some species shows additionally spines (*Acanthocolpidae* and *Stephanochasmidae*) and pores of unicellular glands formed by deeply burrowed epithelial cells. The cuticle is lined with three layers of muscles: circular, diagonal, and longitudinal their contractions producing worm-like movements of the animals. The body is filled with a loose parenchyma surrounding all the viscera.

The digestive system begins with the mouth leading to the muscular pharynx. The pharynx opens into the oesophagus of a length depending on the species. The oesophagus passes into two intestinal caeca in the form of straight or branching blind canals. Two symmetrical ganglias interconnected with transverse commisures are placed between the oral sucker and the pharynx. Each ganglion gives rise to three nerve cords which run toward the anterior and posterior ends of the body and branch off to all the organs. Digenean sensory organs occur as a few sensory bodies, located mostly near the oral sucker. The eye spots are present at the larval stage only, in the miracidia and cercariae, the adults lacking them altogether.

The excretory system consists of protonephridia and canaliculi fusing into two symmetrical lateral excretory vessels; those in turn fuse into a common vessel ending with the excretory pore on the rear end of the body. The digenean excretory system is of key importance in species identification. The digeneans are hermaphroditic, *i.e.*, each individual contains both male and female gonads.

The female system consists, as a rule, of a single ovary; the oviduct leads from the ovary to the ootype, where the fertilisation takes place. The ootype is also connected with vitellaries, their secretion providing the embryo with yolk and participating in the egg shell formation. The eggs, fertilised in the ootype, are passed to the uterus which expells them outside via the genital pore located in the anterior section of the body.

The male system consists of testes, most often paired. Sometimes a single testis (*Asymphylodora*) or multiple testes (*Aporocotyle*) are present. The testes are connected, via the vasa efferentia, with the vas deferens opening to the ejaculatory duct which frequently ends with the penis. The ejaculatory duct opens near the uterine pore. Both gonad systems may contain such additional elements as the seminal vesicle and bursa in the male system and/or the seminal receptacle in the female one. The digenean life cycle is very complex; it involves both metamorphosis (heterogenesis) and change of hosts. The adult parasites live in the intestine, bile and urinary ducts, and blood vessels of vertebrates (fishes, amphibians, reptiles, birds, and mammals including man). The digenean *Transversotrema patialense* is known to occur in the scale capsules of a fish *Brachydanio rerio*. It is thus an ectoparasite whose larvae are tissue parasites.

The digenean life cycle usually involves two intermediate hosts and a single definitive one. The following developmental stages occur: egg, miracidium, sporocyst, redia cercaria, metacercaria, and adult. The miracidium larva hatching from an egg is oval and elongated, its cilia enabling it to swim freely in water. Eggs of some species, sinking down to the bottom, are eaten by snails. In the intestine of a snail species, typical of a given digenean, the eggs hatch into miracidia which wander to the snail digestive gland to develop further. In this case, the miracidium lacks cilia. An X-shaped pigmented eye spot is visible on its head. The larva contains glands opening anteriorly. Their secretion has lytic properties enabling the larva to penetrate the intermediate host's body wall. Miracidia may show beginings of digestive, excretory, and nervous systems.

The miracidium lives freely for some time (usually not longer than 24 hours) and dies unless the first intermediate host, typically a mollusc, is found. In the latter's digestive gland the miracidium loses its cilia and metamorphoses into the next stage, the sporocyst. The sporocyst is a sac-like, poorly differentiated structure containing in its rear part a group of germinal cells developing into daughter sporocysts, rediae, or cercariae. The redia, also sac-like but more advanced in development, has a distinct pharynx and the fore section of the intestine. The redia also has an opening to release the next developmental stages. The redia contains germinal cells, too; give rise to either the daughter rediae or next juvenile stages, the cercariae. Thus the cercariae may be formed in the sporocysts or in the rediae, depending on the digenean species. The cercaria has a well-developed intestine and beginnings of all other organs of the adult. Additionally, this stage contains some larval organs lacking in the adults, namely the X-shaped eye spot and two groups of penetration glands on both sides of the mouth, the glands secreting lytic enzymes which dissolve host's tissues.

Cercariae of many species have additionally a stiff rod near the mouth, the so-called stylet, used to pierce the host's skin. The cercaria, as a rule, is provided with a locomotory organ, the tail, differing in shape and size from species to species. It enables the cercaria to swim freely for some time after leaving the mollusc host's body before the second intermediate host is encountered. For many species it is a fish. The cercaria penetrates its skin to get to the muscles, loses its tail and encysts to metamorphose into the next stage, the metacercaria. Some cercariae enter the fish body swallowed with food. The metacercariae, encysted or not, settle in various organs of the second intermediate host and wait there for their definitive (primary) host—another fish species, an amphibian, reptile, bird, or mammal (including man). In the definitive host's intestine the metacercaria loses its encystment, develops gonads to their final form and metamorphoses into a mature fluke capable of producing eggs.

Not every developmental cycle involves all the stages described above. Some species lack a sporocyst or redia; even a cercaria may metamorphose directly into adult, the metacercaria stage being skipped. The ability to reproduce as a sporocyst and redia is a characteristic feature in the digenean life cycle. As a consequence, a single egg may eventually give rise to very numerous offspring, thus increasing the parasite's chance of finding a proper host and perpetuate the species. There is a diversity of opinions as to the sporocyst and redia reproduction. Many authors regard it as larval parthenogenesis, while others consider it as polyembryony, the first opinion, however, prevailing. More than 15000 digenean species are known from fish. It is difficult to give a precise number, as new species are being continually described.

The pathogenicity of adult digeneans living in a fish intestine is usually weaker than that of the tissue-dwelling larvae. This is particularly evident in the case of the cercariae actively penetrating the fish skin; the damage caused by the skin-piercing metacercariae can be fatal to newly hatched and juvenile fish. The damage in tissues is both mechanical and toxic, the latter being caused by the lytic secretion of the penetration glands. Occasionally, swimmers happening to come across a cercarian "cloud" emitted by molluscs can be affected by the cercariae whose attack produces an itching and burning sensation on the skin accompanied by a rash. The settlement of metacercariae in fish organs is species-specific; they encyst or not, depending on the species.

Diplostomum spathaceum is a very common species in freshwater fish of Europe, Asia, and North America; it is common also in brackish lagoons and bays of the Baltic. The adult digeneans live in the fish-feeding birds (*Larus*, *Sterna*, *Ardea*, *Phalacrocorax*, *Sula*, *Alca*, and

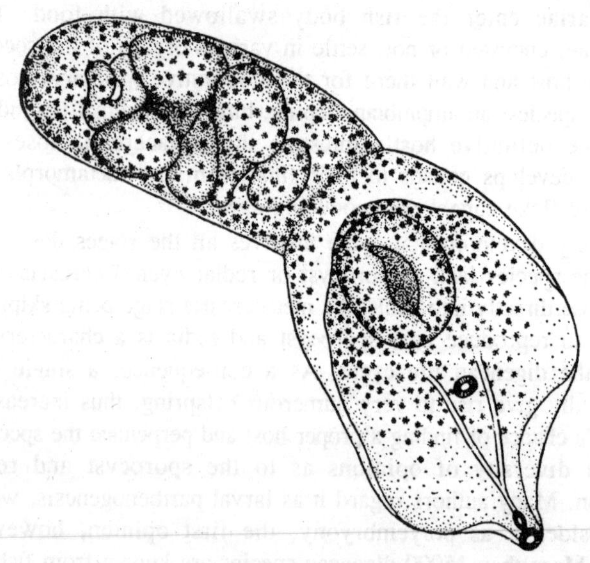

Fig. 2.9. Diplostomum paraspathula.

others). On the coast it is mainly seagulls that are the definitive hosts. The first intermediate hosts are such lymnaeid gastropods as *Lymnaea stagnalis*, *Radix auricularia*, and *Galba palustris*. Within the snails, sporocysts develop with cercariae in them. The cercariae liberate themselves and attack fishes. They pierce the fish skin and then move within the fish body to settle finally in the eye lens where they remain unencysted, awaiting further development in the intestine of a fish-feeding bird. They often occur in masses, several or few hundred individuals in a single lens. They inflict a disease called diplostomulosis. The eye lenses, affected by the metacercariae invasion, are destroyed, become opaque and brittle. The invasion is often accompanied by exophthalmus, the cornea breaking down and the lens falling out of the eye ball. A secondary infection, bacterial or fungal, may ensue, the mold covering then the entire eyeball. The blinded fish moves haphazardly, keeps to the water surface and thus becomes an easy prey of piscivrous birds.

D. spathaceum dwell in many freshwater fishes and also in those living in marine coastal waters, *e.g.*, flounder (*Platichthys flesus*), various gobiids (*Gobius niger*, *Pomatoschistus minutus*, *P. microps*), broad-nosed pipefish (*Syngnathus typhle*) and straight-nose pipefish (*Nerophis ophidion*). Particularly vulnerable to the invasion are salmonid pond cutures, frequently experiencing high losses. It is very difficult to control the parasite. The most efficient method is to eliminate snails, the intermediate hosts for *Diplostomum*, and thus to break the parasite's life cycle. Parasite-borne melanosis is a disease inflicted by metacercariae of various digeneans settling on or just underneath the fish skin. The metacercariae are encysted and surrounded by accumulating black pigment, melanin, forming black patches on the skin. At the same time, the rest of the skin is often depigmented, which makes the black spots still more conspicuous. The condition is induced by several species. Some of them are discussed below.

Posthodiplostomum cuticola (Nordmann, 1822) (= *Neascus cuticola*), a freshwater species common in Poland, brings about melanosis in cyprinids. The disease is particularly common in the vicinity of heron rookeries as the grey heron is the definitive host of the species.

Apophallus muehlingi occurs in coastal waters of the Baltic and is particularly common in the Szczecin Lagoon and Lake Dabie. Its metacercariae attack such cyprinids as bream, rudd, and vimba and settle in the subcutaneous tissue and in the fins. Melanin accumulates around the metacercariae and forms fine black spots. The species' life cycle was studied by Odening (1970). The first intermediate host for *A. muehlingi* is a prosobranch snail *Lithoglyphus naticoides* of the family *Hydrobiidae*, originally a Black Sea and East Baltic species introduced to western Europe.

As a consequence, the parasite's range is widening westward, too. The second intermediate hosts are cyprinids, the seagull *Larus ridibundus* being the definitive host. Mammals may be infected, too. *A. donicus* (= *Rossicotrema donicum*), a species related to *R. muehlingi* and occurring in the same areas, more often affects the percids. Particularly heavy invasions are recorded in perch; very frequently, the whole body and fins are covered with fine black dots. The parasite's life cycle, described also by Odening (1970), proceeds similarly to that of *A. muehlingi*. The first intermediate host is the same snail, *Lithoglyphus naticoides*, while the role of the second intermediate host may be played by piscivorosu birds (mostly seagulls) and mammals (dogs and cats). Malczewski (1962) found the presence of *A. donicus* in farm-bred fur-bearing animals (minks and foxes) fed with fish.

Cryptocotyle lingua occurs in marine fish, causing melanosis symptoms similar to those caused by previously discussed species. As in the previous cases, the affected fish's skin shows black spots with metacercariae inside. The adult parasite, 0.5–2.0 mm × 0.2–0.9 mm in size, lives in the intestine of seagulls (*Larus argentatus*, *L. fuscus*, *L. ridibundus* and others) and also in the intestine of mammals such as seals, dogs, and foxes.

The parasite's eggs get to the water with feces, sink down to the bottom, and are subsequently swallowed by snails, mainly *Littorina littorea*. The hatched larvae migrate from the snail intestine to the digestive gland where rediae develop with cercariae inside. The cercariae leaves the snail and attacks fishes. They settle and encyst underneath the fish skin and metamorphose into metacercariae. These are surroundded by melanophores. A number of different fish (herring, mackerel, and others), mainly in the NW Atlantic and off the Atlantic coasts of Europe, can be infected. The parasite was also recorded off Japan. When abundant, the metacercariae cover the entire fish body, fins, and even the eye cornea, thus blinding the affected fish.

The parasite is most dangerous, particularly for juvenile fishes, when the cercariae migrate within the fish body. The herring fry is particularly vulnerable and frequently dies as a result of the invasion. Thus the parasite has a detrimental affect on the fish population. The fishes infected by *C. lingua* should be barred from consumption for aesthetic reasons and, more important, because cases of human infestation have been recorded. The two *Apophallus* species and *Cryptocotyle* belong to the family *Heterophyidae*, containing numerous species known to be human parasites (see: Anthropozoonoses). Hsiao (1940) described an interesting case of digenean infection in cod (*Gadus callarias*) caught off Race Point (Provincetown). The fish was completely black, its body surface being very rough, hardened, and devoid of mucus.

A closer examination revealed a complete melanism caused by the mass occurrence of digenean metacercariae on the skin. The number of cysts was enormous and the melanocytes occurred in great numbers; they were particularly abundant around the cysts. The cysts conained metacercariae at various stages. On the other hand, neither the muscles nor the body contained the parasites. The cysts occurred within the epidermis; the connective tissue was at least three times thicker than normal. The cysts and melanophores were abundant also in the eyes, thus causing blindness. The author referred to was unable to identify the metacercariae; presumably they were a species of *Cryptocotyle*. Difficult to explain is also the occurrence of an infection on such an immense scale. Numerous digenean metacercariae are not surrounded by a pigment and as a consequence become invisible to the naked eye. They settle in the fish skin, in muscles or in viscera (hert, intestinal serous membrane). Their further development proceeds in the intestine of predatory fish, birds, or piscivorous mammals. One of such species is *Stephanostomum baccatum* whose life cycle is well-known as a result of the studies of Wolfgang (1955). The digenean, 3.34 × 0.75 mm in size, has two rows of spines around the oral sucker. Adults live in the intestine of marine predatory fishes, *e.g.*, *Hemitripterrus americanus*, *Hippoglossoides platessoides*, *Pleurogrammus azonus*, *Myoxocephalus scorpius*. The parasite's range extends from the Atlantic coasts of North America to the shores of Europe, including the Baltic Sea.

The cercariae develop in buccinind snails (*Buccinum undatum* and *Neptunea decemcostatum*), and then actively attack fish of the family *Pleuronectidae* (*Limandla limanda*, *Platessa platessa*, and others), piercing through their skin and mostly settling under it, in muscles of the blind side, and in the fins. They occur in masses, particularly in large fishes, encyst and metamorphose into metacercariae infective for other fishes, their definitive hosts. The fish blood system may contain digeneans of the family *Sanguinicolidae*, consisting of 7 genera with 21 species known to date. They are elongated, fusiform, and lack suckers and pharynx. Their development involves two hosts. Molluscs are intermediate hosts providing a site for the development of cercariae, while marine and freshwater fish are the definitive hosts.

The life cycle of the freshwater *Sanguinicola* species was described by the Polish parasitologist Ejsmont (1926). Most often, the eggs are released into the water through the broken gill epithelium. The eggs hatch into miracidia which actively search for their first intermediate hosts, usually pulmonate snails, and metamorphose in them into sporocysts. The cercariae with forked tails (furcocercariae) develop in the sporocysts and thus release themselves from the snails. Then, using their penetration glands, they make their way through the fish skin and reach blood vessels to metamorphose into adults.

The family *Sanguinicolidae* contains species of the genus *Sanguinicola*, typical of freshwater fish. Marine fish parasites include *Aporocotyle simplex* (Odhner, 1900) and *A. theragrae* (Ichihara, 1970). *A. simplex* was found for the first time and described by Odhner in the gill of dab (*Limanda limanda*) off Sweden. *A. theragrae* is frequently recorded in the Alaska pollock (*Theragra chalcogramma* of the North Pacific. This is a small (up to 7 mm in length and 1—2 mm in width) parasite having a characteristic, H-shaped intestine. Numerous (more than 100) testes fill the space between the intestinal caeca. The single, large ovary is located posteriorly in the body. These digeneans are often encountered extravasated on the gills, intestine, kidney or between the viscera during dissection or processing of fish, hence occasionally the fillets may be accidentally contaminated. Very serious infections are observed at times in *Theragra chalcogramma*. Soviet authors reported on about 50% invasion incidence and about 54 digeneans per host. There are no data on the pathogenicity of *A. theragrae*, but a strong infection of the blood system doubtless has a bearing on fish condition.

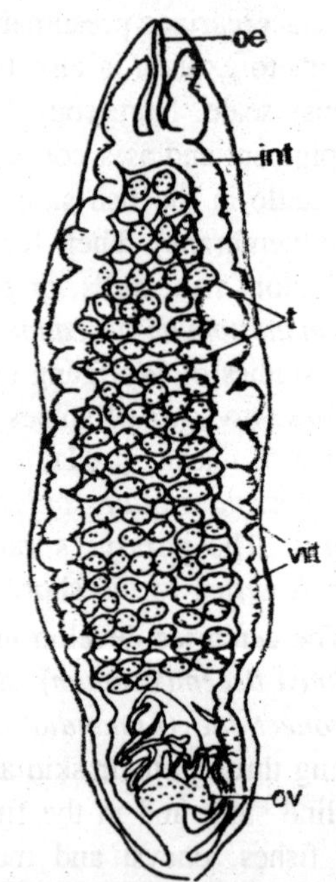

Fig. 2.10. Aporocotyle theragrae: int—intestine; oe—oesophagus; ov—ovary; t—testes; vit—vitellaries.

Various *Phyllodistomum* species dwell in the urinary baldder and ducts of freshwater and marine fish. Their body is tapered anteriorly, the posterior part often being broadened leaf-like, with wrinkled edges. The species belonging to this genus are highly host-specific. The first intermediate hosts in their life cycle are bivalves (*Dreissena*, *Anodonta* and others) harbouring cercariae which later encyst either in the same bivalves in daughter sporocysts or in dragonfly larvae. Thus the second intermediate host is not always present. The definitive host—a fish—is infected by cating metacercariae-containing bivalves or dragonfly larvae.

Most *Phyllodistomum* species parasitize freshwater fish; marines species, however, are known as well, for instance *Ph. marinum* (Layman, 1930) from the urinary duct of *Spheroides borealis*, and *Ph.*

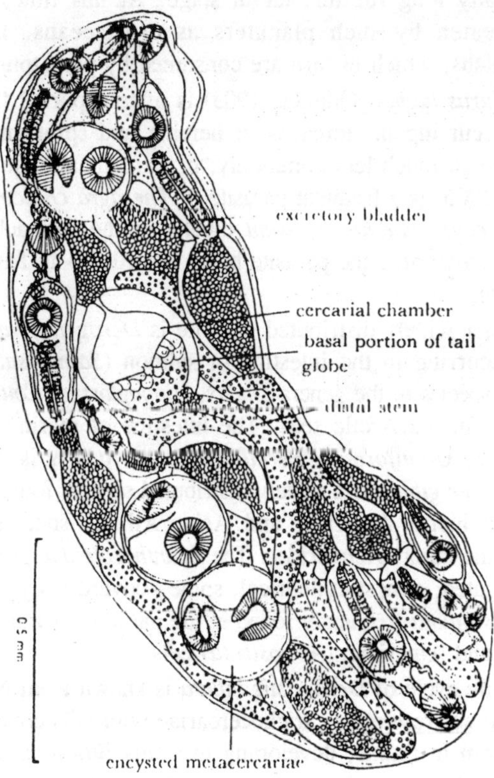

Fig. 2.11. Sac-like daughter sporocyst of Phyllodistomum simile; the adult fluke is found in urinary bladder of brown trout, Salmo trutta.

pacificum (Yamaguti, 1951) from *Caranx equula*. Both species were found in the Sea of Japan. In the same sea, a related species, *Urinatrema hispidum* (Yamaguti, 1934) was described from the urinary bladder of *Hexagrammos otakii*. The gall bladder of *Anarrhichas lupus* often contains the already mentioned digenean *Fellodistomum fellis*. Most digenetic trematodes dwell in the intestine of marine fish.

Digeneans of the family *Hemiuridae* are the commonest species. They usually inhabit the anterior part of the digestive system (stomach). They are small (a few mm in length) parasites whose posterior part is developd as a retractable tail. Their life cycle requires two intermediate hosts. Molluscs are the first hosts and contain cercariae which have characteristic adaptations to the planktonic mode of life in the form of long filamentous processes arranged fan-or umbrella-like. The processes enable the cercariae to float in the water for 3-4 days, which is exceptionally long for this larval stage. At this time, the cercariae may be eaten by such plankters as crustaceans, medusae, and chaetogenaths, which in turn are consumed by plankton-feeding fish.

Hemiurus luehei (Odhner, 1905) is often reported from European waters, occurring in stomachs of herring and sprat and encountered also, although much less commonly, in salmon and trout. *H. levinseni* (Odhner, 1905) is a frequent parasite of *Theragra chalcogramma* in the Pacific. *Lecithocladium excisum* dwells in the stomach of mackerel (*Scomber scombrus*), the parasite being recorded both from the Atlantic and Pacific.

Another widely distributed parasite is *Derogenes varicus* (Muller, 1784), occurring in the intestine of salmon (*Salmo salar*) and many other fish species of the genera *Gadus*, *Hippoglossus*, *Cottus*, and *Clupea* and others in the Arctic seas. The intestine of marine *Clupeiformes*, *Gadiformes*, *Gobiiformes*, *Perciformes*, and others is often inhabited by *Lecithaster gibbosus* a widely distributed species found off the coasts of Europe North America, and Asia. All the species listed above belong to the family *Hemiuridae*. The *Acanthocolpidae* groups trematodes with elongated, almost cylindrical, spine—coverd body, such as that of *Deropristis inflata* frequently found in the intestine of European (*Anguilla anguilla*) and American (*A. rostrata*) eels.

The life cycle of *Deropristis inflata* is known from North America. The cercariae, resembling trichocercariae with tails covered with short spines, form in rediae, developing in snails *Bittinium alternatum* and penetrate polychaetes *Nereis virens* and *N. diversicolor*. They encyst there and metamorphose into metacercariae. Eels are infected after consuming polychaetes with metacercariae. Adult digeneans dwelling in fishes are harmless to man. They are usually removed together with viscera on gutting; however, fillets and carcasses may be accidentally contaminated at a massive invasion, which is undesirable for sanitary reasons.

Subclass: Didymozoida

Members of this subclass, except for the three freshwater fish parasites known to date (*Nematobothrium labeonis*, *Philopinna higai*, and *N. texomensis*), parasitize marine fishes. They are widely distributed in pelagic subtropical and tropical waters of the Atlantic, Pacific, and Indian Oceans and adjacent seas. According to Nikolaeva (1978), they make up about 7% of all the trematodes. Most of them occur in cysts produced by affected fish as a defensive reaction; more seldom they remain unencysted. The cysts may be located in gills, mouth cavity, in eye sockets, under the skin, and in muscles. The muscle-dwelling species are usually found unattached in the muscle tissue. When encysted, they always occur in pairs and exhibit various types of reproduction, from hermaphroditic to dioecious.

When hermaphroditic, some individuals show domination of the male system, while the female system prevails in others (gonochorism, *i.e.*, incomplete hermaphroditism). The cysts are variously shaped, the shape being typical of the genus or even of the species. The *Didymozoida* are elongated, filamentous (*Nematobothrium*) or ribbon-like (*Atalastrophion*). In some species, the body is divided into two sections: the anterior one is usually thin and filamentous or ribbon-like and flattened, while the posterior one is large, swollen, completely filled with gonads and eggs (*Koellikeria*, *Didymozoon*, *Didymocystis*).

The didymozoid life cycle is known only in fragments, hence their taxonomic status is not fully established. Their bean-shaped eggs are very small, only about 10 μm long, and have lids. The eggs hatch into the miracidium, having an anterior ring of about 30 spines surrounding the apical sucker. The whole miracidium is covered by setae. The miracidium development is unknown. It is not known, either, how the eggs and larvae get out of the fish muscles. Presumably, this happens after the fish has died. According to Nikolaeva (1965), the miracidium develops within a gastropod, the first intermediate host. In her opinion, rediae or sporocysts and cystophora-type cercariae develop in snails, similarly to the related family *Hemiuridae*. Crustaceans are the suspected second intermediate hosts providing a site for the development of metacercariae. The metacercariae were being found in cirripeds (*Lepas*

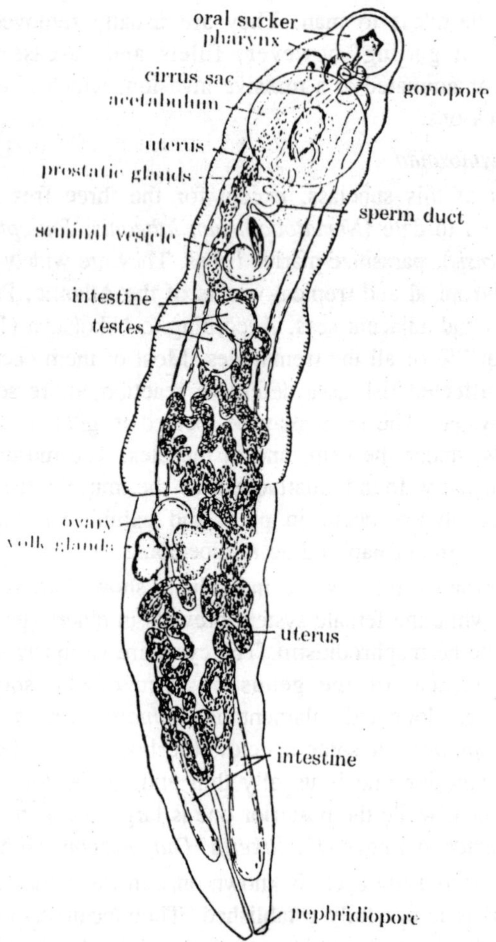

Fig. 2.12. Hemiurus.

sp.) and in copepods *Paracalanus aculeatus*. The metacercariae belong to either of the two following types: the *Monilicaecum* and the *Torticaecum*. Those are collective names for the didymozoid metacercariae. Yamaguti (1971) differentiates between three stages among them:

1. *Premonilicaecum* and *pretorticaecum*, developing in crustaceans, the second intermediate hosts.
2. *Monilicaecum* and *Torticaecum*, juvenile stages dwelling in various fishes, some of which serving as proper hosts and others playing a role of parathenic hosts.

3. *Postmonilicaecum* and *Posttorticaecum*, stages developing into the final form within their proper definitive hosts (fishes).

The subclass groups have about 200 species. They are usually of medium size, but some very large parasites are encountered as well. *Nematobothrium filarina* (Van Beneden, 1858) occurs, incysts the size of a fist, in the gill cavity of meagre (*Sciaena aquila*) and grows to more than 1 m. *Gonapodasmius okushimai* (Ishii, 1935), described from the muscles of *Pagrosomus major* off the Japanese coasts reaches up to 590 cm (the male is much smaller). *Nematobibothrioides histoidii* Noble, 1974 is narrow, ribbon-like, and exceeds 12 m in length.

Unitubulotestis sardae (G.A. and W.G. MacCallum, 1916) (= *Nematobothrium sardae*) up to 3 cm in length, is one of the noteworthy and relatively well-known species. Pairs of individuals live in 7 mm long and 3-6 mm wide, orange cysts located among the gill lobes of the Atlantic bonito (*Sarda sarda*) in the North Sea. The parasite was also recorded in the NW Atlantic. Studies of E. and J. Grabda (1939) and E. Grabda (1947) showed the infection to be quite substantial, up to 64.8%. Single gill arches container, apart from juvenile parasites invisible to the naked eye, up to 10 cysts of various sizes. The cyst walls are very delicate and thin, frequently provided with fine capillary vessels. The cyst colour is imparted by eggs filling the parasites. Each cyst contains two coiled filamentous individuals of *U. sardae*, their maximum length and width reaching 29.7 mm and 1.5 mm, respectively. Both individuals are hermaphroditic, but with a clear domination of male gonads in one and female gonads in the other.

The earliest stage of development is unknown. On the other hand, juveniles 7-14 mm in size and cyst formation were observed. The parasites, first one and then the other, get into gill lobe blood vessels and produce the arteriectasis in the lobe artery, which—in the course of further development of the parasite—leads to the formation of a typical sacculated aneurism (*aneurisna verminosa saccata*). Some cysts may contain blood extravasated from arteries. Undoubtedly, the parasites settling in the blood vessel lumen bring about disorders ingill circulation. However, no anatomo-pathological changes were observed in the gills except for a single case of a strong invasion (10 cysts in a gill arch) in a juvenile 20 cm long fish were clearly pale and anemic.

Koellikeria orientalis (= *Wedlia orientalis*) is a typically dioecious trematode. Its paired individuals occurring in cysts located in various parts of the digestive system of *Germo macropterus*, *Euthynnus pelamys*, and *Thunnus thynnus* in the Pacific. The trematode body is divided into two parts. The anterior section is thin and elongated, slightly widened and flattened around the oral sucker. The posterior section is large, swollen, and almost spherical. The female posterior part is particularly enlarged due to the enormous amount of eggs filling it.

Neodiplotrema pelamydis is an interesting case of two individuals being permanently accreted in a single cyst, similar to the freshwater monogenean *Diplozoon paradoxum*. *N. pelamydis* dwells in the gill cavity of *Euthynnus pelamys* from the Pacific. Both individuals in the cyst are hermaphroditic, their posterior parts being fused. Without doubt, none of the trematode species is indifferent to the fish health and condition. However, those parasites settling in muscles are the main concerns for fish processing. One of those is the largest known trematode, *Nematobibothrioides histoidii* mentioned earlier, occurring unattached in muscles of *Mola mola* off California.

These trematodes interlace fish muscels with numerous loops. Up to 12 individuals per host were found in fish specimens weighing, on average, 16 kg. The fish flesh was literally packed with thin (1.5-4 mm), very long (up to 1200 cm in length) parasites difficult to separate from fish without damaging the latter. Besides the adults, the affected fish contained aggregations of eggs (17 × 15 μm in size, on the average) in its connective tissue, which additionally contaminated the flesh. The *M. mola* behaviour was abnormal: the lazily moving fish kept to the water surface, returning there when pushed deeper with an oar. It should be added that, besides *N. histoidii* described, the *M. mola* individuals examined also showed the presence of numerous parasites of various taxa: protozoans, digeneans, cestodes, and copepods. The abnormal behaviour was thus a net result of the combined effects of all the parasites.

Muscles of *Pagrosomus major* in the Pacific may contain *Gonadopodasmius okushimai* (Ishii, 1935). The species, about 0,5 mm wide, has a long (up to about 6 m) body. Muscles of the Pacific tuna *Thunnus alalonga* sometimes contain *Metanematobothrium guernei* 177 mm long and 1 mm wide. *Didymozoon tenuicolle* was encountered in muscles of *Lampris guttatus* from the Europeen waters. The parasite's "neck" is clearly separated from its "trunk." The total length is 63.8 mm, the "trunk" measuring 47 mm. Wierzbicka (1980) described a new didymozoid, *Neolampididymozoon tenuicolle* from *Lampris guttatus* in the SE Atlantic.

The parasites occur encysted in pairs on the gills and gill membrane as well as in the dorsal fin base muscles. It is interesting to note that the parasite's size depends on its location in host. For the first time in the Atlantic, the same author described the occurrence in *Lampris guttatus* of another species, *Metadidymocystis cymbi*. The parasite occurs unattached among the subcutaneous muscles down to 5 mm and frequently forms large aggregations of hermaphroditic individuals 10-26 mm long. In recent years, muscles of the hairtail (*Lepidopus caudatus*) from off South Africa have often been found to contain trematodes of the family *Didymozoonidae*, not identified any closer. They frequently occur only as aggregations of eggs, 16–18 μm x 30–32 μm in size, typical of the family. The hairtail invasion incidence in 1977 reached 50%. Mackerel (*Scomber scombrus*) caught in the Georges Bank was found to be infected by *Atalastrophion* trematodes, the parasites intertwining with the fish muscle fibres and forming orange-coloured centres. The colour is imparted by bean-shaped tiny eggs, enormous amounts of them filling the parasite's body. Those fillets abundantly infected with the parasites were soft, watery and greyish in colour.

Nikolaeva (1972), too, drew attention to a frequent presence of didymozoid trematodes in muscles of marine fish. She found unencysted and unattached trematodes in many fish species from the Aegean, Red, and Mediterranean Seas and from the Atlantic and Indian Ocean. "Balls" of parasites interlaced the fish skeletal muscles. It was not possible to dissect out all of the trematodes and identify them to species; some could be identified to the generic level only. For example, muscles of the Red Sea *Lutianus bohar*, the Indian Ocean *L. lineatus*, *Letrinus miniatus*, and *Epinnula orientalis* as well as those of *Cypselurus* sp. from the Aegean Sea and *C. furcatus* from the Atlantic were found to contain *Gonapodasmius* trematodes. Muscles of the Indian Ocean *Platycephalus pristis* were infested by *Didymozoon* sp. The fish quality was unquestionably lowered by the presence of these parasites. Although the parasites are harmless to man, attention should be paid to them; badly infected fishes should be rejected. On weak invasion, and when the muscles contain encysted species, the cysts should be removed from fillets.

Class: Cestodes (Tapeworms)

Cestodes are plathelminths whose body is ribbon-like, flattend dorsoventrally, and consists of the scolex, neck, and strobila, the latter made up of a number of sections (proglottids) arranged one behind another. Unsegmented cestodes are known as well (*Caryophyllaeus*, *Khavia*, *Caryophyllaeides*). The class contains exclusively parasites. Adult forms dwell in the intestine of vertebrates. The scolex is an

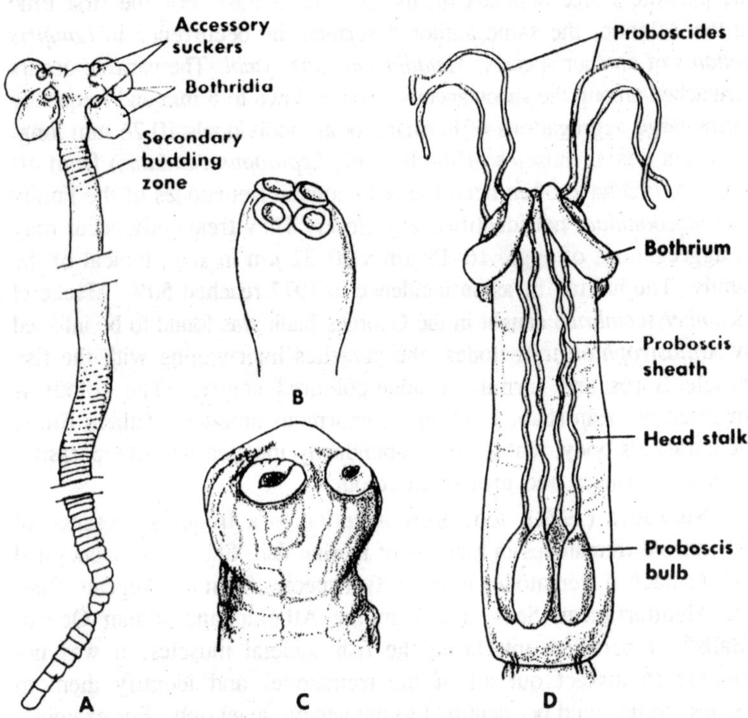

Fig. 2.13. A—Phyllobothrium dornii proliferates in usual manner at rear end of neck; B—Proteocephalus coregoni scolex; C—Corallobothrium giganteum scolex; D—Gilquinia squali.

attachment organ used to fasten a cestode to the host's intestinal mucosa; in this way the parasite overcomes the peristaltic movements and the danger of being removed from the gut.

Consequently, the scolex is provided with various additional holdfast structures such as suckers (acetabulae) and bothria in the form of 2 (*Pseudophyllidea*) or 4 (*Tetraphyllidea*) grooves. More complex are bothridia; they are grooves, having their own muscles and divided by septa into secondary grooves, frequently supported by stalks (*Tetraphyllidea*) Additionally, the scolex may possess various rings of hooks and retractable proboscids covered with numerous rows of fine hooks (*Tetraryhynchidea*). The scolex structure is an important identification character of cestodes. The cestode neck is a growth zone. New proglottids are produced on the borderline between the neck and the strobila. The body is filled with a loose parenchymal tissue, its cells containing numerous calcareous bodies.

The central nervous system is located in the scolex. The system consists of two ganglia interconnected with transverse commisures. The ganglia give rise to anteriorly-directed nerves of the scolex attachment structures and to 10 nerve cords running parallel backwards along the strobila. Two of them are the main nerves, much thicker than the rest, located symmetrically along the sides of the body. All the nerve cords are interconnected with transverse commissures; their branches extend to the organs within the strobila. The excretory system consists of protonephridia and a network of capillary canaliculi opening to the main excretory canals (most often two) located laterally and connected with transverse canals in each proglottid. The excretory ducts open in the terminal proglottid.

Digestive system

Cestodes have no intestine, the nourishment being absorbed by the whole surface of the body. Food absorption is facilitated by the structure of tegument, its surface being thrown into numerous folds (the microtriches) and pierced by numerous pores. With a few exceptions, as for examples *Dioicocestus* sp., cestodes are hermaphroditic, each proglottid having its own set of male and female gonads. In some species the segmentation is obliterated externally, but sets of gonads are replicated several times. As a rule, male gonads develop first, followed by developing female gonads. The male system consists of numerous testes, spherical or oval, scattered in the parenchyma. The

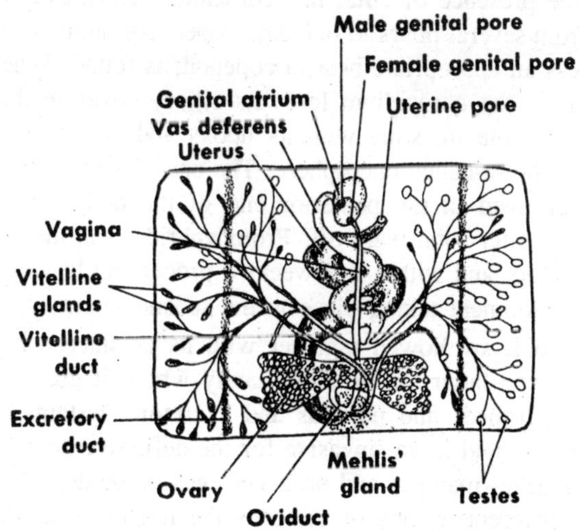

Fig. 2.14. Dibothriocephalus, mature proglottid.

vasa efferentia leave the testes and fuse into a single vas deferens, its terminal part functioning as a copulatory organ (cirrus), opening into the cirrus bursa (*bursa cirri*). The vas deferens often widens in its terminal part and forms the seminal vesicle where spermatozoa are gathered. The female system includes a single ovary, the Mehlis gland surrounding the ootype, vitellaries, and the seminal receptacle. Fertilization of egg cells and egg formation takes place in the ootype, to which open the oviduct, vitellary ducts, and the seminal receptacle. Eggs are collected in the uterus and are subsequently excreted via the gonopore. The vagina is a canal receiving spermatozoa introduced by the cirrus during copulation.

Both the male and female systems form the basis of cestode identification. The cestode development is complex; it involves metamorphosis and change of hosts, a vertebrate always being the definitive host. The larval phase proceeds in one (*Diphyllidea*, *Proteocephalidea*) or two (*Pseudophyllidea*, *Tetraphyllidea*, *Tetrarhynchidea*) intermediate hosts. The first intermediate host in both lines of development is always an invertebrate; in the *Pseudophyllidea* cestodes affecting the fish, it is most often a copepod. The second intermediate host is a fish. The array of definitive hosts covers fishes, birds, and mammals (including man). When in water, eggs of pseudophyllid cestodes hatch to release a spherical and completely ciliated larva provided with 6 hooklets.

Due to the presence of cilia, this coracidium larva can swim for some time (from several hours to one day, depending on the conditions) and dies unless an appropriate host, a copepod, is found. When eaten by a crustacean, the coracidium loses its ciliary cover in the host's stomach, pierces the intestine with its hooks and gets to the host's body cavity as the so-called oncosphere. The oncosphere grows longer, its hooklets grouped in the posterior end of the body set off as a cercomere; the latter soon tears off. Protonephridia and the excretory canaliculi develop, and within 2–3 weeks a procercoid larva emerges, its further development taking place in the second intermediate host.

A procercoid, swallowed together with its crustacean host by a fish, moves to the latter's abdominal cavity where it grows further. The scolex is formed, and nervous and excretory systems develop, before a plerocrercoid larva, invasive for the definitive host, appears. The cestode reaches maturity and produces eggs in the definitive host's intestine. In different groups of cestodes the life cycle as described undergoes modifications with respect to the duration of various stages, advancement in development, larval structure, etc.

Cestode pathogenicity. Both the larval and adult cestodes are found in the fish as pathogens. Plerocercoids of many species, growing in the fish body cavity, exert mechanical pressure on the liver, kidney, and gonads, thus causing malfunctioning of the organs affected. Large plerocercoids of *Ligula* induce sterility in fish, their growth occasionally causing the fish abdominal cover to break down. Toxins excreted by those plerocercoids give rise to fish behavioural disorders, *e.g.*, the *Ligula* plerocercoid-infected bram do not migrate to spawn. The presence of adult cestodes in the host's intestine is not unimportant to the host, since, especially inlarge species, cestodes may cause anaemia and emaciation. Marine fish may contain cestodes belonging to five orders: the *Pseudophyllidea*, *Diphyllidea*, *Tetraphyllidea*, *Tetrarhynchidea*, and *Proteocephalidea*. The division is based mainly on the scolex structure. Of the greatest importance for the fishing industry are the *Pseudophyllidea* and *Tetrarhynchidea*.

Order: Pseudophyllidea

Cestodes grouped in this order have scolex with two bothria: dorsal and ventral; sometimes the scolex is funnel- (*Cyathocephalus*) or fan-shaped (*Caryophyllaeus*). The strobila usually has numerous proglottids, although unsegmented cestodes are encountered, too, viz the family *Caryophyllaeidae*. The segmentation of strobila is usually pronounced and corresponds to internal segmentaion. In some species (*e.g.*, the *Ligulidae*), however, external segmentation does not coincide with the internal one. Life cycle involves three hosts: two intermediate and one definitive. The order contains numerous species dwelling in fresh-water, marine, and migratory fish. Some species are patho-genic for man as well; these will be dealt with in the chapter on anthropozoonoses.

The commonest species include *Bothriocephalus scorpii* (Muller, 1776), a frequent parasite of flatfish, attacking other marine fish as well. It was found in the Southern Baltic in the flounder (*Platichthys flesus*), turbot (*Scophthalmus maximus*), and sculpin (*Myoxocephalus scorpius*). A massive invasion of the cestode is frequently observed; 100% of the North Sea *Rhombus laevis* can be affected. The cestode was also recorded in the Atlantic and Pacific as well as in the Mediterranean, White, and Black Seas.

B. scorpii is a rather large (up to 1 m long) cestode whose adult form lives, attached most often to the pyloric caeca mucosa, in the fish intestine. Its life cycle was followed in the laboratory by Markowski (1935). The first intermediate host is a copepod, *Eurytemora hirundo*; the next host is a gobiid, *Pomatoschistus minutus* (more seldom the straightnose pipefish, *Nerophis ophidion*), the flatfish being definitive hosts. The parasite affects fish condition considerably; a massive invasion may retard fish growth.

The eel in Puck Bay is frequently attacked by *Bothrioce-phalus claviceps* (Goetze, 1782), encountered also in European and American inland waters. Presumably, the eel becomes infected at its freshwater phase of life, which is inferred from the fact that the parasite's first intermediate host is, as observed experimentally by Jarecka (1959), a freshwater copepod *Macrocyclops albidus*. The second intermediate host, *i.e.*, the one passing the infection on to the eel, is not known to date.

The intestine of salmonids (i.a. salmon, trout) may be a suitable habitat for *Eubothrium crassum*. This rather large (up to 60 cm long) cestode with a well-developed scolex and two distinct bothria is very common in European, North Amercian, and Caspian basin waters. The cestode settles most often in pyloric caeca and attaches itself by means of the scolex to the bottom of the caeca, the strobila falling down into the intestine. The infection is usually very heavy, up to several hundred cestodes per host. The *E. crassum* life cycle was described by the Norwegian parasitologist Vik (1963). Typically of the *Pseudophyllidea*, it involves two intermediate hosts: the first is a copepod, either *Cyclops strenuus* or *Eucyclops serrulatus*, while the other is perch (*Perca fluviatilis*), eventually infecting salmon and trout.

There is no detailed data on the *E. crassum* pathogenicity. Even a heavily infected (up to 400 cestodes per host) fish retains good condition and has fat deposits lining the intestine. Mikhailov (1975) is of the opinion that *E. crassum* is pathogenic to the Caspian salmon. A very heavy invasion of up to 3000 cestodes per host was recorded in that species, resulting in the intestinal occlusion and death of the individuals affected. The intestine of cod (*Gadus morhua*), haddock (*G. aeglefinus*), and other Atlantic gadids as well as *G. macrocephalus* and *Theragra chalcogramma* in the Pacific may be inhabited by *Abothrium gadi* van Beneden, 1871. The cestode has a deformed scolex with poorly developed bothria.

The order discussed includes also certain cestodes (*Diphyllobothrium latum*, *D. minutus*, *D. pacificum*, *D. ursi*, and others) whose larvae are intestinal parasites of vertebrates, while plerocercoids occur in fish. They are dangerous for man, infected via fish (see: Anthropozoonoses). The group includes also *Pyramicocephalus phocarum* whose adult forms live in the intestine of seals *Phoca barbata*, *Ph. hispida*, *Cystophora cristata*, *Eumetopias jubatus*, and *Enhydra lutris*. The plerocercoid of the species has a large (up to 4 mm in length and

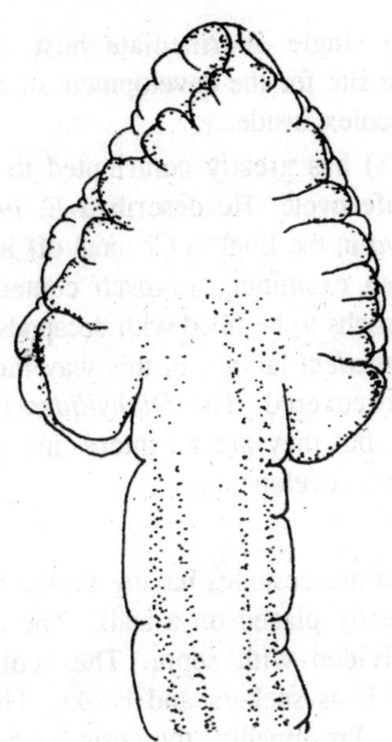

Fig. 2.15. Plerocercoid of Pyramicocephalus phocarum.

3 mm in width as its base) triangular, pyramid-shaped scolex with large, strongly wrinkled bothria and a 5—6 cm long tail.

The *P. phocarum* plerocercoids occur in numerous marine fish such as *Gadus morhua*, *Theragra chalcogramma*, *Myoxocephalus verrucosus*, and *Eleginus gracilis*. Particularly frequent are *P. phocarum* invasions in the Alaska pollock (*Theragra chalcogramma*). The plerocercoids settle in the fish abdominal cavity, between the pyloric caeca, on the intestinal serous memberane, on gonads, and even in muscles. *P. phocarum* tapeworms occur in Atlantic and Pacific fishes as well as in those in the White and Barents Seas. They are particularly abundant in the vicinity of seal colonies, seals being their definitive hosts. Those settling in the liver destroy it. Cases of human infections are common.

Order: Diphyllidea

This order groups cestodes which have a very specific scolex with two bothridia (dorsal and ventral) in the anterior part and two groups of large hooks set on the top of the rostellum. The posterior part of the scolex resembles a handle or a stalk with vertical rows of fine hooks. The very short strobila consists of a few proglottids only. The order contains only one genus, the *Echinobothrium*, with 11 species living in the intestinal spiral valves of rays (*Raidae*). The *Echinobothrium* life cycle involves a single intermediate host, a crustacean or a gastropod, providing a site for the development of a cysticercoid larva with a fully formed scolex inside.

Ruszkowski (1928) has greatly contributed to the knowledge on the *Echinobothrium* life cycle. He described *E. benedeni* from *Raja asterias* and *R. punctata* in the English Channel off Roscoff, and studied its development. When examining stomach contents of the infected rays, he found the stomachs to be filled with decapods *Hippolyte varians*, which contained *E. benedeni* larvae; in this way the intermediate host of the parasite was discovered. The *Diphyllidea* cestodes are of no economic importance, but they are an interesting group with respect to their morphology and development.

Order: Tetraphyllidea

The *Tetraphyllidea* are cestodes having scolex with 4 bothridia of various shapes, frequently placed on a stalk. The attachment surface is large and often divided with septa. The scolex has additional attachment devices such as suckers and hooks. The tetraphyllid life cycle is poorly known. Presumably, the cestodes have, similar to the *Pseudophyllidea* and *Tetrarhynchidea*, two intermediate hosts and a definitive one. A marine crustacean is the first intermediate host the procercoid grows in. The plerocercoid stage develops in teleost fishes, the mature cestodes living attached to the spiral valves of sharks.

Scolex pleuronectis Muller, 1780 (= *S. polymorphus*) is the commonest tetraphyllid larva. It occurs in the intestine, abdominal cavity, gall bladder, and muscles of salmonids, flatfish, and others. It is frequently encountered in the Barents, White, Black, and other European seas. Presumably, *S. polymorphus* is a collective name of various tetraphyllid larvae. Other larva, *Pelichnibothrium caudatum* occurs in the intestine and pyloric caeca of marine fishes and anadromous salmonids of the river Amur basin. The larva was also found in Pacific cephalopods. Adult tetraphyllids occur in the lancetfish (*Alepisaurus ferox*) in the Atlantic, as well as in the blue shark (*Prionace glauca*), sunfish (*Lampris regius*), and occasionally in tuna (*Thunnus thynnus*) off Japan. Both the adult and larval *Tetraphyllidea* are of no importance to the fishing industry.

Order: Tetrarhynchidea

Synonym: Trypanorhyncha

The order groups relatively small tapeworms, living, as adults, in the stomach and spiral valves of sharks. The larvae are very common in marine teleosts. The key character of these cestodes is the structure of the scolex. The scolex is relatively large, with 2 or 4 bothridia anteriorly and 4 or—less frequently—2 very contractile and retractable proboscids, each being retracted and everted by means of special muscles in the form of 4 bean-shaped structures (bulbus) connected with the retractor muscles drawing the proboscids into their canals in the scolex parenchyma. The proboscids are covered by numerous rows of hooks of a shape and size depending on the species and on the location on the proboscid.

The scolex, owing to its complex structure, can be thus divided into three parts: bothridial (*pars bothridialis*), vaginal (*pars vaginalis*), and muscular (*pars bulbosa*). The tetrarhynchid life cycle is still poorly known. To date, it is only the Polish parasitologist J.S. Ruszkowski who, in his work on *Grillotia erinaceus* (Van Beneden, 1858) observed and described the first stage of the cycle, which made it possible to infer the full course of it. The *Tetrarhynchidea*, similarly to the *Pseudophyllidea*, require two intermediate hosts (a crustacean and a teleost) and a definitive one. During his work at the Biological Station in Bergen, Ruszkowski (1934) cultured eggs collected from mature cestodes found in the intestine of *Raja oxyrhynchus*.

Subsequently, he infected various planktonic organisms with typical coracidia hatched from those eggs. The following copepods succumbed to the infection: *Acartia longiremis*, *Pseudocalanus elongatus*, *Paracalanus parvus*, and *Temora longicornis*. While in copepods, the coracidia lost cilia and made their way to the body cavity to metamorphose there into procercoids provided with a cercomere with 6 hooklets. The further development takes place in teleosts and involves plerocercoids having a fully developed scolex with retractable proboscids and bothridia. The scolex is located anteriorly in a special receptacle (*receptaculum scolecis*). This larva does not fully correspond to the pseudophylli-dean plerocercoid, hence various authors call it *plerocercus* or *blastocystis*.

In the intestine of a shark or a ray the scolex is everted, the strobila grows and the cestode takes on its final adult form. The larval *Tetrarhynchidea* are very common in marine fish and often occur in masses causing high losses; particularly harmful in this respect are the muscle-dwelling species. They are not dangerous to man, but the relatively large, well-visible larvae make fish muscles look repulsive. For this reason, the infected fishes are processed into fish meal rather than marketed for consumption. The already mentioned species *Grillotia erinaceus* is common in cod, hake, halibut, and other teleost fish in the North Atlantic as well as North, Irish, and Norwegian Seas, and in the North Pacific and adjacent waters. The larvae settle in abdominal cavity organs, on the peritoneum, and in muscles, particularly along the backbone. The larvae have 2 bothridia and 4 proboscids. The adult cestodes live in the spiral valve of *Raja oxyrhynchus*.

Larvae of *Guilquinia squali* were found in the eyes of the North Sea whiting (*Odontogadus merlangus*). The larvae, frequently with their scolices everted and provided with 4 bothria and 4 proboscids, lie unattached in the hyaline body of the eye. A single eye may contain up to ten 3.5-8.0 mm long and 0.7-1.6 mm wide larvae. The whiting becomes infected with those larvae at a young age (before its third year of life), invasion incidence being much lower in older fish, which

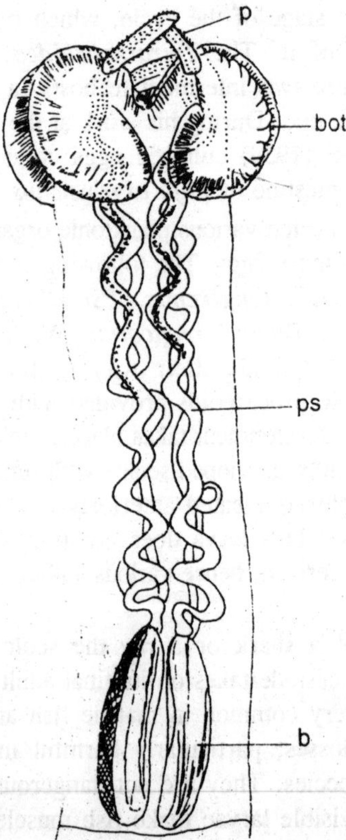

Fig. 2.16. Guilquinia squali from intestine of Squalus sucklii: b–bulbus; bot–bothridia; p–proboscid; ps–proboscid sheath.

is probably related to a switch in their diet at that age. The adult cestodes were recorded in the North Atlantic, Mediterranean Sea, and in the Pacific off American coasts, mainly in *Squalus acanthias* and *S. sucklii*, less frequently in other sharks.

The *Nybelinia surmenicola* larvae form small spherical or oval white cysts 2-3 mm in diameter. They are common in viscera and muscles of the Alaska pollock (*Theragra chalcogramma*), a Pacific gadid. They may also be encountered in other fish such as *Ophiodon elongatus*, *Merluccius productus*, *Pleurogrammus azonus*, *Sebastes flavidus*. *Cleisthenes herzensteini*, *Myozocephalus brandti*, and *Hippoglossoides elassodon*. Moreover, the larvae were found in squids *Ommatostrephes sloani pacificus*. The *Nybelinia* plerocercoids show a characteristic fold (a kind of mantle), the so-called velum which surrounds the posterior end of the body. The scolex has 4 bothridia and 4 proboscids.

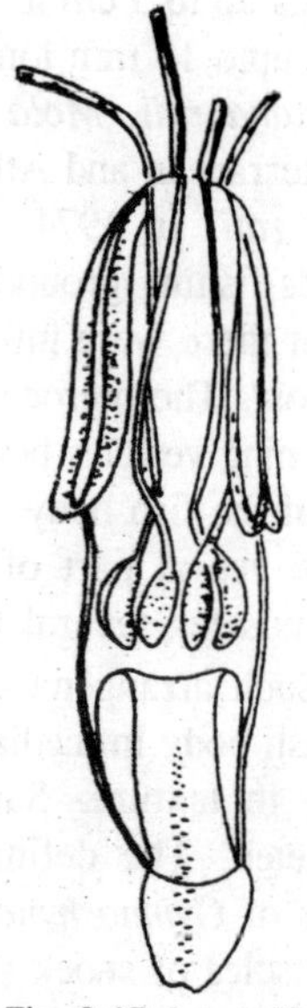

Fig. 2.17. Larva Nybelinia surmenicola.

The *N. surmenicola* larvae are most commonly recorded in *Theragra chalcogramma*, the invasion incidence reaching occasionally 100% at the intensity of up to 205 larvae per host. The larvae occur in different organs: stomach wall, intestinal serous membrane, pyloric caeca, gonads, and muscles. They are particularly abundant, occasionally even as grape-like structures, in the abdominal cavity near the anus. At this site they are also most common in muscles.

The *N. surmenicola* plerocercoids are common in the North Pacific, in the Okhotsk and Bering Seas as well as in the Sea of Japan, and on the shelf off Asia and America. Although the cestodes are harmless for man, any strong invasion is a reason to reject the Alaska pollock affected due to their repulsive appearance. In the Soviet Union, where considerable amounts of the Alaska pollock are landed, a special processing technique is employed to get rid of the parasites and use the fish for consumption. The ventral part containing most parasites is cut off, the dorsal part—much less infected—being retained.

Plerocercoids of *Gymnorhynchus gigas* are large larvae, each having a very long (up to 50 cm) larval tail. The anterior vesicular section, 1.5 to 2 cm long, contains a retractable scolex which, when everted, reaches up to 5 cm in length (together with proboscids), the proboscids being upto 18 mm long. The *G. gigas* larvae are found in the muscles of *Brama raii*, *Mola mola*, *Pagellus centrodntus*, and also in other Mediterranean and Atlantic fish species. They are particularly common in *B. raii*. In 1974, a heavy infection was recorded in the Canary Islands fishing grounds off NW Africa. More than 90% of the *B. raii* caught there were infected, the invasion intensity reaching 17 parasites per host. The plerocercoids gathered near the backbone, their scolex-containing vesicles being directed posteriorly and grouped in the ventral part of the fish body. The larval tails piereced posteriorly and grouped in the venral part of the fish body. The larval tails pierced all the fish muscles several times.

Such arrangement of the tails suggested larval migrations within the fish body immediately after entering it. The affected muscles were loose in texture. Sanitary requirements did not allow them to be marketed. The definitive host of the species is not known. Larval forms of *Gymnorhynchus thyrsitae* (Robinson, 1959) are very common in muscles of snoek (*Thyrsites atun*) and hairtail (*Lepidopus caudatus*) off New zealand and on the South African shelf. *L. caudatus* caught in 1974 off South Africa was infected in 95% 1-18 parasites being found per host, while up to 90% of *The. atun* contained up to 18 cestodes per fish. Off New Zealand, the invasion intensity in *The. atun* muscles reached 286 larvae per host.

The larvae have vesicular blastocysts measuring 5×3 mm and ribbon-like larval tails more than 10 cm in length. The blastocyst contains the scolex, about 5 mm long, provided with 4 large flat bothridia and 4 very long (more than 3 mm) proboscids protracting from the apical part of the bothridia. Hooks covering each proboscid are diverse in size and shape. The proboscid is naked at its base, the armed part beginning with an asymmetrical ring of large hooks. Infected parts of the fish muscles are loose and moistened. The *G. thyrsitae* larvae occur mainly in the abdominal muscles, but can also be found in the dorsal ones in the entire fillet. Although not infective for man, the larvae seriously affect fish quality, a heavy invasion disqualifying the fish as a marketable product.

Molicola horridus Goodsir, 1841 is a close relative of *G. gigas*; some authors even regard *Molicola* as synonymous with *Gymnorhynchus*. The anterior part of the *Molicola* plerocercoid is vesicular, too, the vesicle containing the scolex; the larval tail is ribbon-like and very long. The larvae are very common and abundant in the liver of *Mola*

mola; very often the entire liver is tightly packed with packed with them. Adults occur in the intestine of *Isurus glaucus*.

Large stout larvae of *Hepatoxylon trichiuri* (synonym: *H. squali*) occur commonly in the Atlantic, Pacific, and Indian Ocean fish. The tapeworms are particularly frequent in the West African shelf and off New Zealand. They occur among the abdominal cavity organs in snoek (*Thyrsites atun*), hairtail (*Lepidopus caudatus*), hakes (*Merluccius* spp.), *Genypterus blacodes*, *Macruronus magellanicus*, and other teleosts. They were also found in such sharks as *Isurus oxyrhynchus*, *Prionace glauca*, and *Squalus acanthias*. The larvae reach 84 mm in length and 6-9 mm in width, the 20-30 mm long larvae being found most commonly. The scolex, 7-9 mm in length, is indistinctly separated from the strobila. Two bothridia are embedded in the scolex rather than sticking out above it and externally form two grooves (dorsal and ventral) with slightly thickened edges. Four short, almost spherical proboscids are covered with strong, 0.18-0.26 mm long hooks.

The larvae, mostly non-encysted, settle in the fish abdominal cavity, their proboscids adhering to the intestinal peritoneum, or are surrounded by connective tissue cysts. They can also be found in the liver, gonads, and muscles. Presumably, *Isurus glaucus* is the principal definitive host of the tapeworm, as adults were also found in other sharks (e.g. *Carcarias glaucus*, *Pristiurus catulus*, *Somniosus microcephalus*, *Lamna cornubica*, and *Squalus acanthias*. The larval *Tetrarhynchidea* are particularly common in commercially caught fish species off the African shelf.

Order: Proteocephalidea

Synonym: Ichthyotaeniidea

Tapeworms in this order have a scolex with 4 circular lateral suckers; in some species the fifth sucker occurs on the top of the scolex. These cestodes are intestinal parasites of fish, reptiles, and amphibians. Similarly to the *Diphylloidea*, the life cycle involves a single intermediate host, usually a copepod crustacean. The entire larval development, from the oncosphere to the fish-invading plerocercoid proceeds in the intermediate host. Most of the *Proteocephalidea* parasitize freshwater fishes. They are highly host-specifc.

Both European and North American eels often contain *Proteocephalus mascrophthalmus*, a rather small (up to 20 cm in length), very delicate and easily disintegrating tapeworm. Its intermediate hosts, as found experimentally by Doby and Jarecka (1966), are freshwater copepods *Cyclops strenuus*, *Cyclops* sp., *C. abyssorum*, *Acanthocyclops vernalis*, and *Diacyclops viridis*. The copepods infect eel during its fresh-water phase of life. A few *prosobothrium* species only were found in sharks of the genera *Squalus*, *Zygaena*, *Carcharias*, and *Prionace* in the Mediterranean Sea and in the Sea of Japan. The *Proteocephalidae* are of no economic importane.

CLASSIFICATION

Phylum platyhelminthes is divided into three classes. The classification followed here is from *Hyman* (1951).

I. Class-Turbellaria

II. Class-Trematoda

III. Class-Cestoda

CLASS - TURBELLARIA

Turbellaria is derived from Latin word turbella means a *stirring*. The animal produce turbulance in water by body cilia. They possess following characters:

1. The members of this class are broad, flat and leaf-like in shape.
2. They are lead a free living life in fresh or salt water, in moist soil, some members may be commensal or parasite.
3. Body unsegmented, dorso-ventrally flatenned and covered with cilia.
4. The anterior end is differentiated into head.
5. Epidermis cellular or syncytial contains rod-like structures called rhabdites. The rhabddites are curved structure capable of discharging a chemical secretion.
6. Epiderrmis and mesenchyma are provided with unicellular glands of two types (a) cyanophilous which secrete adhesive secretion and (b) eosinophilous.
7. Turbellarians are abundently provided with adhesive organs which are of two types (a) glandulo-epidermal adhesive organs and (b) glandulo-muscular adhesive organs.
8. Alimentary canal incomplete i.e. with single opening, the mouth. Mouth is mid-ventral, sometimes near the centre of body. Intestine is blind or absent.
9. Excretory system with protonephridia.
10. Sense organs consist of tangoreceptòrs and chaemoreceptors.
11. Mostly bisexual or hermaphrodite or unisexnal with very few exceptions.
12. Reproduction by asexual, sexual and by regeneration.
13. Cleavage spiral. Development direct without larval stage.

Classification of Turbellaria

The members of this class are found almost all places. It includes more than 1500 species class Turbellaria is divided into 5 orders.

1. Acoela
2. Rhabdocoela
3. Allocoela
4. Tricladida
5. Polycladida

Order 1 - Acoela

1. They are minute, delicate turbellarians with a ventral mouth.
2. Mouth and pharynx are present. Pharynx simple when present but usually absent and so the intestine is also absent.
3. Parenchyma syncytial not differentiated into mesoderm and endoderm.
4. Protonephridia absent.
5. Food is taken in solid form and digested by endodermal cells.
6. Gonoducts, gonads and yolk glands are absent.
7. Exclusively marine forms living under stores algae etc. at the sea bottom. Some live in the intestine of holothurians. Examples: Convoluta, Hoplodiscus

Order 2 - Rhabdocoela

1. Small Tubellarian usually less than 3mm in length.
2. Mostly free-living, marine of fresh water.
   Some of them inhabit in the respiratory cavities of various fresh water crustaceans, viz. *Temnocephala*.
3. Body is elongated, cylindrical or compressed with simple pharynx and simple unbranched sack-like intestine.
4. Yolk-glands may or may not be present.
5. Yolk with two to many testes, whereas the female with 1 or 2 ovaries.
6. Nervous system consists of two longitudinal nerve cords. Eyes are present in most of them.

   Order Rhabdocoela is divided into four sub-orders:

Sub-order (i) Natandropora

1. Exclusively fresh water forms.
2. Pharynx simple.
3. Single median protonephridium.

4. Single compact testis mass, penis unarmed.
5. Yolk-gland is absent.
6. Sexual fission with chains of zooids.

   Examples: *Catenula, Rhycoscolex, Stenostomum*

Sub-order (ii) Opisthandropora

1. Fresh water or marine forms.
2. Pharynx is simple.
3. Excretory system consists of paired protonephridia.
4. Compact testes, penis armed with a stylet.
5. Yolk glands absent.
6. Asexual reproduction with chains of zooids.

   Examples: *Microstomum, Macrostomum*

Sub-order (iii) Lecithophora

1. Fresh water, marine or terrestrial, commensel or parasitic.
2. Pharynx bulbose.
3. Protonephridia paired.
4. Separate ovaries and yolk glands.
5. Exclusively sexual reproduction.

   Examples: *Graffilla, Anaplodium, Merostoma, Gyratrix*

Sub-order (iv) Temnocephalida

1. Freshwater ectocommensal forms.
2. Anterior end having 2-12 tentacles.
3. Posterior end with 1 or 2 adhesive discs.
4. Pharynx dolii form.
5. Gonopore simple.

   Examples: *Temnocephala, Monodiscus*

Order 3-Allocoela

1. Mostly marine, few freshwater and brakish water forms. Some are ectoparisite or ectocommensal in habit.
2. The pharynx is bulb-like or plicate.
3. Intestine is straight or branched.
4. Excretory system consists of paired protonephridia with two or three branches and nephridiopores.
5. Nervous system with 3 or 4 pairs of longitudinal nerve cords provided with transverse commissures.
6. Reproductive system consists of numerous testes and a pair of ovaries. Penis papilla mostly present.

Order Allocoela is divided into four Sub-orders.

Sub-order (i) Archiophora

1. Marine animals.
2. Pharynx plicate.
3. Female reproductive system primitive.

   Examples: *Proporoplana*

Sub-order (ii) Lecithoepitheliata

1. Marine, freshwater or terrestrial forms.
2. Pharynx simple or bulbose.
3. Penis with a cuticular stylet.
4. Female duct simple or none. Typical yolk glands absent, nutritive cells may surround the ova.

   Examples: *Hofstenia, Prorhynchus.*

Sub-order (iii) Cumulata

1. Fresh water or marine forms.
2. Pharynx plicate or bulbose.
3. Intestine mostly without diverticulae.
4. Penis unarmed.
5. Separate ovaries and yolk glands or *germoritellaria*.

Examples: *Hydrolimax, Plagiostomum.*

Sub-order (iv) Seriata

1. Mostly marine and a few are fresh water forms.
2. Pharynx plicate.
3. Intestine with lateral diverticula.
4. Statocysts mainly present.
5. Ovaries and yolk glands are separate.

   Examples: *Monocelis, otoplana*

Order 4 Tricladila

1. Marine, fresh water or terrestrial forms.
2. Large turbellarians, measuring 2-60 cm in length.
3. Pharynx plicate usually backwardly directed.
4. Intestine three-branched sacs.
5. Eyespots usually present.
6. Two ovaries, 2 or numerous testes.
7. Female reproductive system provided with one or two copulatory bursa.

8. Penis with penis papilla, common genital aperture.

Order Tricladida is divided into three suborders.

Sub-order (i) Maricola

1. Exclusively marine forms.
2. A pair of eyes and auricular groores.
3. Typical penis papilla sometimes armed with stylet.
4. Rounded copulatory bursa is generally present.
5. Only sexual reproduction.

Examples: *Bdelloura, Ectoplana*

Sub-order (ii) Paludicola

1. Mostly freshwater rarely brackish water forms.
2. Two to many eyes or eyeless.
3. Bursa with bursal canal, anterior to penis.
4. Reproduction mostly asexual.

Examples: *Dugesia, Kenkia, Dendrocoelum*

Sub-order (ii) Terricola

1. Terrestrial tropical forms.
2. Two or many eyes.
3. Bursa, if present, lies behind the penis.
4. Reprodution may be asexual.

Examples: *Geoplana, Dolicoplana*

Order 5-Polycladida

1. Size modrate, from 2-20 mm. Dorsoventrally flat leaf-like body; they are exclusively marine.
2. Intestine with many branches.
3. Eyes and nerve cords are present.
4. There are numerous ovaries and testes.
5. Genital apertures separate.

It includes two suborders:

Sub-order (i) Acotylea

1. Pharynx usually vertical curtain like.
2. No suckers are present behind the female pore.
3. Tentacles nuchal type.
4. Eyes concentrated at anterior margin, not in pairs.

Examples: *Shylochus, Notoplana*

Sub-order (ii) Cotylea

1. Sucker present behind the female pore.
2. Pharynxtubular.
3. A pair of marginal tentacles bearing eyes or with a cluster of eyes in place of tentacles.

   Examples: *Thysanozoon, Yungia*

CLASS-II TREMATODA

1. They are ecto or endoparasites and commonly called 'flukes.'
2. Body is usually unsegmented, dorso-ventrally flatenned and leaflike.
3. Size varies from 0.5 mm to 1 or several cm. long.
4. Body is covered over by thick cuticle and no cilia.
5. Suckers and sometimes hooks are present. Oral sucker present at anterior end for feeding ventral sucker for attachment.
6. Epidermis absent. Rhabdites are absent.
7. Digestive system consists of mouth, muscular pharynx and intestine usually with two main branches. Anus is absent.
8. In some cases there is a system of parenchymal vessels showing the presence of primitive circulatory system.
9. Excretory organs consists of protonephridia.
10. There are three pairs of longitudinal nerve cords. Sense organs are poorly developed.
11. Bisexual or hermaphrodite except blood fluke (*Schistosoma*).
12. Single ovary and many testes. Cross fertilization takes place. Development direct or indirect.
13. Life-cycle simple or complicated with more than one hosts.

There are about 6000 species in class Trematoda. This class is divided into three orders:

Order 1-Monogenea

Order 2-Aspidobothrea

Order 3-Digenea

Order 1–Monogenea

1. They are ectoparasite mostly in cold-blooded animals (fishes).
2. The life-cycle is simple and passed in one host only, monogenetic.
3. Oral sucker absent, when present is poorly developed.
4. A pair of adhesive structures are present at anterior end.
5. A large disc-shaped structure, the opisthaptor, is found on the posterior end. It contains hooks and is lined with chitin.

6. Paired excrecory pores are present in the anterior part of the body.
7. Male and female genital pores are separate.
8. One or two vaginae are present. The uterus is small and contains few eggs.
9. Development is indirect through a larval stage called onchomiracidium.

Order Monogenea is divided in two suborders:

Sub-order (i) Monopisthocotylea

1. Opisthaptors are present without suckers but with hooks.
2. Genito-intestinal tract is absent.
3. Posterior adhesive organ is single or divided by septa.
4. Eye-spots usually present.

Examples: *Gyrodactylus, Monocotyle, Heterocotyle, Amphibdella*

Sub-order (ii) Polyopisthocotyle

1. Opisthaptor contains many suckers and also hooks.
2. Oral suckers two, opening into oral cavity.
3. Genito-intestinal tract present.
4. Eye-spots are absent.

Examples: *Polystoma, Diplozoon, Microcotyle.*

Order 2–Aspidobothrea

1. Endoparasite in the digestive tract of fishes and reptiles.
2. Oral sucker is absent.
3. The ventral sucker is divided by septa into many small parts. The hooks are absent in it.
4. Single excretory pore.
5. Only one testis is present.
6. Development direct, without alternation of hosts.

Order Aspidobothrea is divided into two families:

Family (a) Aspidogastridae

1. Sucker enormous circular, oval or elongated, divided by septa into one, three or four longitudinal rows of depressions termed alveloli.
2. Protonepliridia have separate bladders and separate or a common nephridiopore situated near the posterior end.
3. Testis single.
4. Laurer's canal may be present.

5. Endoparasite in the mantle, pericardial and renal cavities of clams and snail and in digestive system of fish and turtle.
   Examples: *Aspidogaster, Cotylaspis, Cotylogaster*

Family (b) Stichocotylidae

1. It consists of only one genus which lives in bile passage or spiral valve of fishes.
2. The life cycle is complex one, involving at least two hosts.
   Example: *Stichocotyle.*

Order 3–Digenea

1. They are endoparasite in invertebrate and verte-brates.
2. Two to four hosts in the life-cycle.
3. Two suckers, without hooks, mouth within angerior sucker.
4. There is only one excretory pore present in the posterior region of the body.
5. There is a common gonopore for male and female reproductive systems. Vagina is absent.
6. Uterus is long branched with many eggs.
7. Their larval forms reproduce asexually before metamorphosis.

Order Digenea is divided into many families; some important ones are as:

Family (a) Bucephalidae

1. They are commonly called gastrostomes. The mouth is located in the middle of the body.
2. Anterior end bears a simple or lobate oral sucker or has a generally lobed hood-like extension.
3. Intestine sacciform.
4. Acetabulum absent.
5. Genital pore posterior.
6. The cercaria larvae known as oxhead, have a pair of long tails.
7. The primary host is fish and secondary host is clams.
   Examples: *Bucephalus, Bucephalopsis.*

Family (b) Paramphistomidae

1. Body thick and fleshy with a powerful posterior sucker.
2. Cirrus pouch is present.
3. Oral sucker is absent.
4. Gonopore, situated in the anterior body half and provided with in some genera with a genital sucker.

5. Nephridiopore dorsal infront of acetabulum.
6. Uterus long having ascending limb. Vitelline glands follicular.
7. These are the endoparasites of gut and bile ducts.
   Examples: *Paramphistomum, Gastrothylax*

Family (c) Opisthorchidae

1. These are distomes of small to large size.
2. They live in the bile passage of fish-eating reptiles, birds and mamnials.
3. Body flat and semitransparent.
4. Excretory bladder 'V' shaped.
5. Testes posterior Genital pore infront of acetabulum.
   Examples: *Opisthorchis, Amphimerus, Metorchis*

Family (d) Fasciolidae

1. Commonly known as liver flukes.
2. Body leaf-like and acetabulum near the oral sucker.
3. Gonopore just infront of the acetabulum.
4. Testes, ovary, vitelline glands all are branched.
   Examples: *Fasciola hepatica, Fasciplopsis buski*

Family (e) Echinostomatidae

1. Body slender, elongated with spiny or scaly cuticle.
2. Acetabulum large behind the oral sucker.
3. Collar of spines is present in cercarial and metacercarial stages.
4. It is commonly found in the intestine of birds.
   Examples: *Echinostoma, Echinoparyphium, Parochis, Himasthla*

Family (f) Schistosomatidae

1. These are blood flukes that inhabit the hepattic portal system and pelivc veins of birds and mammals.
2. Sexes are separate.
3. Female lodges in a gynaecophoric canal of male.
4. Testes numerous and ovary tubular.
   Example: *Schistosoma*

Besides above families the others are:

Family - Zoognidae. Examples: *Zoogonus, Nassa.*
Family - Allocreadidae. Examples: *Bunodera, Plagioporus.*
Family - Azygiidae. Examples: *Azygia, Otodistomum.*
Family - Hemiuridae. Examples: *Halipegus, Derogenes.*

Family - Didymozoonide. Examples: *Didymocystis*
Family - Gorgoderidae. Examples: *Gorgoderina, Gorgodera*
Family - Plagiorchidae. Examples: *Plagiorchis, Styphlodora.*
Family - Dicrocoelidae. Examples: *Dircrocoelium, Eurytrema.*
Family - Heterophyidae. Examples: *Stictodera, Heterophyes.*
Family - Microphallidae. Examples: *Spelotrema, Maritrema.*
Family - Troglotrematidae. Examples: *Paragonimus, Troglotrema.*
Family - Cyclocoelidae. Examples: *Cyclocoelus, Typhlocoelum.*
Family - Clinostomatidae. Example: *Clinostomum.*
Family - Notocotylidae. Examples: *Notocotylus, Catatropis.*
Family - Strigeidae. Examples: *Ophiosoma, Cyanthocotyle*
Family - Brachylaimidae. Examples: *Brachylaima.*
Family - Sanguinicolidae. Example: *Sanguinicola.*
Family - Spirorchidae. Examples: *Spirorchis, Hapalorhynchus*
Family - Heronimidae. Example: *Heronimus.*

CLASS-III CESTODA

1. They are found as endoparasites in the intestine of vertebrates and are called tapeworms.
2. Body is flat, elongated and ribbon-like, measures from 1 mm to 10 metres in length.
3. Body is without epidermis, rhabdites, cilia but a cuticle is present.
4. Body is usually divided into a scolex or head, neck and few to many proglottids.
5. Scolex is provided with hooks or suckers or both.
6. Scolex is followed by neck which forms new proglottids by transverse budding called strobilization.
7. Mouth and digestive tract are absent.
8. Excretory system consists of protonephridia with flame - cells.
9. Nervous system consists of ganglionated nerve ring in scolex and two pairs of lateral nerve cords in proglottids.
10. Each proglottid contains one or two sets of male and female reproductive organs. Thus they are hermaphrodite.
11. Several testes in the form of follicles and single-lobed ovary.
12. Uterus presents a great variety of structures.
13. Fertilization internal and self.
14. Life-cycle is complicated with hooked embryo, and is passed in more than one host.

Class Cestoda has about 3400 species. The class is divided into two sub-classes:

Sub-class (A) Cestodaria or Monozoa

Sub-class (B) Eucestoda or Merozoa

Sub-class (A) Cestodaria

1. Endoparasites in the intestine and coelomic cavities of fishes and reptiles.
2. Body is formed by one segment and without a scolex.
3. No alimentary canal.
4. Single set of reproductive organs.
5. Larva with ten hooks, known as Pycophore.

Sub-class Cestodaria is divided into two orders.

Order 1 - Amphilinida

1. They are endoparasites in the body cavity of ganoid fishes.
2. Frontal gland and protrusible pharynx are present in the anterior part of the body.
3. Suckers are absent.
4. A proboscis is present.
5. Male and female pores present in the posterior part of body.
6. Uterus is coiled and opens near the anterior end.

Example: *Amphilina*

Order 2 -Gyrocotylidea

1. They are endoparasites of chimaeroid fishes.
2. Anterior end has cup-like sucker.
3. Eversible proboscis at anterior end.
4. Male and female pores are situated at the anterior half of the body.
5. Uterus not coiled.

Example: *Gyrocotylus.*

Sub-class (B) Eucestoda

1. Elongated, flat body is divided into many proglottids.
2. Body is divisible into scolex, neck and strobila.
3. Scolex with suckers, spines and hooks for attachment to host.
4. Each proglottid with male and female reproductive organs.
5. Larva with six hooks.

Sub-class Eucestoda is divided into following orders.

Order 1 - Tetraphyllidea

1. Endoparasite, exclusively in the intestine of elasmobranch fishes.
2. Scolex provided with four bothria (sessile suckers) often with hooks.
3. Testes lie infront of ovaries.
4. Cirrus armed with hairs, spines or hooks.
5. Common genital atrium marginal.

   Examples: *Phyllobothrium, Acanthobothrium.*

Order 2 - Lecanicephaloidea

1. Intestinal parasites of elasmobranch fishes.
2. Scolex divided by a transverse groove in two parts upper with disc and lower with four suckers.
3. Vitellaria in two lateral bands.

   Examples: *Lecanicephalum, Polypocephalus*

Order 3 - Proteocephalidea

1. These are endoparasites of freshwater fishes, amphibians and reptiles.
2. Scolex mobile with four lateral sucker, an apical sucker in some.
3. Ovary bilobed, Uterus with many branches and pores.
4. Vitellaria scattered.
5. Common genital atrium marginal.

   Examples: *Proterocephalus, Ophistaenia*

Order 4 - Diphyllidea

1. Parasitic in the intestine of elasmobranch fishes.
2. Proglottids only 20 or less.
3. Scolex with two bothria and a spiny head stalk.
4. Vitellaria scattered.

   Example: *Echinobothrium*

Order 5 - Trypanorhyncha

1. Parasitic in the spiral valve of digestive tract of elasmobranch fishes.
2. Strobila is segmented and is of modrate size.
3. Scolex with two or four sessile bothria and four protrusible spiny proboscis.
4. Testes extend beyond ovary posteriorly.

   Examples: *Haplobothrium, Tetrarhynchus.*

Order 6 - Pseudophyllidea

1. Endoparasites in the intestine of fishes, birds and mammals.
2. The larva lives in crustaceans.
3. Generally six shallow groove-like suckers and hookless bothrior are found on the scolex.
4. Proglottids may be absent or present.
5. Neck is not developed.
6. Many yolk glands are present.

   Examples: *Amphicotyle, Dibothriocephalus*

Order 7 - Nippotaeniidea

1. Endoparasite in the intestine of Japanese fresh water fishes.
2. Scolex with single terminal sucker.
3. Proglottids few with, each with one set of genitalia.
4. Vitellaria compect.

   Examples: *Nippotaenia, Amurotaenia*

Order 8 - Taenioidea

1. Endoparasites in the intestine of birds and mammals.
2. Elongated body, with proglottids.
3. Scolex bears four suckers often with an apical rostellum armed with hooks.
4. Excretory system with four longitudinal vessels and flame cells.
5. Complete sets of reproductive organs in each proglottids. Single compact yolk gland.

   Examples: *Taenia, Echinococcus.*

Order 9 - Aporidea

1. It is endoparasite of swan (birds).
2. Scolex with four suckers, rostellum armed.
3. No segmentation, ootype, yolk glands and sex-ducts or pores.

   Examples: *Nematoparataenia, Gastrotaenia*

POLYCHOERUS

Phylum	—	Platyhelminthes
Class	—	Turbellaria
Order	—	Acoela
Genus	—	*Polychoerus*

It is found in littoral zone of ocean shores in excessive winter. It is bringht orange or orange red in colour. Body flat, oval, changeable shape. The anterior end is round while posterior end is not ched bearing tail filament. Entire body is covered with cilia. Statocyst is present in median like. Digestive tract is absent. Definite gonads are wanting. Yolk glands are absent. Definite oviduct absent. The seminal bursa opens ventrally by female gonopore. Penis is present.

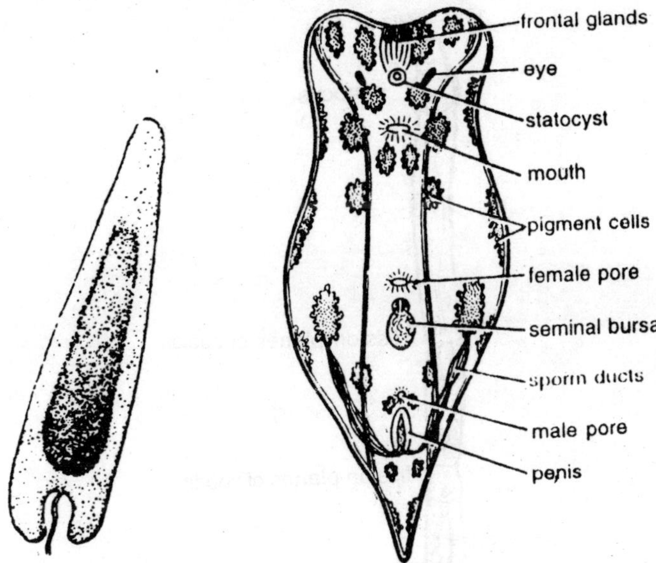

Fig. 2.18. Polychoerus. *Fig. 2.19. Convoluta.*

CONVOLUTA

Phylum	—	Platyhelminthes
Class	—	Turrbellaria
Order	—	Acoela
Genus	—	*Convoluta*

Convoluta is found in sea water. Living under stones, among algae and most commonly on the bottom mud. Small worm with side of the body are held curved ventrally without any projections. But the body has pigment cells. The anterior of cephalic end has two eyes, a statocyst and cluster of frontal glands. The mouth is located near the anterior half of the ventral surface. A digestive cavity is lacking. The excretory system is completely absent. Hermaphrodite but protandry is common. The eggs are enclosed in a thin gelatinous capsule and are struck to some objects. The eggs are developed directly into young worms that escape the jelly.

Centula

Phylum	—	Platyhelminthes
Class	—	Turbellaria
Order	—	Rhabdocoela
Suborder	—	Notandropora
Genus	—	*Centula*

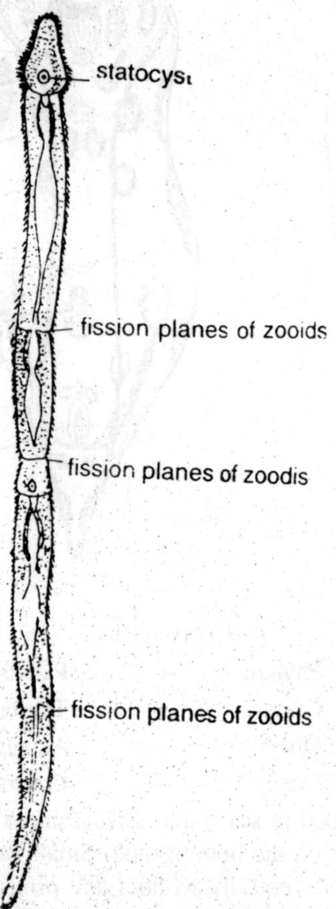

Fig. 2.20. Centula.

It is a freshwater Turbellaria. The pharynx is simple. Single median protonephridium. Nervous system consists of four pairs of longitudinal nerve cords and a statocyst. Asexual reproduction by fission with chains of zooids. No sexual reproduction.

MICROSTOMUM

Phylum	—	Platyhelminthes
Class	—	Turbellaria
Order	—	Rhabdocoela
Genus	—	*Microstomum*

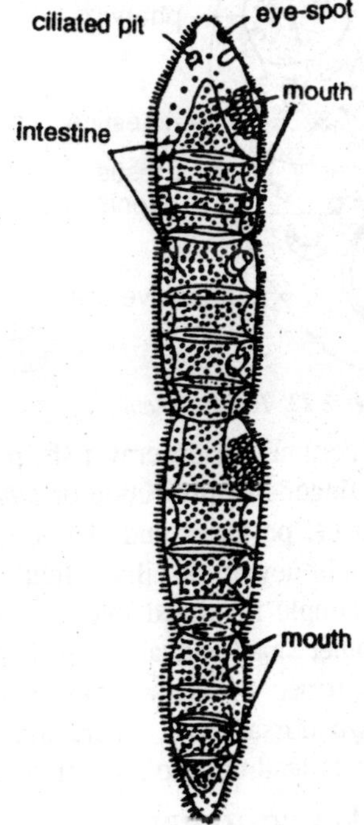

Fig. 2.21. Microstomum.

It is a microscopic, freshwater form. It swims by means of cilia present on the body surface. Its epidermis contains nematocysts derived from hydroid food. Its is found that Microstomum attack Hydra only when it need of nematocysts. The digestive cavity is simple, unbranched and straight. It reproduce asexually by repeated transverse fission forming a linear chain of 4, 8 or 16 zooids. Each zooid separates to become an individual.

TEMNOCEPHALA

Phylum	—	Platyhelminthes
Class	—	Turbellaria
Order	—	Rhabdocoela
Genus	—	*Temnocephala*

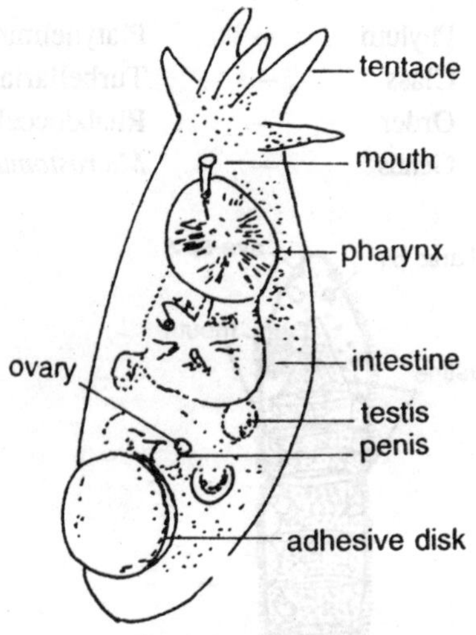

Fig. 2.22. Temnocephala.

It is found as an ecto commensal on cray fish, prawns and other crustaceans. It contains 12 finger-like projection or *tentaces* at anterior end and an adhesive disc at the posterior end. The border of syncytial epidermis is apparently cuticle devoid of cilia. Mouth anterio-ventral. Pharynx thick walled and simple lobulated intestine present. It feeds upon diatoms, rotifers and other small animals. A pair of nephridiopores is present on the dorsal surface of the anterior part of the body. Cerebral ganglion bears two dorsal eyes. There are four testes and single ovary. Their gonoducts leading into a common genital atrium.

Plagiostomum

Phylum	—	Platyhelminthes
Class	—	Turbellaria
Order	—	Allocoela
Genus	—	Plasgiostomum

It is present in both fresh water and marine. Mouth is anterior. Pharynx is bulbose. Intestine with short diverticula. Gonopore posterior. Sperm is characteristic in having witng-like lateral expansions.

Bipalium

Phylum	—	Platyhelminthes
Class	—	Turbellaria
Order	—	Tricladida
Genus	—	Bipalium

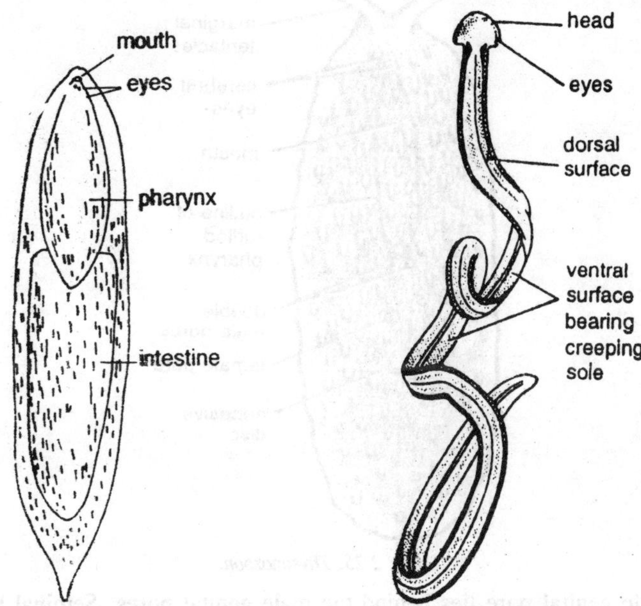

Fig. 2.23. Plagiostomum. *Fig. 2.24. Bipalium.*

Bipalium is a long terrestrial turbellarian living the humid soil on the floor of tropical forests. It is cosmopolitan in distribution. The animal measures 20-50 cm in length. It consists of an expanded lunate head and cylindrical long body. Numerous eyes are present on the margin of the head and sides of body. There are stripes on the dorsal surface and a creeping sole on the ventral surface. The stripes are purple, black, yellow, olive and grey in colour. Reproduction generally asexual. It never becomes sexual in temperate climate and it propogates b fragmentation. Bipalium adventium breeds sexually.

THYSANOZOON

Phylum	—	Platyhelminthes
Class	—	Turbellaria
Order	—	Polycladida
Genus	—	Thysanozoon

It is found in a less cold water of sea. It has an oval body with dorsal surface beset with papillae, each receiving a branch form intestine. Anterior end bears a pair of marginal tentacles and numerous cerebral eyes. Pharynx is tubular. Glandulo-muscular adhesive organ is present on the dorsal surface behind the female gonopores. A single

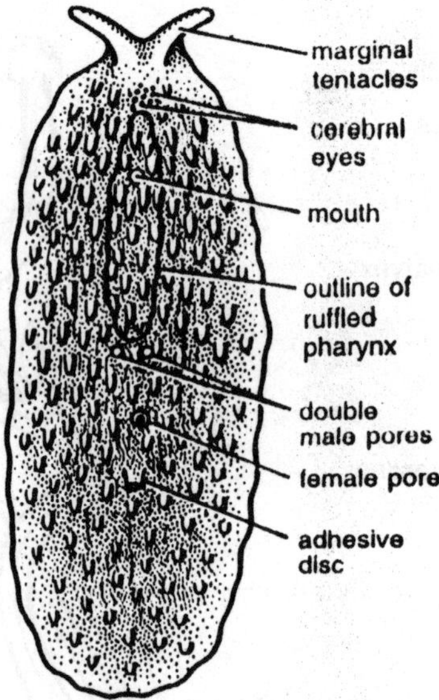

Fig. 2.25. Thysanozoon.

female genital pore lies behind the male genital pores. Seminal bursa is absent.

GUNDA

Phylum	—	Platyhelminthes
Class	—	Turbellaria
Order	—	Tricladida
Genus	—	*Gunda (Procerodes)*

It is a marine animal. The auricle and a pair of eyes or ocelli are present in the anterior end. Intestine has regularly arranged intestinal diverticulae. The gonads alternate with the diverticulae giving the appearance of segmentation. Incapable of sexual reproduction. *Lang* used the feature of intestine as a classical example in support the gonocoel theory of coelom formation by Haeckel.

APIDOGASTER

Phylum	—	Platyhelminthes
Class	—	Trematoda
Order	—	Monogenea
Genus	—	*Apidogaster*

It is an endoparasite in the pericardial and renal cavities of fresh water mussel and in the gut of fishes and turtles. Body is elongated

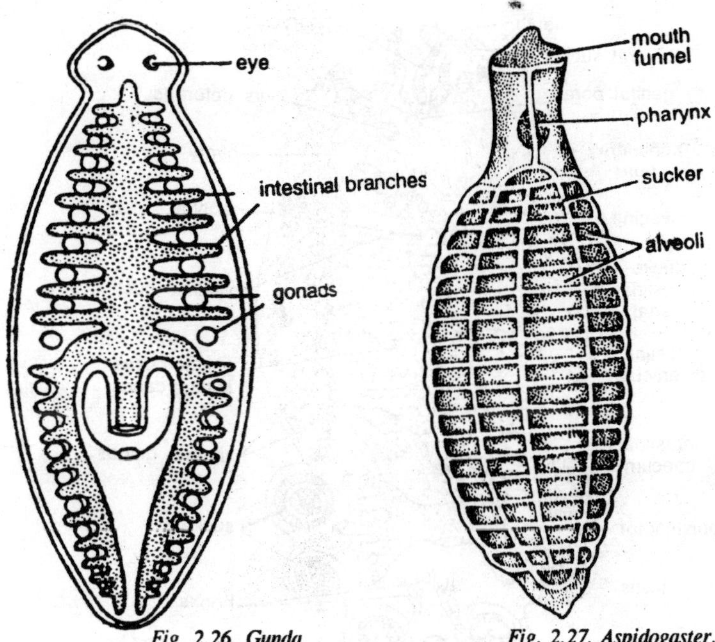

Fig. 2.26. Gunda. *Fig. 2.27. Aspidogaster.*

and dorsoventrally flatenned with anterior narrow, end. The narrow anterior end has a subterminal mouth which is devoid of oral sucker. A large sucker appears which occupies great space in the posterior region of body. The sucker is sub-divided into four longitudinal rows of sucking cups or alveoli. Gut is simple and straight. Excretory system consists of protonephridia with excretory bladders. Bisexual life-cycle simple without change of host.

POLYSTOMUM

Phylum	—	Platyhelminthes
Class	—	Monogenea
Order	—	Monogenea
Genusm	—	*Polystomum*

It is an endoparasite in the urinary bladdar of frogs, and turtles. Body is leaf like and dorso-ventally flattened. The posterior region has a disc-shaped opisthaptor with six cuplike suckers and 2 to 4 large chitinous hooks. Mouth is surrounded by an oral sucker. Intestine is bifurcated. Hermaphrodite with single ovary and numerous tests. Breeding season starts in spring. Eggs are produced in the urinary bladder of frong and discharged into the water along with the urine of

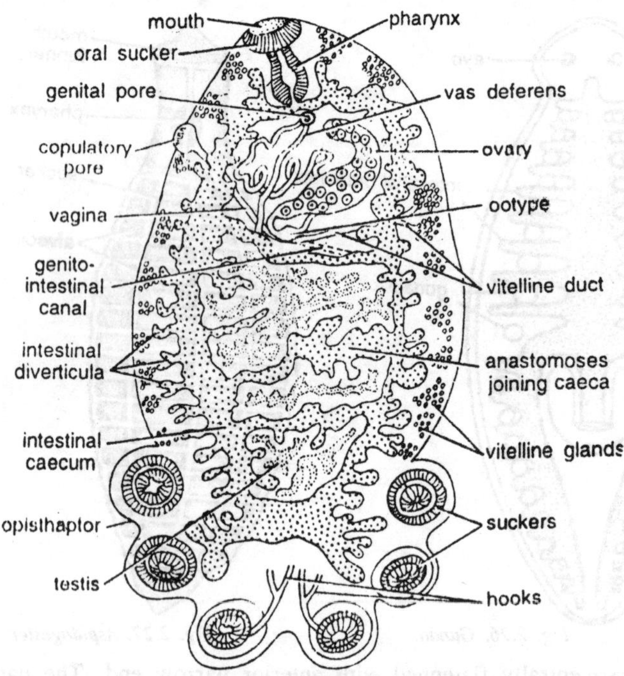

Fig. 2.28. Polystomum.

the host. Hatching takes place after 4 to 5 weeks and larva attaches to the gills of tadpole and later develop into adult in the alimentary canal of metamor-phosed tadpole.

Diplozoon

Phylum	—	Platyhelminthes
Class	—	Trematoda
Order	—	Monogenea
Suborder	—	Polyopisthocotyla
Genus	—	*Diplozoon*

Diplozoon is an ectoparasite, found attached to the gills of freshwater fishes, feeding on their blood. Two individuals are always found attached permanently together in the form of 'X'. Opisthaptor bears eight suckers arranged in two rows of four each. Digestive system consists of mouth surrounded by oral sucker, pharynx and intestine. Intestine is not bifurcated. Reproductive system consists of testis and ovary along with their usual ducts. The sperm duct, uterus and vagina of one individual cross with those or other and thus ensures cross fertilization.

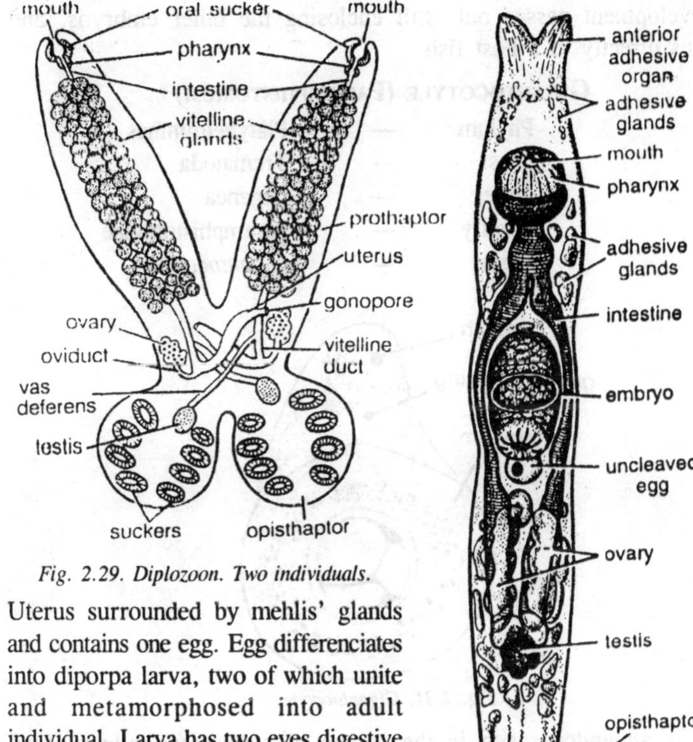

Fig. 2.29. Diplozoon. Two individuals.

Uterus surrounded by mehlis' glands and contains one egg. Egg differenciates into diporpa larva, two of which unite and metamorphosed into adult individual. Larva has two eyes digestive system & a pair of suckers.

GYRODACTYLUS

Phylum — Platyhelminthes
Class — Trematoda
Order — Monogenea
Suborder — Monopisthocotyla
Genus — *Gyrodactylus*

Fig. 2.30. Gyrodactylus.

It is an ectoparasite on the skin and gills of freshwater fishes. Body is minute and elongate. Anterior end is provided with adhesive glands and organs. Oral sucker is absent. Eyes are absent. Opisthaptor is disc or wedge-shaped and bears one pair of anchors and eight pairs of hooklets. Intestine is sac-like forked into two branches without diverticula. Genito-intestinal canal is absent. Genital pore is median. Development is intra-uterine *i.e.*, viviparous. A remarkable feature is that a second embryo is produced within the first one, a third within the second and a fourth within the third. The first embryo, on completing its development passes out, still enclosing the other embryos, and attaches directly to a host fish.

GIGANTOCOTYLE (PARAMPHISTOMUM)

Phylum	—	Platyhelminthes
Class	—	Trematoda
Order	—	Digenea
Family	—	Paramphistomidae
Genus	—	*Gigantocotyle*

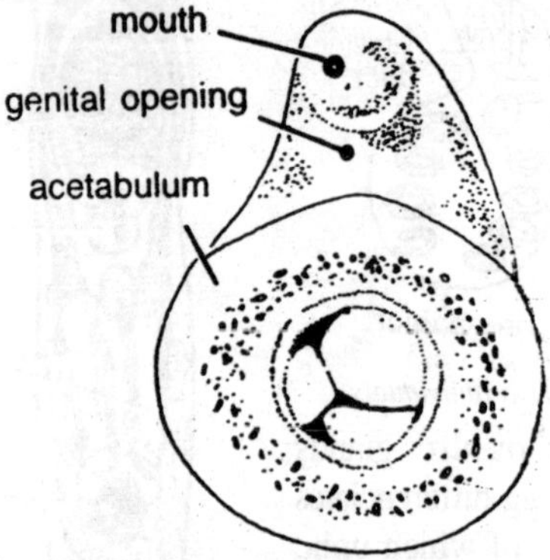

Fig. 2.31. Gigantocotyle.

It is an endoparasite in the rumen of cattles and bile-duct of sheep, goat etc. Body is dorsoventrally flatenned and measuring 12-14 mm in length. Mouth lics at the anterior end and without an oral sucker. Acetabulum is very prominant and lies at the posterior end. Digestive system consists of mouth, highly musculr pharynx and bifurcated, unbranched intestine. Parenchyma has a sort of delicate tubes forming much branched lymphatic system. Testes are lobed and occur near the middle of the body. Ovary is small, lobed and lies posterior to testes. Vitellaria are scattered on the sides of animal. Genital pore is just behind the bifurcation of intestine. Uterus is slightly folded and eggs are very large.

GASTRODISCOIDES HOMINIS

Phylum	—	Platyhelminthes
Class	—	Trematoda
Order	—	Digenea
Family	—	Paramphistomidae
Genus	—	*Gastrodiscoides*
Species	—	*hominis*

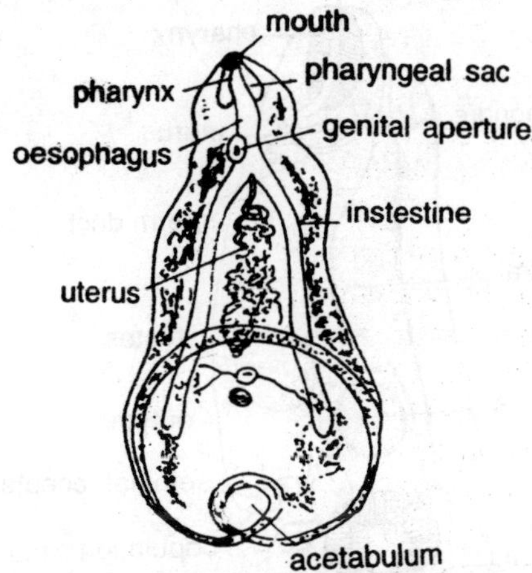

Fig. 2.32. Gastrodiscoides.

It is an endoparasite in the colon and caecum of man. If is found in India and Indo-China. It measures 5-7 mm in length. Body is differentiated into anterior and posterior part. Terminal mouth is present an anterior part. Oral sucker is absent. Posterior part bears a prominant acetabulum on ventral side. Gonopore is in the anterior half of the body on ventral side. Excretory pore lies on the dorsal side infront of acetabulum. The secondary host is snail. It causes inflammation and diarrhoea.

Gastrothylax

Phylum	—	Platyhelminthes
Class	—	Trematoda
Order	—	Digenea
Family	—	Paramphistomidae
Genus	—	*Gastrothylax*

It is an endoparasite in the rumen of cattles. It is elongated animal possessing a deep ventral pouch which extends upto the ventral sucker and it is supposed to be elongated genital chamber. Digestive system includes mouth, pharynx, oesophegus and simple non-diverticulated intestine. Two rounded testes near the base of the pouch. Female reproductive organs consist of single ovary, uterus, Ootype, vagina and seminal recepticle. Single excretory pore is present. Posterior end is divided with large acetabulum or adhesive sucker.

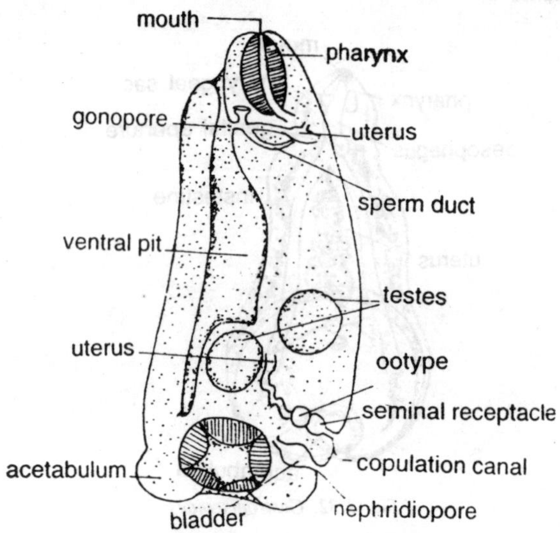

Fig. 2.33. Gastrothylax.

FASCIOLOPSIS BUSKI

Phylum	—	Platyhelminthes
Class	—	Trematoda
Order	—	Digenea
Family	—	Fasciolidae
Genus	—	*Fasciolopsis*
Species	—	*buski*

It is a large intestinal fluke of man, dogs, pigs etc. It is found in India and China. It measures about 70 mm in length and 20 mm in width. Oral sucker and acetabulum both are well developed. Mouth surrounded by oral sucker leads into small pharynx. Oesophegus is short and leads into non-diverticulated caeca. Gonads are present in posterior region of body. Intermediate host is snail. Cercaria larval encyst on tubers of water caltrop or singhara. The singhara is eaten raw and thus infection occurs. It causes inflammation and hemorrhage of the intestine.

PARAGONIMUS WESTERMANNI

Phylum	—	Platyhelminthes
Class	—	Trematoda
Order	—	Digenea
Genus	—	*Paragonimus*
Species	—	*westermanni*

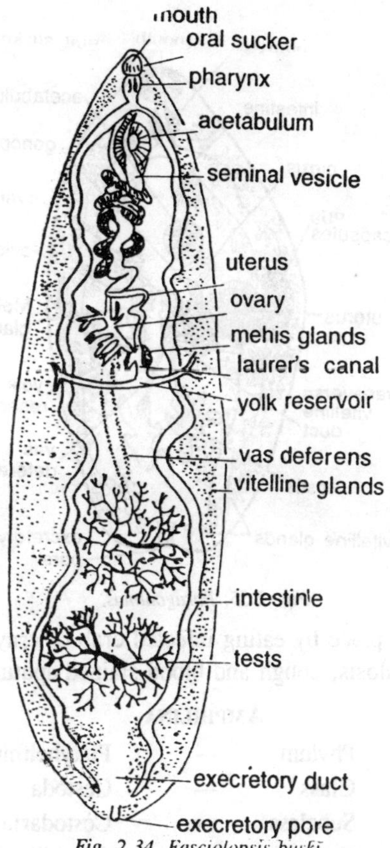

Fig. 2.34. Fasciolopsis buski.

Paragonimus westermanni or Lung fluke is parasite in the lungs of mammal including man.' The body of the fluke is oval with well developed oral sucker surrounding the mouth and a ventral sucker in the centre of the body. The internal organs are very much like that of Fasciola hepatica. The cyst like pockets are formed around the parasite which rupture and liberate the eggs into bronchial tubes to be executed with sputum. The cyst likc pockets are formed around the parasite which rupture and liberate the eggs into bronchial tubes to be executed with sputum. The miracidia live only few hours after hatching. The cercariae are long, have small knob like tail and spiny cuticle. They do not swim but creep in leech-like manner. They pierce the cuticle of crayfish where they become encysted inside the tissues and gradually develop into mature infective metacercariae. The infection of the new

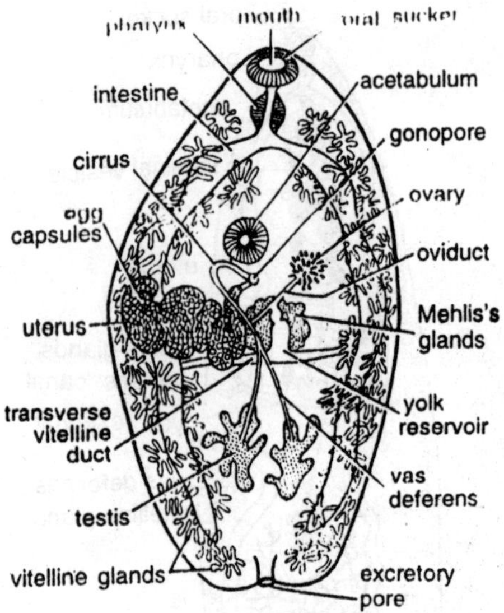

Fig. 2.35. Paragonimus.

host usually takes place by eating infected crab or crayfish. The cause or help in tuberculosis, cough and blood stained sputum.

AMPHILINA

Phylum	—	Platyhelminthes
Class	—	Cestoda
Subclass	—	Cestodaria
Order	—	Amphilinida
Genus	—	*Amphilina*

It is an endoparasite found in the coelom of fish, Acipenser. Body flat and leaf-like. Scolex is absent. Anterior end is provided with weakly developed protrusible proboscis. Testes are scattered and cirrus is armed. Vagina lies behind the ovary and opens on the left side. Uterus is coiled. The intermediate host is a crustacean, Gammarus, which is eaten by Acipenser.

GYROCOTYLE

Phylum	—	Platyhelminthes
Class	—	Cestoda
Subclass	—	Cestodaria
Order	—	Gyrocotylidea
Genus	—	*Gyrocotyle*

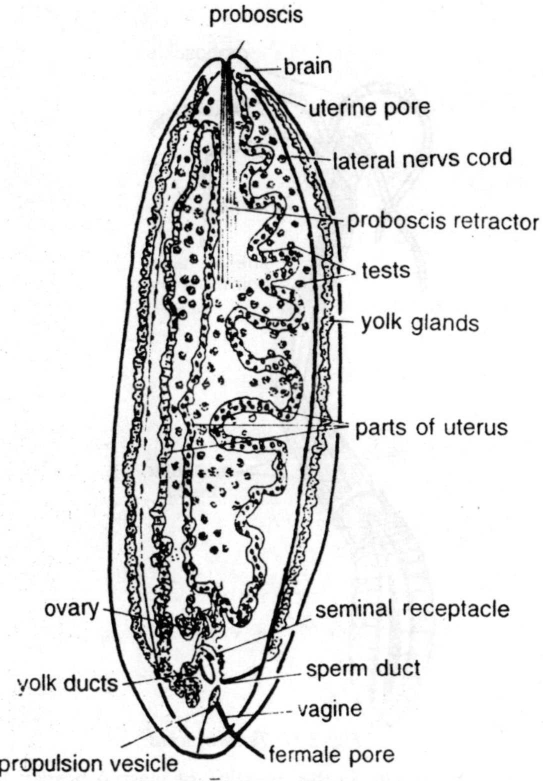

Fig. 2.36. Amphilina

These are endoparasite in the intestine of chimaeroid fishes. Body is elongated and dorsoventrally flatenned. Anterior end bears a large opening that leads into a highly muscular protrusible mass, the proboscis. Posterior end bears a rosette-shaped adhesive organ, surrounding a funnel shaped depression. The margins of body are generally ruffled. Bisexual Male reproductive system comprise scattered testes, sperm duct and penis papilla. Female reproductive system comprises ovary, oviduct and uterus. Follicles of yolk gland scattered throughout.

TETRARHYNCHUS

Phylum	—	Platyhelminthes
Class	—	Cestoda
Subclass	—	Eucestoda
Order	—	Trypanorhyncha
Genus	—	*Tetrarhynchus*

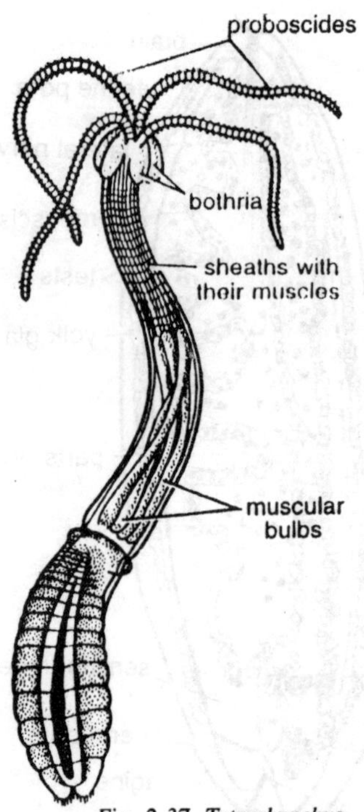

Fig. 2.37. Tetrarhynchus.

It is an endoparasite in the intestine of elasmo-branch fishes. Body is long divisible into scolex and proglottids. Scolex is long and differentiated into a long proximal part containing the proboscis apparatus and a distal part bearing the bothria or sucker. Scolexbears four suckers each with a eversible proboscide armed with spines. The proboscides are retractile in muscular sheaths which end in muscular bulbs. Each proglottid contains a complete set of reproductive organs.

DIPHYLLOBOTHRIUM (DIBOTHRIOCEPHALUS)

Phylum	—	Platyhelminthes
Class	—	Cestoda
Subclass	—	Eucestoda
Order	—	Pseudophyllidea
Genus	—	*Diphyllobothrium*
Species	—	*latum*

It is found in the intestines of man. It is world wide in distribution. It is the largest and most pathogenic cestoda of human beings,

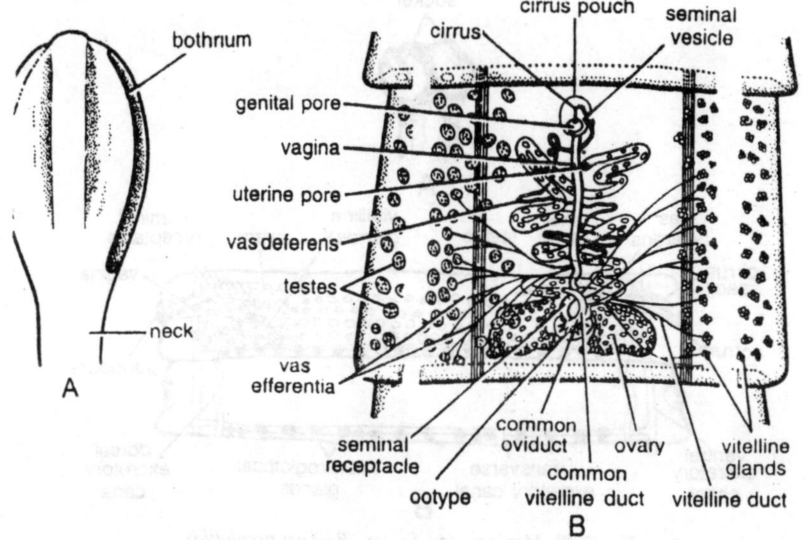

Fig. 2.38. Diphyllobothrium. A—Scolex; B—A mature proglottid.

measuring 20 metres in length. Body is differentiated into scolex, neck and proglottids. The number of proglottids is 3000-4000. Scolex is fusiform and bears two slit-like suckers or bothria. Mature proglottid is broader than long. Each proglottid is hermaphrodite and contains complete sets of male and female reproductive organs. Testes are numerous. Vasa efferentia unite to form coiled sperm duct. Ovary is trilobed. Vagina runs posteriorly forming seminal vesicle. Uterus very much twisted. Gonopores ventral. Life-cycle involves two intermediate hosts, one is *cyclops* (a copipod) and other a fish. It causes bothriocephalus anaemia, erythropenia and haemorrhage.

MONIEZIA

Phylum	—	Platyhelminthes
Class	—	Cestoda
Subclass	—	Eucestoda
Order	—	Taenioidea
Genus	—	*Moniezia*
Species	—	*expansa*

It is found in sheep, cattles and other ruminants. Body is divisible into scolex, neck and proglottids. The scolex is small with prominant suckers. The rostellum and hooks are absent. Each proglottid is broader than long and contains a double set of reproductive organs. Genital

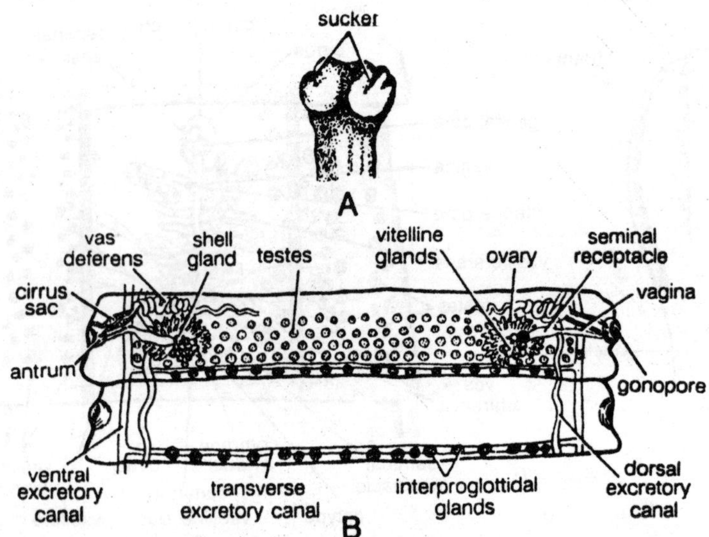

Fig. 2.39. Moniezia. A—Scolex. B—Two proglottids.

ducts are dorsal to osmoretgulatory canals. Testes are numerous scattered throughout the centre of proglottid. Cirrus in cirrus pouch. Ovaries are in the form of an open fan. Uteri form networks. On one side vagina is dorsal to cirrus pouch and on other ventral. The intermediate host is mite (*Galumna*). It is not so pathogenic but sometimes may cause diarrhoea or anaemia.

Hymenolepls Nana

Phylum	—	Platyhelminthes
Class	—	Cestoda
Subclass	—	Eucestoda
Order	—	Taenoidea
Genus	—	*Hymenolepis*
Species	—	*nana*

It is commonly called the dwarf tapeworm of man. It lives in the intestine. It is world-wide in distribution. It measures 25-40 mm in lengths and has about 100-200 proglottids. The rostellum is well-developed and retractile. It bears a single circlets of 20-30 hooks. Neck is long and slender. There are three testes with varying arrangement. Cirrus pouch is large with external and internal seminal vesicle and with an accessory sac. Life-history is simple without any intermediate host. The inner membrane of the hexacanth bears long,

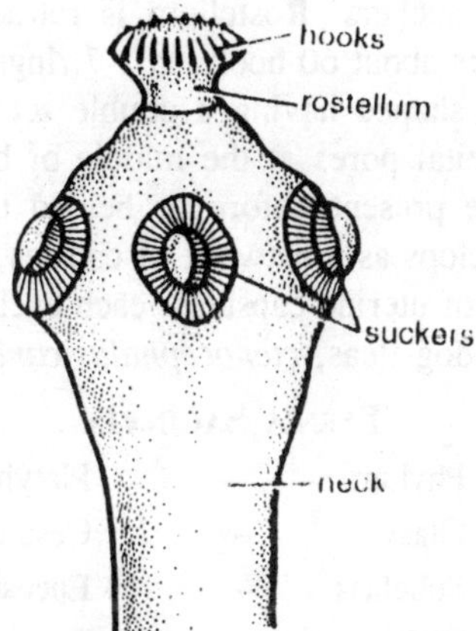

Fig. 2.40. Scolex of Hymenolepsis nana.

wavy filaments. Larvae develop in the intestinal villi while the adult lives in the intestine. It causes abdominal pain and diarrhoea.

DIPYLIDIUM CANINUM

Phylum	—	Platyhelminthes
Class	—	Cestoda
Subclass	—	Eucestoda
Order	—	Taenioidea
Genus	—	*Dipylidium*
Species	—	*caninum*

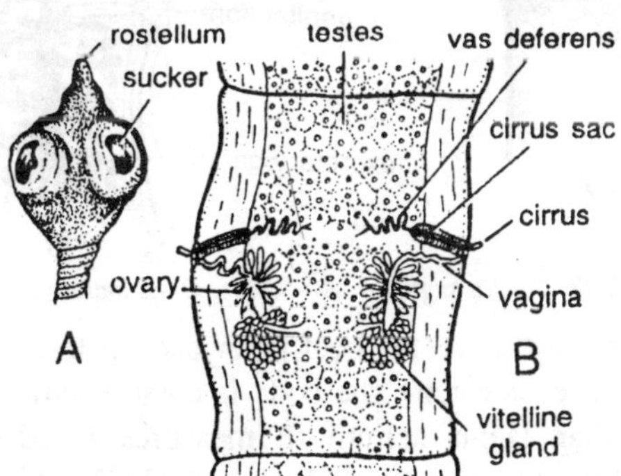

Fig. 2.41. Dipylidium caninum. A—Scolex. B—Mature proglottid.

It is commonly called dog-tapeworm. It has world wide distribution. It lives in the intestine. It measures 15-40 cm in lengths and has under 200 proglottids. The scolex is somewhat triangular and bears four deeply cupped suckers. Rostellum is retractile into a rostellar sac in scolex. It bears about 60 hooks in 3-7 rings. Mature proglottids are long and barrel shaped having a double set of male and female sex organs. The genital pores at the middle of both lateral margins. Numerous testes are present before or behind the female genitalia. The uterus first develops as a network of canales, but latter breaks up into a large number of uterine capsules, each enclosing 3-30 embryos. Intermediate host is dog-fleas, *ctenocephalus canis* and *pulex irritans.*

Taenia Saginata

Phylum	—	Platyhelminthes
Class	—	Cestoda
Subclass	—	Eucestoda
Order	—	Taenoidea
Genus	—	*Taenia*
Species	—	*saginata*

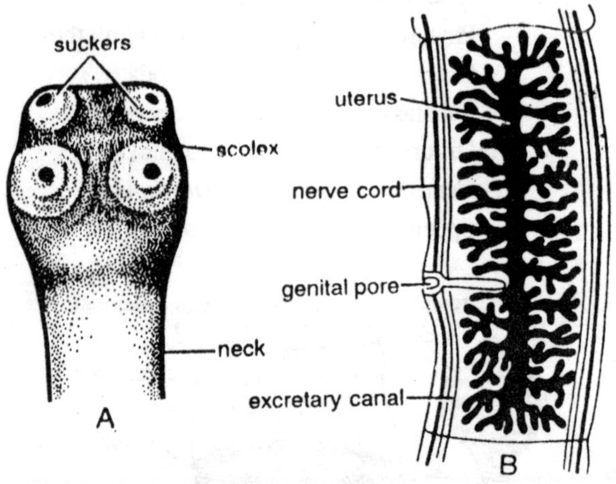

Fig. 2.42. Taenia saginata. A—Scolex. B—A mature proglottid.

It is commonly called the beef tapeworm. It lives in the intestines of man where beef is eaten. The body is dorso-ventrally flatenned and much larger than taenia solium. It measures 15-20 feet in length. Body is divisible into scolex, neck and proglottids. Scolex bears four large suckers, adhesion and devoid of rostellar hooks. About 1000 proglottids are present. Testes are about 300-400. Ovary in the form of two lobes. Ootype is present surrounded by Mehli's gland. Life-history is similar to *T. solium.* Intermediate hosts are cattles.

3

Parasitic Physiology

It is not yet possible to present a satisfactory general account of nematode biochemistry and physiolgy, for the simple reason that nearly all the information available relates to the animal-parasitic species only. As many of these species are highly specialized, it would be presumptuous to supose that details of their physiology would apply without serious modification to the less specialized plant-parasitic and freeliving forms. Nevertheless, the historical development of parasitology has been such that a reviewer attempting to deal with nematodes generally has no alternative but to accept the existing knowledge of physiological processes in the animal parasites as his basic material and from this to compare and predict, however modestly, whatever he feels may be applicable to the free-living and plant-parasitic forms. Those who work with nematodes are in a particularly fortunate position to study the development and nature of parasitism. In no class of animals is the gradation of adaptations from the free-living to the facultagtive and oblighate modes of parasitism, whether plant or animal, more readily apparent or more widespread. It is therfore unfortunate that the freeliving species, which are the obvious starting point for physiological study, have received so little attention. However, there are indications, among which discussions in the present context are not the least, that biologists are now aware of the problem facing them.

It is hardly necessary, therefore, to defend the attempt which is made in the following pages to present the subject of nematode physiology as a whole, nor to apologize for the repeated emphasis on the need for the broadly conceived studies of all nematodes, regardless of their ecological classification as free-living, plant-parasitic, or animal-parasitic species. There are many approaches to any discussion of nematode physiology, not all of which are entirely suitable to a given occasion. In the present account of the plan which has been adopted deals first with the processes of digestion, absorption, and composition of the perienteric fluid. These lead naturally to the discussion of assimilation of the various foodstuffs, to the composition of the organs and tissues, to the biosynthesis of tissue components, when known, and to the consideration of organ and tissue function.

A general account of respiration, both aerobic and anaerobic, is followed by a more detailed account of the metabolism of carbohydrates, fats, and proteins. This is succeeded by a discussion of excretion, water balance, permeability, and egg-hatching. No attempt will be made to deal comprehensively with all aspects of nematode physiology, nor to make mention all the literature. Large and complex subjects such as immunity in animal hosts, resistance in plant hosts, and the effect of alterations in the host's physiology upon the establishment of infections have been regretfully excluded. References to most of the literature dealine with nematode physiology are to be found in reviews which are readily available.

Digestion and Absorption

Nematodes feed on a wide variety of substrates, including bacteria, fungi, algae, protozoa, other nematodes and soil microfauna, and living and decaying higher plants and animals. Whether their buccal armament features stylet, teeth, or lips, the ingestion of food appears to depend ultimately upon the powerful sucking action of the esophageal muscles. The food of many species, notably those in which the diameter of the stylet is usually too small to admit the passage even of bacterria, is primarily fluid, whereas in others such as the predators, the food is solid. Considering the complexity and diversity of the natural food, it is probable that in one or another species most of the known digestive enzymes are to be found, although there is little concrete evidence supporting this supposition.

Similarly, it is not unreasonable to suppose that the nutritional requirements may very widely from one species to another, and here it is gratifying to know that some success has been achieved by studying nutrition in axenic cultures. This phase of nematode physiology will not be discussed as it is given special attention by Dougherty. Digestion of foodstuffs by nematodes may be extracorporeal, intracorporeal, or a combination of the two, but is always, so far as is known, extracellular. There appears to be no authentic description of intracellular digestion following phagocytic engulfment of food particles, such as occurs in more primitive metazoans.

Extracorporeal digestion has long been thought to occur in nematodes whose food must enter the body by way of a hollow stylet whose dimensions are such that even bacteria would traverse it with difficulty. This stylet is of course employed in piercing the hard parts of the food material, and as a rule it receives the duct of the dorsal esophageal gland somewhere near its base. Students of plant-parasitic nematodes have grown accustomed to the obvious inference that the dorsal gland produces a secretion which is used to predigest the plant food. It is curious to find, therefore, that Linford (1937) appears to be the only investigator who has actually described the outward flow of the secretion, the organisms observed being *Aphelenchoides* sp. and *Meloidogyne* spp. In these species secretion is followed by extracorporeal liquifaction of the affected tissues, and then by ingestion of the liquid.

Linford suggested also that the predaceous *Aphelenchoides* produced a paralysant which immobilized its prey while digestive secretions were pumped into it. This early study by Linford did not include any observations on the actual nature of the secretion, and twenty years later not much has been added to our knowledge. Recent work by Myuge (1957a) and Zinovev (1957) has revealed that the bulb nematode (*Ditylenchus allii*), the potato tuber nematode (*D. destructor*) and the root-knot nematode (*Meloidogyne* spp.) all secrete protease and amylase into the distilled water in which they are suspended. *D. allii*, which is specific for the onion, does not penetrate the onion cell, the fluids of which are toxic to it, but feeds on the inter cellular substances, and may therefore be supposed to secrete protopectinases or pectinases which digest the middle lamella. This species secretes only 14% as much amylase as *D. destructor*, which feeds on potato starch. Moreover, the thermophilic *Meloidogyne* spp. secrete increasing amounts of enzymes as the temperature is raised to 24°, whereas the secretion of the non-thermophilic *D. allii* does not change.

These observations of the Russian workers seem to provide the only clear evidence supporting the specific enzymic nature of the salivary secretions of plant-parasitic nematodes. In passing, however, it is worth noting that Tracey (1958) was able to detect both chitinase and cellulase in homogenates of three species of *Ditylenchus* (*D. destructor*, *D. dipsaci*, *D. myceliophogus*). Whether these enzymes play a part in the extracorporeal digestion of the cell walls of higher plants and fungi has not been shown. Nor is it possible to distinguish clearly between the secretions which may be responsible for tissue penetration and those which are directly concerned with the predigestion of foodstuffs.

There is no reason to believe, in fact, that the dissolution of a cellulose cell wall in the process of nematode penetration into tissues does not result in the creation of soluble sugars which are of dietary value, and in the case of *D. allii* already mentioned, which digests the middle lamella and thus produces maceration of the onion cells without actually penetrating them, it is tempting to suppose that the products of protopectin digestion are used in part at least as nutrients. This uncertainty regarding the function of extracorporeal digestion is even more striking when one turns to a consideration of the animal-parasitic nematodes.

Many non-migratory parasites produce extensive cytolysis of the host tissues which has been thought to be due to secretions of the esophageal glands. A striking example is to be found in the dog kidneyworm, *Dioctophyme renale*, which has a large and highly dendritic dorsal gland opening for forward in the esophagus. Similarly, Thorson (1956) found that esophageal extrascts of the dog hookworm (*Ancylostoma caninum*) contain a protease, and that the sera of dogs refractory to further infection with hookworm stongly inhibit this protease. Thus, it appeared that the protease was part of the glandular secretion, and that it possessed antigenic activity. It is in the migratory larvae of various species, however, that the difficulty inherent in distinguishing between enzymes responsible for tissue penetration and predigestion of foodstuffs are clearest.

Lewert and Lee (1954, 1956) have shown that several species of skinpenetrating filariform larvae cause definite changes in the acellular ground substance of the host dermis, and that these effects are associated chiefly with the production of a collagenase by the larvae. It is not difficult to conceive that the extracorporeal digestion of collagen which seems essential for effective penetration of the parasite, produces amino acids or soluble peptides which may also be eminently suitable as larval food. The whole subject of extracorporeal digestion, therefore, is one which will need much more study before its general features are clearly outlined. Intracorporeal digestion is probably more common in nematodes than the extracorporeal type, and might be expected to occur even in those species in which a preliminary external solubilization of foods occurs.

Many nematodes feed by ingestion of microorganisms, decaying organic matter, intestinal contents, or tissue particles. Saprophytic

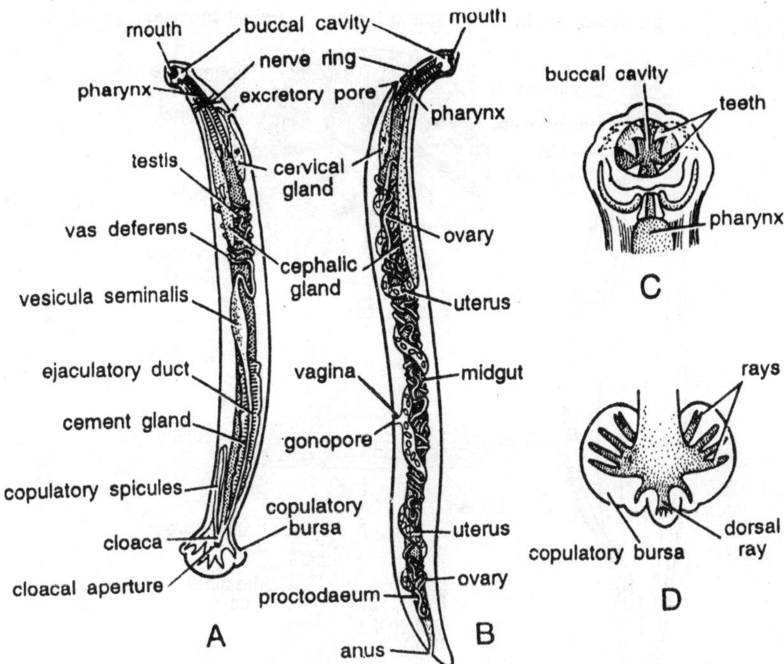

Fig. 3.1. Ancylostoma. A—Adult male. B—Adult female. C—Anterior end. D—Posterior end of male.

species, including the free-swimming intestinal parasites, may of course feed in part on predigested material, but there is good evidence that solid particles are also ingested and digested. Hobson (1948) has thorougly reviewed the evidence in the case of intestinal parasites.

As for free-living saprophytes, McCoy (1929) showed that the larval stages of *Ancylostoma caninum*, which normally develops in the host feces, can be grown successfully in pure cultures of various coliform bacteria, on which they feed; and Overgaard Nielson (1949), in his extensive study of free-living nematodes observed that saprophytic species feed on the microorganisms associated with the decaying plant tissues, rather than on the tissues themselves. Thus, the existence of intracorporeal digestion in many, and perhaps most, nematodes seems assured. The convincing observations which have just been cited have not yet been well-supported by detailed biochemical studies.

Hirsch and Bretshneider (1937), and Janicki (1939) described granules in the foregut of *Ascaris lumbricoides* and *Dioctophyme renale* which were regarded as being secretory, although the discharge of the

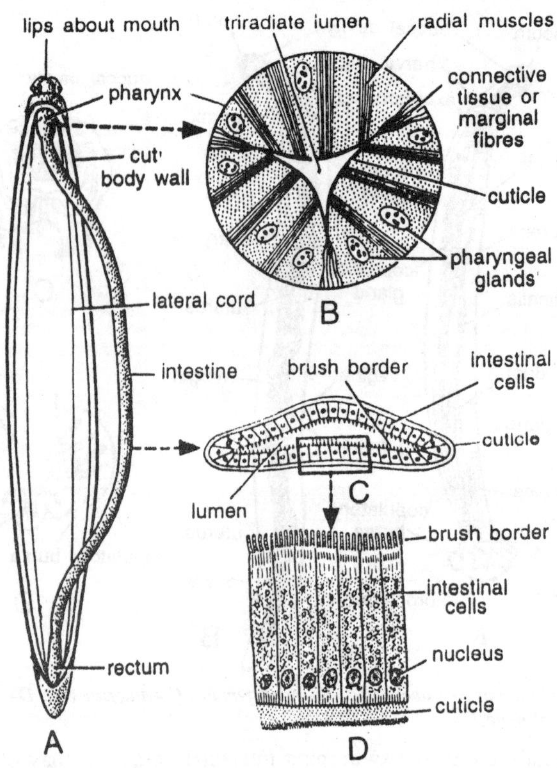

Fig. 3.2. Ascaris (digestive system). A—Alimentary canal. B—Pharynx in T.S. C—Intestine in T.S. D—A portion of intestinal wall showing brush border.

granules into the intestine was not seen. The first thorough study of digestion in nematcdes was made by Rogers (1941a), using the combined intestine and intestinal contents of two species, *Strongylus edentatus* and *A. lumbricoides*. Rogers found that amylase protease and lipase were present in both species, and that they were well adapted in their properties to the natural habitats of the two parasites, that is the horse caecum and the pig jejunum and ileum, respectively.

The investigation of *Ascaris* digestive enzymes was later continued in more detail by Carpenter (1952), but with reference only two enzymes present in the esophagus and intestinal cells. Carpenter's experiments, which were exceptionally complete, revealed that the maximum enzyme activity occurred in the foregut, which also contained most of the socalled secretory granules. Most of the enzymes, which included an amylase; lipase, protease and four different peptidases, were present also in the esophagus and to a lesser extent in the mid- and hindgut. Savel (1955b) reported that more protease and peptidase activity were to be found in *Ascaris* intestine than in other tissues, and several workers, *e.g.* Rogers (1940a), have found that blood feeders digest hemoglobin, as shown by the formation of hematin in the intestinal lumen. In summary, there can be no doubt that nematodes are likely to contain an array of digestive enzymes. Some of these may be secreted to the exterior by the dorsal esophageal gland. Less certainly, the esophageal glands may contribute secretions which pass to the intestine, where they are augmented by secretions from the foregut.

In some nematodes, extracorporeal digestion is essential; in some it may be useful but unessential, and in many others it may play no part in the normal digestive processes, which are then entirely intacorporeal. Considering the doubts which still exist concerning the mechanism of absorption of digested foodstuffs in higher animals, it is not surprising to find that absorption by nematodes is scarcely understodd at all. It sems certian, however, that absorption occurs by way of the intestine, and not through the cuticel. Intestinal cells of enmatodes are apt to contain a number of inclusions, some of which have been reported to consist of fats, carbohydrates, proteins, or minerals, which tend to disappear during starvation. It is not known whether such inclusions are synthesized in the intestine from absorbed foods.

A number of workers, including Rogers (1947), Yamao (1951) and Chowdhury (1955) have pointed out that the nematode intestine contains an abundance of acid or alkaline phosphatases, which have often been thought to be essential for the absorption of simple sugars against a concentration gradient. This hypothesis has never been strongly supported by experimental evidence, and it may be significant that in several nematodes the tissues contain very little glucose, but contain instead considerable quantities of the disaccharide trehalose. As trehalose is not a common dietary carbohydrate there exists an as yet unexplored possibility that in nematodes, as in the desert locust *Schistocercagregaria*, the concentration gradient in the intestine normally favours the absorption of glucose, the absorbed glucose being efficiently converted to trehalose in intestinal or other tissues.

Hemolymph

All nematode tissue are bathed in the perienteric fluid, or hemolymph, which fills the body cavity completely and is always under pressure owing to the tonicity of the longitudinal muscles of the body wall. Obviously, the hemolymph must be used to transport solutes from one tissue to another, whether these be the products of intestinal digestion and absorption, mobilized storage products, or metabolic wastes. On the other hand, hemolymph is essentially cell-free, and it would be surprising, in the light of modern knowledge, to find that an appreciable synthesis of substances such as proteins or fats occurred in it. While there is as yet no evidence on this point, it is nevertheless true that *Ascaris* hemolymph, which has been studied is some detail, is a very complex solution.

Nearly all our present information concerning hemolymph stems from studies on *Ascaris* and *Parascaris*, and as it is improbable that knowledge concerning more normally-sized nematodes can be secured in detail, its composition in *Ascaris* warrants close consideration. In *Ascaris*, hemolymph is a clear, pink, slightly sticky solution, somewhat acid (pH 6.2 to 6.4), with a rather high osmotic pressure (equivalent to 0.192 M sodium chloride) which is, however, consistently lower than the osmotic pressure of the host gut fluids. This high osmotic pressure is the reflection of a large ash content (0.9%) in which the ions of sodium, potassium, calcium and magnesium predominate. There is much less chloride than is required to balance these cations, the anion deficiency being made up in large part by volatile and non volatile organic acids and possibly by bicarbonate. Although the resemblance between the ion content of *Ascaris* hemolymph and the host fluids is remarkable this should not, for reasons which will appear later, be regarded as a general feature of nematode hemolymph.

Proteins, carbohydrates, and fats are all present. Proteins, comprising 4.9% of the fresh weight and 66% of the total solids, consist primarily of albumins and gloulins which differ little in their amino acid composition from those of mammalian plasma. Hemoglobin is present in small amounts and several enzymes have been identified, among which are invertase, maltase, amylase, esterase, and protease. Non-protein nitrogen consists chiefly of free amino acids, peptides, urea and ammonia. As already mentioned, it is improbable that synthesis of proteins occurs in the hemolymph itself, although no evidence to show their origin in other organs or tissues has been reported. Possibly the intestine, whose cytology is complex may be the source of the hemolymph proteins.

Carbohydrates are abundant in hemolymph, consisting of glycogen (0.40%) and trehalose (0.77%). Glucose and other free carbohydrates occur only in traces. Fats are less abundant; but significant amounts of phospholipids (0.18%) and triglycerides (0.13%) are present. Although it is probable that the composition of the hemolymph of other species in general is similar to that of *Ascaris*, there may also be important differences. Thus, the ionic composition might be expected to reflect the conditions under which a particular species lives, and might change from one developmental stage to another, depending upon the extent of the ecological variation encountered.

Similarly glycogen has been thoroughly identified in *Ascaris*, where it occurs in most tissues and less thoroughly in many other species; and trehalose occurs in the tissues of a number of nematoides. The species examined, however, have all been animal parasites. The food of free-living and plant-parasitic forms is so varied that there is no reason to expect that soluble sugars such as trehalose, or polysaccharides like glycogen, will be found to be the main carbohydrates in all nematode. The dietary carbohydrate of many species, in fact, probably contains very little of the glucose which is required for the synthesis of glycogen and trehalose. Investigation of the carbohadrates of free-living and plantparasitic nematodes is, therefore, a matter of importance.

Assimilation and Chemical Composition

Assimilation is the means by which simple food-stuffs or products of digestion of foodstuffs are converted into the complex constituents of the organism, and in the widest sense includes biosynthetic processes in general. It presupposes, therefore, some knowledge of the complex molecules which are its end result. In the present discussion it is proposed to combine, where possible, a review of the chemical composition of nematodes with the means by which complex molecules are synthesized in these animals. This arrangement is not burdensome, as for the most part there is no exhaustive information concerning either chemistry or biosynthetic mechanisms. It is worth remembering that nematodes possess a few biological characteristics which much strongly affect the quantitative aspects of assimilation. Their growth rate, for example, is not exceptionally fast, and indeed in many parasitic species is rather slow.

At various stages in the life cycle, moreover, metabolism may be greatly depressed, as in larvae of plantparasites living temporarily in the soil, filariform larvae of animal parasities, and surviving infective eggs of plant or animal parasites. In all nematodes the amounts of energy required for motion and locomotion are probably small. On the other hand the needs of the female reproductive organs are impressively high. Every day, for example, a mature female *Ascaris* lays eggs and accompanying secretions which amount to some 10% of her own weight; and in 22 days a minute free-living *cephalobus elongatus* weighing approximately on microgram was observed to lay eggs weighing about 6 μg. This seemingly disproportionate amount of energy expended on egg production is, of course, related to the large numbers of eggs laid, but is also due to the fact that the nematode egg is cleidoic.

So far as is known the embryonating eggs absorb no nourishment from the environment, and as a result all their caloric requirements must be provided by the adult female in the form of yolk materials which are deposlted in the mature ooyte. Moreover, in a numebr of plant and animal species the fully embryonated and infective egg survives for long periods in the absence of a suitable hatching stimulus, and must therefore contain additional caloric reserves for this purpose. As the yolk material is comprised primarily of proteins, carbohydrates, and fats, the adult female must possess adequate systems for the extensive biosynthesis of all three chemical groups.

Carbohydrates

As already mentioned, information concerning carbohydrates in nematodes is largely confined to the animal parasites, whose metabolism might be expected to be centered around glucose and glycogen. It is not safe to assume that these compounds are of equal importance in free-living or plant-parasitic species. Until recently, glucose in animal parasites was estimated by means of relatively unspecific reducing agents. In one investigation, however, it has been determined by oxidation with glucose oxidase, which is extremely specific for glucose. The total glucose in seven species ranged widely from 0.01% of the dry weight in the filarial worm, *Litomosoides carinii*, to 0.7% in the fowl ascarid, *Ascaridia galli*. In a parallel series of analyses, trehalose, which is a dimer of glucose, ranged from a high value of 2.18% of the dry weight in codworm (*Porrocaecum decipiens*) larvae to 0.0% in *L. carinii*. In all two species (*A. galli*, *Heterakis gallinae*) there was more trehalose than glucose and no other low molecular weight carbohydrate seemed to be present in appreciable amount.

The distribution of glucose and trehalose in the different organs and tissues is known only in *Ascaris* in which, comparatively, the intestine and cuticle contain more glucose, and the muscle, hemolymph and reproductive tissues an overwhelming amount of trehalose. Reference has already been made to the well-characterized *Ascaris* glycogen which has its counterpart in many other animal-parasitic species and which differs little in detailed structure from mammalian glycogen. Intestine, muscle and reproductive organs are rich in glycogen, whereas cuticle and hemolymph are not.

Sometimes glycogen is stored in immense amounts, as in *P. decipiens* larvae, where it comprises 55% of the dry weight, and *Ascàris* muscle, which contains 70%. These amounts, however, are exceptional, and many species contani much less. Apart from *Ditylenchus dipsaci*, whose solids contain 10% of unidentified carbohydrate no plant parasites have been examined. Glycogen synthesis in mammals occurs by phosphorylation of glocose to glucose-1-phosphate (hexokinase phosphoglucomutase), followed by polymerization of glucose-1-phosphate (phosphorylase) with liberation of phosphate. Adenosine triphosphate is also required. Although clear evidence is lacking, this may also be the pathway followed in nematodes, for several tissues of *Ascaris* may contain phosphorylyse, hexokinase and ATP.

In some species, most notably in *L. carinii* addition of glucose to a saline medium containing serviving worms reduces the rate of utilization of tissue glycogen or leads to an actual synthesis of glycogen. The mechanism by which trehalose is synthesized in plants and animals which normally contain it has not yet been learned. Trehalose is of course formed by linkage between the first carbon atom in each of two glucose molecules, whereas glycogen, including *Ascaris* glycogen consists primarily of 1→4 linkages between glucose units. Clearly, trehalose cannot be directly converted to glycogen by simple polymerization, but must, if it contributes to glycogen synthesis, first submit to hydrolysis or rearrangement from 1→1 to a 1→4 type of linkage. It is not necessary of course, for either carbohydrate to be formed directly from dietary glucose.

Many amino acids are freely convertible to 3-carbon compounds from which glucose may be synthesized and there is also evidence that in embryonating *Ascaris* eggs higher fatty acids give rise to similar 3-carbon compounds from which both trehalose and glycogen are synthesized. The latter represents a conversion of fat to carbohydrate which has not been conclusively demonstrated in other animals, although it is well-known in germinating seeds, which bring about the conversion by adaptation in the tricarboxylic acid cycle. Apart from the carbohydrates discussed above, which are primarily concerned in energy metabolism, others of a more or less complex nature might be expected to occur in nematodes. These include the ribose and deoxyribose of nucleic acids and the amino sugars and uronic acids of glycoproteins and mucoproteins, which have not been much studied, although a number of reports have appeared describing the histochemical distribution of nuclei acids and glycoproteins in nematode tissues and eggs. Two structural carbohydrates, however, have been investigated in some detail, and therefore deserve mention.

The first of these is ascarylose, a distinctive hexose occur ring in the glycosides which are components of the yolk in *Parascaris* and *Ascaris* oocytes, and later make an important contribution to the vitelline membrane of the fertilised eggs. These glycosides will be discussed later. Ascarylose itself is 3, 6—di-deoxyaldodexose a unique carbohydrate not known to occur naturally except in nematodes. No details of its biosynthesis are available. The second structural carbohydrate requiring mention is chitin, which occurs in many nematodes only in the hard part of the shell. No actual isolation has been reported, but the results of chemical and histochemical tests, and enzymic hydrolysis on eggs of many plant- and animal-parasitic species are convincing. Free-living species have not been examined, and the presence of chitin in some animal-parasitic species seems doubtful. It cannot be claimed, therefore, that chitin is universally present in nematode egg-shells.

As the hard shell, like the vitelline membrane lying close to its inner surface, is a primary egg envelope, the chitin in it must be synthesized from yolk material. This synthesis and extrusion of chitin or its precursors from the fertilized oocytes of *Parascaris* and *Ascaris* has been described cytologically, and it is probable that glycogen and trehalose in the yolk provide the glucose from which the N-acetylglucosamine units of chitin are formed. The conversion of glucose-6-phosphate to glucosamine in rat liver proceeds in the presence of glutamine as a source of amino groups. *Ascaris* oocytes contain considerable glutamine or related substance, as well as a source of acetate needed for the acylation of glucosamine to N-acetylglucosamine. There is no question, therefore, that the oocyte possesses appropriate substrates for chitin synthesis, although the way in which the synthesis takes place has not yet been studied.

Lipids

In some respects the nematode lipids, first studied in detail by Flury (1912), are better known than the carbohydrates, although as usual, detailed studies have been made only on *Ascaris* and *parascaris*. Generally speaking, they can be said to resemble the lipids of higher animals, but with important differences. Before discussing these in detail, however, the results of analyses for total lipids in several species are worth noting. Viewed generally, the amounts of lipids in five species of animal parasites, ranging from 1% to 8% of the dry weight, are not exceptional. In contrast, the values for three species of *Ditylenchus* and for the vinegar eelworm, *Turbatrix aceti*, are extremely high. Unfortunately, details of the lipids in *Ditylenchus* and *Turbatrix* have not been secured.

Investigations of *Ascaris* lipids, however, have established clearly that, if the over-all concentration is not unduly high, the distribution in various organs and tissues is interesting. Thus, lipids comprise 17.5% of the fresh weight of eggs, 6.0% of the ovaries, and considerably less than 1% in other tissues. Clearly, the reproductive organs are the most important sites for lipid deposition. In all tissues there are appreciable quantities of triglycerides, phospholipids and unsaponifiables, although the relative concentration of phospholipids in the reproductive tissues and egg is low. The composition of the triglyceride and phospholipid fractions is not dissimilar to that of higher animals and need not be discussed further, except to say that the menas by which *Ascaris* synthesizes these substances are unknown.

The unsaponifiable fraction, on the other hand, is particularly interesting, for it contains unique substances of great importance in the parasite's economy. In higher animals, most of the unsaponifiable fraction is composed of sterols, and chiefly of cholesterol. *Ascaris* unsaponifiables contain cholesterol, and also a second, saturated sterol, but the bulk of this fraction consists of a substance isolated many years ago by Flury (1912) and Faure-Fremiet (1913) and named by them ascaryl alcohol. Quite recently Fouquey *et. al.* (1957) established the structure of this compound which is not a true lipid, but a series of closely related glycosides with the solubility characteristics of true lipids.

The generic name, ascaroside, which was suggested for these glycosides should now be accepted as correct. One component of ascarosides, ascarylose, has already been mentioned in the discussion of carbohydrate chemistry. This hexose is combined in glycosidic linkage with a high molecular weight, aliphatic, straight-chain diol to form ascaroside B, the main ascaroside in *Parascaris* ovaries. Ascarosides A and C contain the same hexose, but differ slightly with respect to the higher alcohol. Ascarosides occur in *Ascaris* muscle and cuticle and in *Eustrongylides* males, but are concentrated mainly in the oocytes and in the fertilised eggs, and it is in these cells that their function is clearly understood.

Ascarosides first appear in the developing oocytes as refringent granules which are finally discharged from the fertilized eggs to become the main component of the innermost, or vitelline, memberane. The process of shell formation in *Ascaris* and *Parascaris*, which lies beyond the boundaries of the present discussion, has been reviewed recently; and the function of the vitelline membrane will be mentioned later. At this point it is appropriate to suggest, however, that although ascarosides have been identified only in *Ascaris* and *Parascaris*, it is likely that they, or closely related substances, are constituents of other nematode eggs. Thus, a vitelline membrane is present, and when more closely examined is found to be soluble in lipid solvents, highly impermeable, chemically resistant and to melt at about 70°C. These are the outstanding properties of *Ascaris* and *Parascaris* membranes, and are sufficiently unusual to suggest similarities in chemical structure.

However, much more work will be necessary in ordre to establish the validity of this supposition. It is clear from this brief discussion that in many, perhaps in most, nematodes there is an extensive synthesis of lipids, some of which are incorporated in the somatic tissues, but most of which, in the female, are required for the production of eggs. It is probably safe to assume that as in other animals, the phospholipids and sterols are structural components of cells, concerned in the formation of cellular and subcellular membranes. In the embryonating fertilized egg, for example, these lipids, unlike the triglycerides, are apparently not metabolized. The part played by ascarosides in the somatic tissues is not clear. Triglycerides, on the other hand, serve as energy stores, as will become apparent later.

Proteins

The proteins of nematodes are of exceptional interest but have been studied by modern analytical methods only in a few instances. The total protein in *Ascaris* makes up eight to nine per cent of the wet weight and the amino-acid composition generally resembles that of its host. The haemolymph of *Ascaris* contains 4.9 per cent proteins of which 2.8 per cent are albumins and 2.1 per cent globulins. The fibrous proteins of the cuticle have been most thoroughly studied but complete agreement has not been reached on their composition. X-ray and chemical analysis have inducated the presence of keratin, collagen and tanned protein although there are differences between species. There is little evidence in nematodes that protein is stored for subsequent use in energy metabolism. It is true that nitrogen catabolism during prolonged starvation *in vitro* is fairly extensive, as will appear later, but the large amount of protein deposited in *Ascaris* oocytes is apparently not used at any stage of embryonation for energy. Although protein granules in the larval intestines of *Agamermis decaudota*, *Ditylenchus dipsaci* and *Meloidogyne javanica* disappear during development to the adult stage, it is not clear whether these are mobilized for incorporation into other tissues, or are catabolized.

In general then, it seems likely that the bulk of nematode proteins are structural, or potentially structural, tissue components. Mention was made previously of the large amounts of albumins and globulins in *Ascaris* hemolymph, whose origin and function are unknown, but which must surely originate in some tissue other than the cell-free hemolymph itself. A third hemolymph protein is hemoglobin, which in *Ascaris* comprises about 2% of the total. Muscle and epidermis, contain a second, and somewhat different, hemoglobin. In animal parasites, the presence of hemoglobin is the rule, for von Brand (1952) listed 13 species in which it occurs, and several others have been added since. More important than these minor statistics is the fact that no species has yet been shown to lack this protein. It is, therefore particularly necessary to examine freeliving and plant-parasitic species, concerning which no report has yet appeared. Hemoglobin is widespread amongst parasitic nematodes, only three species having been reported to lack haemoglobin it is localized differently in different species. There is evidence that it plays a respiratoryy role in some, but not all species. The haemoglobin in the body wall may also serve to transfer oxygen from the cuticle to the body fluid.

Hemoglobins of *Heterakis gallinae* and trichostrongyles have characteristic α-and β-absorption bands which differ little in position from those of the host, whereas the differences in *Ascaris* hemoglobin are distinct. The fundamental distinction between host and parasite proteins, however, lies in the extremely high affinity of the parasite proteins for oxygen, exen at very low partial pressure of this gas. As a result, no dissociation of oxyhemoglobin occurs except at very low tissue concentrations of oxygen. After quantitative study of this phenomenon, both Rogers and Davenport are agreed that an oxygen transporting function of hemoglobin in *Ascaris* and *Nippostrongylus muris* is improbable. As no alternative function for hemoglobins has been suggested it is tempting to think that these are simply pigments derived by absorption of modified host hemoglobins.

A serious objection here is that some species, such as *Ascaris*, do not feed on the host tissues, and as a result have limited access to

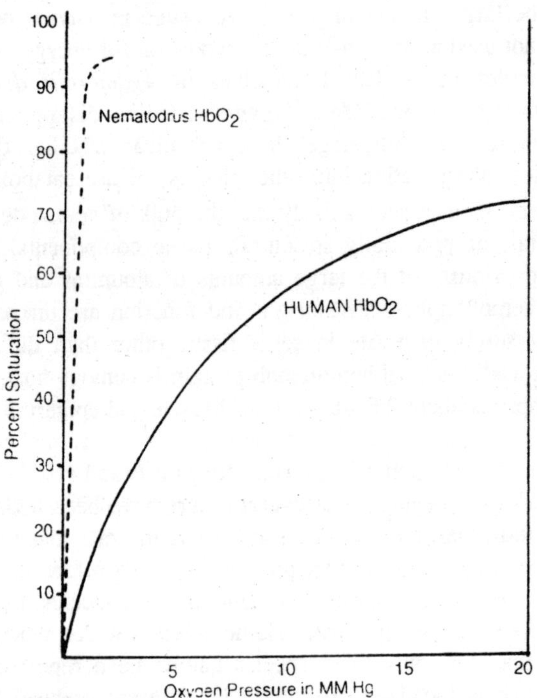

Fig. 3.3. Comparison of the affinity of Nematodirus and human hemoglobin for oxygen.

host hemoglobins. There has been no chemical investigation of the proteins in nematode muscle or epidermis. In recent years, however, a good deal of interest has been taken in the cuticle, which is largely protein, and is of well-recognized importance both in the protection of nematodes against their normal environment, and in their destruction by nematocides.

The organization and composition of the cuticle is therefore worth considering in some details, particularly since it has the same fundamental structure in all nematodes. The cuticle is, so far as is known, a secretion product of the underlying epidermis, but is nevertheless highly organized. It is soft and flexible, and contains about 75% water. When isolated, it can be stretched very considerably in the direction of the longitudinal axis, but very little in the lateral axis. Most of its solids are proteins. However, small amounts of various lipids and carbohydrates are to be found in it whose nature differs little from those occcurring in the muscle. Whether these contribute to its physiological functions or have merely "strayed" into it during the secretion of proteins has not been determined. Recent work by Bird and Deutsch (1957) and Bird (1958) has considerably increased our understanding of the structural complexity of the cuticle. For present purposes, a simplified diagram provides a suitable basis for the discussion of cuticular proteins. Here the basement membrane which lies next to the secreting epidermis is overlaid (in *Ascaris*) with three fibre layers. In many species there are apparently only two fibre layers. These are crossed at an angle of about 70° with the long axis of the worm, and if there is a third layer it lies parallel to the first.

An apparently structureless layer lying externally to the fibre layers was called the homogeneous layer by Chitwood (1936) who was the first to make detailed chemical studies of the various parts. This is succeeded by a cortex of varying complexity. Bird (1958) has produced evidence, based on electron microscopy, that in *Ascaris* culticle the cortex contains a radiating network of pore canals which extend into the external cortex, almost to the cuticular surface. The surface itself is covered with an osmiophilic layer less than 1000 Å thick which may be a lipoid. If whole *Ascaris* cuticle is extracted with water about 20% of its proteins dissolve without seriously modifying the cuticular structure. The extract contains at least five different proteins, all of which seem to be albumins of similar, low molecular weight. Their initial distributionin the cuticle is unknown, although the homogeneous layer would seem to be the most probable locus. This layer also contains sulphur-rich fibrous proteins which resemble fibroin or elastin in their chemical and histochemical properties. It is the fibre layers and the external cortex which have been most studied, partly because of their great intrinsic interest, but equally because they are relatively easy to isolate.

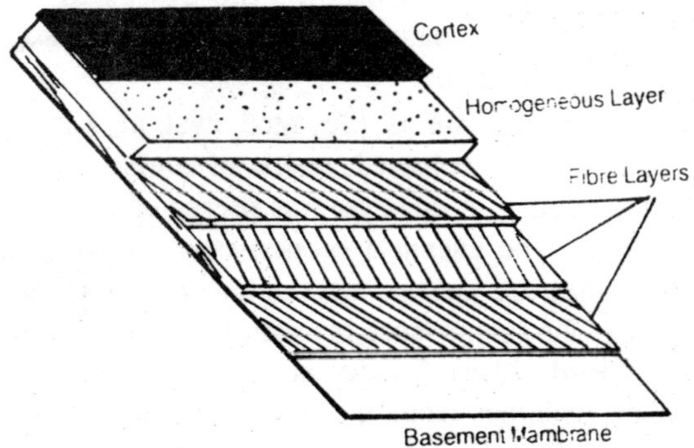

Fig. 3.4. Schematic representation of Ascaris cuticle.

The fibre layers have the staining properties and X-ray diffraction pattern characteristic of collagen which is of course the characteristic protein of tendons and other connective tissues. Amino acid analyses show that hydroxyproline, an amino acid peculiar to collagen, is present in the fibre layerse and in the basal lamella. The collagenous nature of these layers is therefore demonstrable by several techiques.

The nature of the protein in the external cortex, which is responsible for the protection of nematodes against potentially damaging environmental influences, is less certain. Early work by Flury (1912), since confirmed established the presence of large amounts of sulfurbearing amino acids, which, considered in relation to its great chemical resistance, suggested a keratin-like structure. Such a structure is not supported by the X-ray diffraction patterns which are those of a collagen, although the presence of hydroxyproline was not reported in amino acid analyses of this protein. To further confuse the picture, evidence has been presented for the presence in the external cortex of polyphenols and polyphenol oxidase, substances which are commonly associated with tanning processes in the arthropod cuticle.

Bird (1957, 1958) suggested that these phenolic substances necessary for tanning are transported to the external cuticle through the pore canals which, as already mentioned, were revealed by electron microsocpy. As tanning seems always to be associated with the development of colour, and as the nematode cuticle is not highly coloured, it is probable that if tanning occurs, it is moderate in extent. This view is supported by the fact that well-tanned proteins are resistant to the action of the proteolytic enzyme papain, whereas the adult nematode cuticle is not. Papain and related enzymes have, in fact, been fairly widely used for this reason as antihelminthic agents against intestinal parasites. It seems desirable, therefore, to remain uncommitted or the chemical nature of the external cortex until additional studies have been made.

The one possible exception is the aged cyst of *Heterodera rostochiensis*, which is deeply coloured, and is of course, a modified cuticle. Evidence for tanning in this cyst was presented several years ago by Ellenby (1946). The chemical properties of other nematode cuticles, in so far as they are known, are similar to those of *Ascaris*. Chitwood (1936) noted the resemblance between the cuticles of *Ascaris* and *D. dipsaci*, and Bird (1956) detected no important differences in protein amino acids obtained from the entire cuticles of *Ascaris*, *Toxocara mystax*, and *Strongylus eguinus*. Similarly, histochemical examination of *T. canis*, *Subulura*, *Haemonchus*, *Uncinaria*, *Chabertia*, and *Echinuria* suggested the presence of collagenous fibre layers, an elastin-containing homogeneous layer, and a polyphenol containing cortex. It seems clear, also that the exteroal critical layer lines the surface of the stomodaeum and esophagus in nematodes generally and that tanning may occur in this lining.

The discussion of cuticle proteins has up to this point been concerned only with adult nematodes. Much less work has been done with the larval cuticle, which is naturally harder to obtain in quantities sufficient for chemical analysis. Monne (1955*b*) reported that the larval cuticles of the lungworms *Metastrongylus* and *Dictyocaulus* were microscopically structureless, but stained in a manner suggesting the presence of tanned protein and were not attacked by papain, which, as

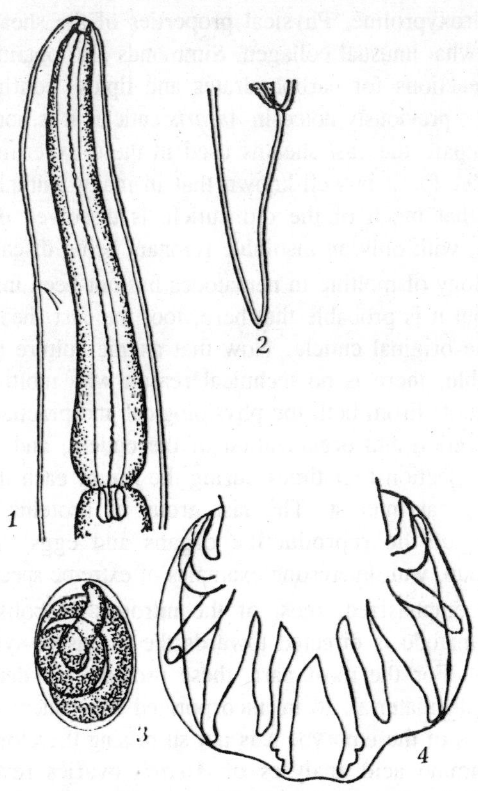

Fig. 3.5. Dictyocaulus, a lungworm of cattle and other mammals. 1–Anterior end. 2–Posterior end of female. 3–Egg with larva. 4–Bursa of male.

noted above, is a fairly efficient agent for the digestion of the adult cuticles. Despite their fragility, therefore, the larval cuticles may be sufficiently well tanned to offer considerable resistance to an adverse environment. Bird and Rogers (1956) in a study of the cast shealths of third stage *Haemonchus*, *Trichostrongylus* and *Nippostrongylus* larvae found no aromatic amino acids in the proteins, and no evidence for quinonetanning, although considerable hydroxyproline was present.

Interpretation of their results is uncertain, as sodium hypochlorite, a strong oxidizing agent, was used to induce exsheathment of the larvae, and the destructive effect of this solution on cuticular proteins must be considered, as the authors realized. A more physiological procedure was employed by Simmonds (1958), who found that fourth-stage *Nippostrongylus* larvae cast off the remains of the third stage cuticle when held overnight in saline solution. Proteins of the sheaths obtained in this way contained an abundance of the aromatic amino acid tyrosine and also of hydroxyproline. Physical properties of the sheaths. were those of a somewhat unusual collagen. Simmonds also obtained strong histochemical reactions for carbohydrates and lipids, confirming the chemical analyses previously noted in *Ascaris* cuticle. It is not possible, naturally to compare the cast sheaths used in these experiments with the whole cuticle, for it is well-known that in many animals such as the arthropods that much of the old cuticle is dissolved during the molting process, with only an insoluble remnant being discarded.

The physiology of molting in nematodes has not been investigated histologically, but it is probable that here, too, the cast sheath is only a remnant of the original cuticle. Now that axenic culture of several species is possible, there is no technical reason why molting cannot be studied in detail. From both the physiological and practiacal points of view the secretion and organization of the cuticle, and its partial dissolution and rejection four times during the life of each individual, are matters of great interest. The last group of proteins requiring mention is that of the reproductive organs and eggs. This is a heterogenous group, with interesting examples of extreme specialization.

As already emphasized, most of the nitrogen anabolism in the adult female nematode is directed towards the extensive synthesis of ovarian proteins. For the most part, these proteins are deposited in the oocyte as yolk material, to be incorporated subsequently into the structural proteins of the embryo. It is not surprising therefore, to find that extensive amino acid analyses of *Ascaris* ovaries revealed the presence of most of the common acids in well-distributed amounts.

During embryonation the yolk protein is converted to the many different proteins necessary for the various cells and tissues. No waste nitrogen accumulates in the egg as a result of this conversion showing that the formation of cell protein from yolk protein is remarkably efficient, and that the yolk protein is not used by the egg as a source of energy. Many years ago van Beneden described the appearance in *Parascaris* oocytes of refractile granules which are now known to consists primarily of proteins containing more than 20% proline.

The clear-cut segregation of these prolamine-like proteins from the yolk material is due to the fact that following fertilisation, the granules are ejected from the cytoplasm, and, according to Yanagisawa (1954, 1955) become a part of the hard shell. Thus, as has long been suspected, the hard shell, although chitinous, also contains a specialized protein. Similar gram-positive inclusions have been described in a number of animal-parasitic nematodes, and in all cases they appear to be transformed after fertilization into constituents of the hard shell. The spermatogonia of many nematodes contain similar gram positive refringent granules which ultimately fuse to form the prominent acrosome of the sperm. This protein, called ascaridine, has been isolated from *Parascaris* sperm. It contains no phosphorus, sulfur, or purines, but two amino acids, aspartic acid and tryptophane, account for 35% and 15% of the total nitrogen, respectively.

After the sperm penetrates the surface of the oocyte, it moves slowly towards the female pronucleus, and as it does so the acrosome, and hence the ascaridine, gradually disappears. Ahead of it, in the egg cytoplasm, appears a dense and local concentration of ribose nucleic acid. Thus there is some connection between the disappearance of ascaridine and the appearance of newly-formed RNA in the fertilized egg, but whether this represents a direct contribution of ascaridine to RNA synthesis has yet to be determined. If, as seems clear, there is a continual synthesis of protein in the ovaries, then the sources of the amino acids contributing to this synthesis are of interest.

Dietary amino acid transferred directly from the intestine to the ovaries are of course an obvious source, and may normally be the most important one. It is also true, however, that *Ascaris* ovaries contain active transaminases which are able to convert α-keto acids derived from carbohydrate metabolism to the corresponding amino acids, with the equilibrium favouring the formation of amino acids. Moreover, Pollak has shown that ovaries can, in the presence of suitable co-factors, condense pyruvate with ammonia to form the amino acid alanine, which he regards as the key amino acid in ovaries. Thus, although most details have still to be worked out, there is evidence that formation of ovarian proteins occurs in the ovaries themselves, and that the required amino acids may be supplied by synthesis, as well as from the diet. This concludes the discussion of nematode chemistry, and the means by which these animals synthesize some of the complex molecules which are of importance to them. If the discussion has been brief, the references supplied will be found to cover most of the existing literature.

Emphasis has been placed on those aspects of nematode chemistry which are most likely to be of general importance, regardless of the species on which the particular investigation was made. In so doing, there is an ever-present danger of fostering an unwarranted complacence, and it cannot be too strongly urged that many more studies are needed, particularly of the free-living and plant-parastitic species. Until these are made, there is no possibility that a discussion based on fact as distinguished from a discussion based on hopeful generalizations, can be presented. In the succeeding pages, which will deal for the most part with processes of a catabolic nature, it will quickly become apparent that there is an equally great need for strengthening and broadening the foundation on which present knowledge rests.

Respiration

As a rule oxygen is needed for the efficient metabolism of organic compounds. When aerobic respiration is not possible, the anaerobic metabolism which replaces it is much less efficient, owing to failure to oxidize organic compounds completely to carbon dioxide and water. One result of this anaerobic respiration, or fermentation as it is more often called, is that the end-products of incomplete combustion must be eliminated from the body before they attain toxic concentrations. Successful adaptation to an anaerobic or semianaerobic existence therefore involves the development of mechanisms for the excretion of low molecular weight alcohols, organic acids and so forth as waster products. In some animals, and certainly in many nematodes, fermentation continues even when oxygen is being used, so that some molecules are oxidized completely to carbon dioxide and water simultaneously others are oxidized less completely and are excreted chiefly as organic acids. The general physiology of adaptations to aerobic and anaerobic environments is dealt with by von Brand elsewhere in this book.

Here it is only necessary to mention some details which may be of direct concern to the experimenter, or which may influence the mechanism by means of which tissue oxidations occur. No nematode yet studied has failed to use oxygen when given the opportunity, *i.e.*, no obligatively anaerobic species has been discovered. Some, *e.g.* *Ascaris*, are poisoned by high concentrations of oxygen, so that it may eventually be desirable to classify these as microaerophiles, a term used by microbiologists to describe the same phenomenon in bacteria and other microorganisms. Generalizations on the subject of oxygen utilization should certainly be deferred, however, until more studies have been made, particularly on plant-parasitic species and the many free-living species which inhabit the essentially anaerobic fresh-water and ocean muds.

Oxygen utilisation by nematodes varies greatly from one species to another, and has not been successfully correlated with weight, volume, or habitat. The upper and lower limits recorded for a number of animal-parasitic and free-living species are set out in Table 15.6, which is based on a summary by von Brand (1952) and on the original investigation of Overgaard Nielsen (1949). It is immediately apparent that the range of Qo_2 values is wide, and that there seems to be no great difference between free-living and parasitiic forms in the ability to use oxygen. The results, of course, were obtained using isolated animals suspended in a saline solution in air, and do not necessarily, or even probably, represent the rate of oxygen consumption in a more normal environment. The represent, in fact, only the *capacity* for metabolizing oxygen.

Direct comparison of the animal parasites with the free-livin species is also complicated by the large difference in the temperature at which the Qo_2 values were determined. In such analyses, a number of factors in addition to temperature may have an important bearing on the results obtained. As a rule, moderate variations in pH or osmotic pressure have not been found to be highly critical. The ionic composition of the suspending medium, on the other hand, may profoundly affect the respiration. Again, the usual Warburg method for determining oxygen consumption, depending as it does upon the continuous removal of carbon dioxide from the gas atmosphere, may result in Qo_2 measurements different from those found under more natural conditions when carbon dioxide is present. Thus, the oxygen consumed by *Nippostrangylus muris* larvae in their early, freeliving stages is unaffected by the presence of 5% CO_2, whereas the Qo_2 of infective larvae is doubled reflecting, perhaps, an adaptation of the infective larvae to the high CO_2 concentrations in the animal tissues in which they will complete their development. Age can also be important.

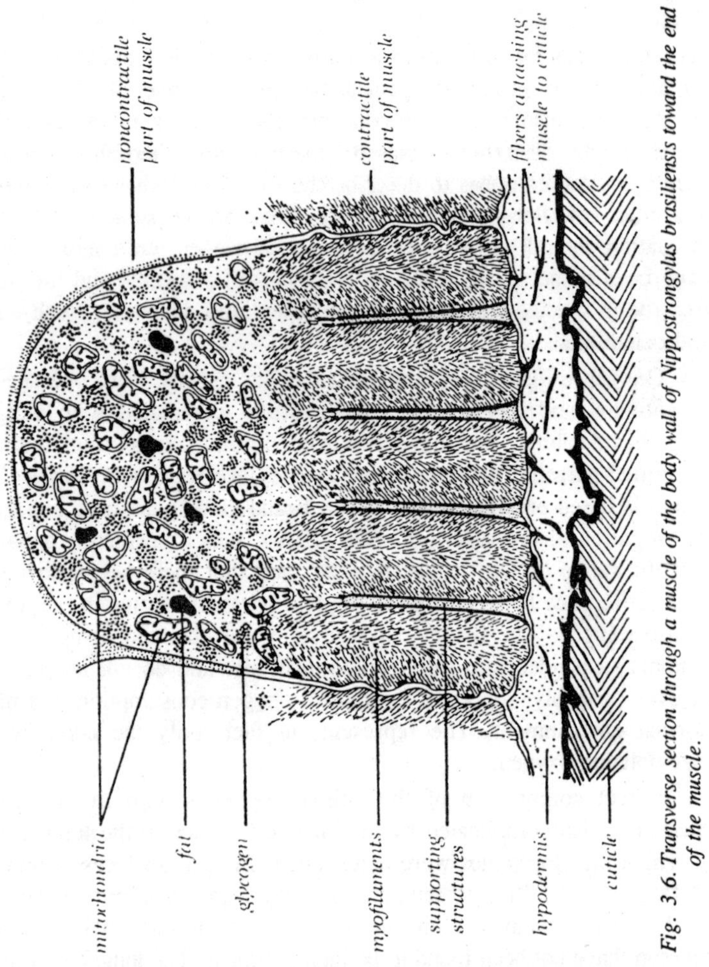

Fig. 3.6. Transverse section through a muscle of the body wall of Nippostrongylus brasiliensis toward the end of the muscle.

Rogers (1948) observed a considerable decrease in the Qo_2 of infective *N. muris* larvae after twelve days, and the Qo_2 of free-living *Panagrellus redivivus* adults feel about 50% after 24 hours starvation. Although these observations on *Nippostrongylus* and *Panagrellus* were all made in nutrient-free saline, the results must be interpreted differently. Decrease in the Qo_2 of *Panagrellus* can be attributed to starvation, and would not be expected to occur in a nutrient solution. In *Nippostrongylus*, however, the decrease is physiological *i.e.* the infective larvae do not feed, and yet can survive for many days in the soil. The ability to survive must entail a severe restriction in the basal metabolism, which is reflected in decreased oxygen consumption.

The importance of recognizing this experimentally uncontrollable condition can be illustrated by considering the changes which take place in the respiration of *Ascaris eggs* during embryonation. Here it has been shown that the Qo_2 falls at first, possibly as an oxygen debt is repaid, but then rises rapidly during the period of cell-division, reaching a peak when the embryo has become vermiform. A sharp decline then sets in, and by the time the egg is infective (about 20 days) the Qo_2 is once again small.

Over a period of months the respiration continues to decline, and is in fact scarcely measurable, although the egg remains infective. Clearly, Qo_2 measurement on these eggs would be of little value if recorded without reference to physiological age. It would be surprising, too, if other eggs failed to display a similar variation, and in more general terms, one might predict that variable oxygen requirements are a physiological feature of all species where eggs, larvae, or adults alternate between metabolically active and quiescent states.

A number of investigations have revealed that nematode Qo_2 measurements may or may not be highly dependent upon oxygen tension. This is not a matter requiring discussion here, except to point out that two factors are of primary importance. These are (1) the size of the animal (2) the nature of its oxygen-activating enzymen system. It has been shown that at oxygen tensions less than 16 mm of Hg the rate of diffusion of oxygen into the tissues will limit the Qo_2 of a nematode more than 0.05 mm in diameter. If the diameter is 0.14 mm, the limiting tension is 32 mm, and so forth. Assuming that diffusion is not

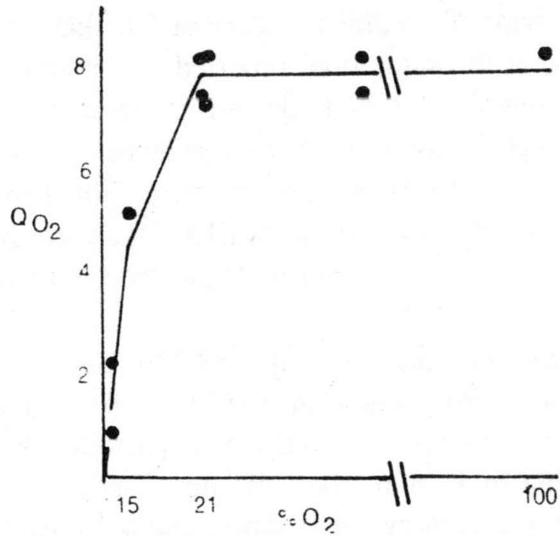

Fig. 3.7. Oxygen consumption of Nippostrongylus brasiliensis at varying concentrations of oxygen in the gas phase.

a limiting factor, however, the Qo_2 may still vary according to the oxygen tension if the terminal oxygen activating respiratory enzymes have a relatively low affinity for oxygen. Such is the case in *Ascaris* which seems to depend upon flavin enzymes for oxygen activation.

Other species, such as largvae of *Trichinella spiralis*, contain the cytochrome oxidase system, which has an extremely high affinity for oxygen. In these, small size and suitable enzymes create a situation in which respiration is independent of oxygen tension even at 15 mm Hg. Parasites removed from their normal environment in salime are apt to be very active. *Heterakis gallinae*, whose movements in the fowl caecum must be restricted owing to the thick consistency of the faecal mass, move very actively during the course of Qo_2 measurements in saline, Rogers (1949a), in his pioneering efforts to assess the physiological availability of oxygen to a number of intestinal parasites, suggested that the Qo_2 observed *in vitro* might, on account of the parasites' muscular activity, be considerably higher than the normal *in vivo* requirement.

Moreover, repeated observations on several species have led to the conclusion that in the absence of oxygen hitherto active nematodes become quiescent motility being restored upon admission of oxygen. Thus, there appears to be a connection between oxygen consumption and motility which may be of practical importance to the experimenter if motility under unnatural conditions is abnormally high. Without denying the apparent necesity of oxygen for motility, there is at the same time reason to believe that only a small part of the oxygen consumed is associated with motility. Santmeyer (1956), in observing that the Qo_2 of surviving *P. redivivus* decreased to half its original value after 24 hours, at the same time reported that motility was not affected. Thus, high motility and high Qo_2 are not necessarily directly related. Similarly, in his extensive report on soil nematodes, Overgaard Nielsen (1949) failed to observe a positive correlation between Qo_2 and motility and found that when motility was abolished by anaesthetization of several species with urethane the Qo_2 was reduced only by about 5%.

Another pertinent observation was reported by Bueding (1949b), who found that in *Litomosoides carinii* a 97 to 98% suppression of oxygen consumption by cyanide immobilized the parasites, whereas an 85 to 95% suppression by cyanine dyes did not. Thus, there is good evidence that oxygen is necessary for motility, and some evidence that the amount of oxygen used for this purpose is only a small part of the total. This leaves the initial problem, whether abnormal motility of isolated specimens is a serious practical problem in Qo_2 measurement, essentially unresolved. In addition to measuring oxygen consumption, many workers have determined carbon dioxide production during respiration, and from the results have calculated the molar ratio CO_2 : O_2 *i.e.* the respiratory quotient (R.Q.), Carbon dioxide measurements are in themselves legitimate, but the interpretation of R.Q. values even in animals whose metabolism is comparatively well-known is apt to be hazardous, and in nematodes has sometimes led to unwarranted conclusions concerning the kinds of substrates being metabolized. Any interpretation of this sort presupposes that all CO_2 arises from the complete oxidation of the substrate, and that CO_2 fixation into tissue components is not extensive.

In nematodes, neither of these suppositions seem justified. As has already been said, oxidative metabolism is often associated with fermentation processes which themselves give rise to CO_2, and in adults of *H. gallinae*, as well as in eggs and muscle of *A. lumbricoides* there is evidennce for a fairly extensive CO_2 fixation. Thus, while measurement of CO_2 output should not be discouraged, interpretation of respiratory quotients should be made with caution. From the foregoign account it can be said that all nematodes studied so far can use oxygen, that the rate of utilization is dependent upon the physiological stage of development, and that the oxygen consumed is not merely the reflection of an abnormal motility under experimental conditions.

To these conclusions one may add another, namely, that the species studied with two dubious exceptions must have oxygen for their continued well-being. The exceptions are *Ascaris*, which while surviving in saline does not seem to favour aerobic over anaerobic conditions, and *Turbatrix aceti*, the free-living vinegar peelworm, some individuals of which have been reported to survive for six months in culture without oxygen. However, experiments of this kind supply no information concerning the possible oxygen requirement for normal growth and maintenance of tissues and production of eggs. Other species including many intestinal parasites quickly succumb to anaerobic conditions, although oxygen tensions in their normal environment are low. It is known that many parasites survive temporary anaerobiosis, and von Brand (1946) has pointed out that flooding of soils must result in oxygen deficiencies to which nematodes must adapt themselves if they are to survive. There has been general agreement for some years that the metabolism of nematode eggs is primarily, and even obligatively, aerobic.

Oxygen requirements of *Ascaris* eggs, which have already been mentioned, may be taken as typical, although no other species has been examined in such detail. These eggs survive anaerobically for some days, but fail to develop unless oxygen is present. Similar observations have been made on a number of other animal-parasitic species. It is interesting to note, however, that no decisive experiment seems to have been performed with the developing eggs of parasites or free-living forms which under natural conditions can have access to very limited supplies of oxygen. An example is to be found in the common pinworm, *Enterobius vermicularis*, which lives in the vermiform appendix and at various levels of the host's caecum, migrating from these loci to the anal and perineal regions where the eggs are laid.

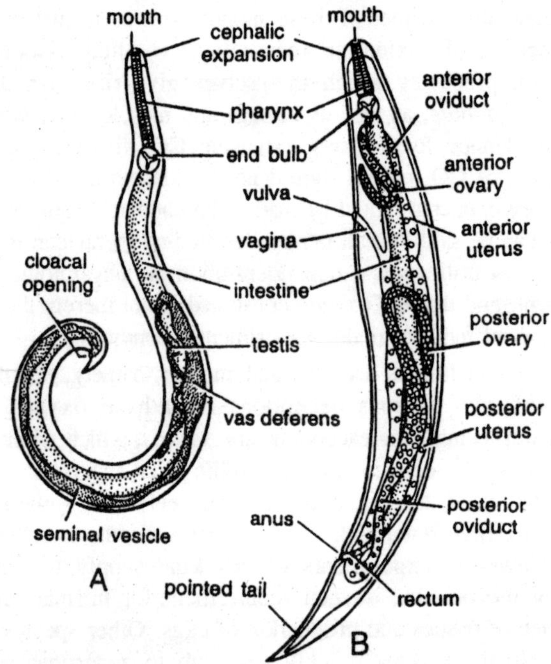

Fig. 3.8. Enterobius vermicularis. A—Adult male. B—Adult female.

Despite the essentially anaerobic character of the natural habitat, the eggs when laid are usually fully embryonated. There is need, therefore, for careful experimental examination of these and similar eggs before generalizations concerning the need of oxygen for development can be accepted. Oxidative energy needs in animals can be satisfied by proteins, fats or carbohydrates, which in turn may come from the diet, or from body depots. There is no evidence for or against the oxidation of dietary amino acids by nematodes living under more or less natural conditions, nor is there much evidence for any extensive catabolism of tissue proteins. Surviving animal parasites excrete modest amounts of nitrogenous end products but under conditions which can scarcely be called normal, although exactly what is meant by normal is not yet understood.

Recent work by Pollak (1957a) has, however, furnished a striking example of how a simple change or addition to the suspending medium can effect metabolism. Pollak found that after two days survival in a balanced saline solution, the ammonia and free amino acid content of *Ascaris* ovaries decreased to less than half its initial value. If, however, the medium contained ammonium ions, the ovarian concentrations of both ammonia and free amino acids was maintained, owing to the ability of this organ to sysnthesize amino acids from keto acids and ammonia. As already mentioned, considerable protein is synthesized in the ovaries and leaves the body in the eggs.

During subsequent embryonation this protein is not used as a source of energy, but is concerned solely with the formation of new tissue protein. The only adult nematode which has been studied appears to have no oxidative fat metabolism, for after several days of starvation *in vitro*, von Brand (1941) could detect no change in the total body fat of *Ascaris*. Under similar conditions fat globules in the intestinal cells disappeared, but it seems likely that these globules represent such a small fraction of the total body fat that their disappearance would not be detected by gravimetric methods. In *Trichinella spiralis* larvae, on the other hand, aerobie *in vitro* survival experimennts resulted in a significant loss of body fat.

Histochemical evidence suggests that the surviving infective larvae of several species rely on the metabolism of stored fat for their energy requirements, which are no doubt small, but may extend over a period of several months. In these, and in the infective egg of *Ascaridia galli* loss of infectivity seems to be associated with the reduction of stored fats to a low level. Such observations should naturally be extended to include many more species, using more precise methods. The most thoroughly studied instance of oxidative fat metabolism is that which occurs in embryonating *Ascaris* eggs.

The unembryonated eggs contain a great deal of fat, chiefly as triglycerides which disappears slowly during the first few days of embryonation and then much more rapidly until after 18 to 20 days the egg has become infective. Thereafter the still considerable fat stores decrease much more slowly as the egg enters the semi-dormant respiratory period previously described. Curiously, however, the maximum rate of fat utilization occurs at a time when the Qo_2 is decreasing, the evidence showing that only part of the disappearing fat is completely oxidized, the remainder being partially oxidized and then used for carbohydrate synthesis. It is tempting to speculate that fat metabolism in adult nematodes is primarily anabolic, with synthesized fats being deposited chiefly in the oocytes, and that fat catabolism occurs chiefly in the eggs and larvae, particularly in eggs which are capable of prolonged survival following embryonation, or in infective larvae which feed little or not at all. Present evidence supports such a view, but is insufficient to permit strong endorsement of it.

As for the mechanisms by means of which nematodes oxidize fats, nothing at all is known. In contrast to proteins, which do not contribute much to energy metabolism, and fats, whose catabolism seems to be restricted to certain developmental stages, carbohydrates are extensively oxidized. As already pointed out, glycogen is stored, sometimes in very large amonunts, in a number of animal-parasitic nematodes. During starvation this glycogen is rapidly depleted in part by fermentative and in part by oxidative processes, the relative proportions depending upon the characteristics of the particular species. This problem will be discussed in more detail later. Here it is sufficient to say that most of the evidence linking oxygen with carbohydrate metabolism is convincing, but indirect. However, in the filarial worm (*Litomosoides carinik*) Bueding (1949b) showed conclusively that at least during the early stages of survival the disappearance of glycogen was linked with the Qo_2. Under natural conditions of course, dietary carbohydrate may be expected to contribute directly to the energy metabolism, any excess being employed for resynthesis of depleted glycogen reserves.

Resynthesis of glycogen *in vitro* has been demonstrated in *Litomosoides*, when excess glucose, fructose, or mannose were supplied in the saline medium. In *Ascaris*, sorbose can also be used for this purpose. Glycogen is stored in *Ascaris* chiefly in the bulbs of the muscle cells and in the oocytes of the ovaries, and durinng survival both of these tissues are depleted. Fertilized eggs, with which the uterus is filled, cannot be depleted by starvation as they are not permeable to carbohydrates. During embryonation of *Ascaris* eggs, there is an extensive disappearance of carbohydrate (both glycogen and trehalose) during the period of active cell division and increasing Qo_2. There can be little doubt that these carbohydrates are a major source of ennergy for the complex biochemical conversions and synthesis taking place at this time.

Later, while the embryo is becoming infective, both glycogen and trehalose are resynthesized, so that the infective egg contains just as much of these two carbohydrates as the unembryonated egg. As already indicated, the resynthesis is the result of conversion of egg triglycerides to glycogen and trehalose. It is remarkable that during subsequent survival of the infective egg these large carbohydrate reserves are not used, although, as has been mentioned, there is a gradual disappearnace of fats. The hatched larva therefore begins its new life with a plentiful supply of carbohydrate, whose fate is not yet known. Nothing is known about the changes occurring in reserve carbohydrates of other nematode eggs, all of which seem to be cleidoic and therefore dependent upon food reserves of some kind. Eggs of most species, of course, hatch spontaneously instead of developing further to an infective, resistance stage, so that it need not be expected that the resynthesis of carbohydrate will be found to be a characteristic feature. However, the eggs of many plant-parasitic nematodes (*Heterodera*, *Meloidogyne*) are comparable in their development to *Ascaris* eggs and in these it will be of interest to compare the details of carbohydrate biochemistry.

Oxidative Mechanisms

In most animals the first phase of carbohydrate metabolism results in the conversion of glycogen to glucose to pyruvate by way of the well-known Embden-Meyerhof reactions. A number of animal-parasitic nematodes have been found to conform without exception to this metabolic pattern. In many animals alternative pathways for glucose oxidation exist, which have not yet been searched for in nematodes, but which are generally of a secondary nature. The Embden-Meyerhof reactions, of course, do not require oxygen, but merely produce the pyruvate which is then either fermented, or oxidized to carbon dioxide and water. At this point in the metabolic processes, intermediates formed in the catabolism of amino acids and fats also enter the metabolic pool. The general features of intermediary metabolism up to this point are so well-known that they need not be discussed here in detail. A great many animals proceed with the oxidation of pyruvate by converting it to a two carbon fragment which is then converted to two moles of CO_2 by participation in one complete turn of the Krebs tricarboxylic acid cycle.

Hydrogen atoms (protons and electrons) which are liberated during these oxidations are simultaneously oxidized by oxygen to form two moles of water. Oxygen must itself be activated before it can participate in these reactions, the entire conversion of protons, electrons, and activated oxygen to water usually proceeding through the agency of the cytochrome–cytochrome oxidase system. This concept of biological oxidations is widely accepted, and certainly applies generally to the tissues of higher animals. In lower animals, however, where fewer investigations have been made, it is clear that other oxidative pathways may sometimes be more important, although their precise nature is obscure. The limited knowledge of oxidative metabolism in nematodes certainly leads to the conclusion that some species employ the Krebs cycle-cytochrome oxidase system in tissue oxidations, and that some do not. How general this division is cannot be stated, as only a few animal parasites have been studied.

The Krebs cycle has been identified in whole or in part in *Nematodirus* spp., *Ascaridia galli*, *Neoaplectana glaseri* and *Trichinella*

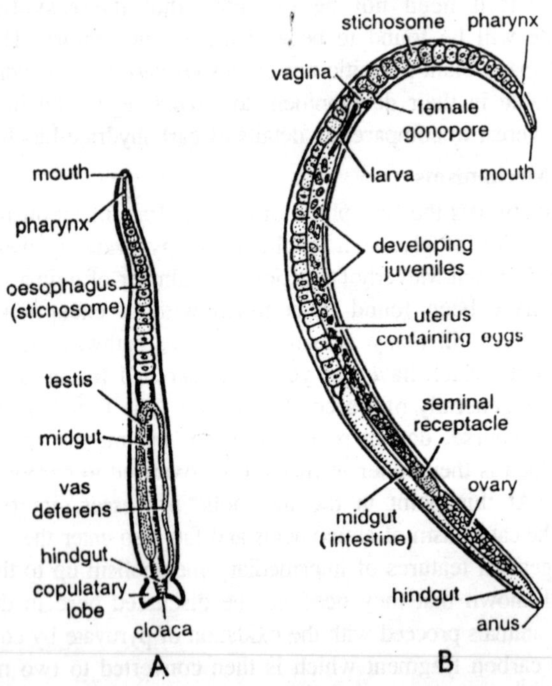

Fig. 3.9. Trichinella spiralis. A—Adult male. B—Adult female.

spiralis (larvae) and unequivocal evidence for the occurrence of cytochrome–cytochrome oxidase in *Trichinella* has been reported. On the other hand neither *Litomosoides* nor *Ascaris* possess a cytochrome system, nor does *Ascoris* make use of the Krebs cycle. Several methods can be used for the identification of the cytochrome system in tissues. historically, the first of these was based on the spectroscopic detection of cytochrome c, and *Ascaris* muscle was one of the first tissues in which cytochrome *c* was reported. Later work, however, has shown that enzymic assay is a more reliable procedure, and it is on this basis that Bueding and Charms(1952) rejected the idea that cytochromes were concerned in biological oxidatios by *Ascaris*. A third method, which is indirect, relies upon the effect of respiratory inhibitors.

The inhibitor most commonly used is cyanide, which is well-known to repress the activity of ferric enzymes almost completely at concentrations approximating 10^{-3}M. Unfortunately, enzymes such as peroxidases are also affected so that cyanide is by no means specific for the cytochrome system. A better inhibitor is carbon monoxide, whose effect in the dark, if reversed by light, appears to be quite specific for ferrous enzymes, of which the components of the reduced cytochrome system are the only known examples. Even this method can lead to puzzling results. Thus, embryonating *Ascaris* eggs are highly sensitive to carbon monoxide and cyanide, but give negative results for the cytochrome system by both enzymic and spectroscopic tests. The only conclusion to be drawn here is that a system is present which at least closely resembles the cytochrome system in its properties.

Nematodes do not seem to differ from other animals in their requirement for high-energy phosphate compounds. Oxidative phosphorylations occur in the species studied by Rogers and Lazarus (1949b) and Chin and Bueding (1954) and several phosphate esters have been tentatively identified. Some details of phosphorus metabolism may be modifications of the pattern normal to higher animals. *Ascaris*, for example, does not appeares to contain appreciable reserves of high energy phosphates such as creatine or arginine phosphates although filariform larvae of *Strongyloides ratti* probably do contain such substances. A reasonably explanation of this difference between species might be that *Ascaris*, which leads a sheltered life in the intestine, has lost the capacity for storing high energy phosphorus compounds which in other, less-sheltered species, are still necessary to the conduct of a normal life. However, more observations on the phenomena of oxidative metabolism in nematodes must be forthcoming before it will be possible to discern a general pattern, if, indeed, such exists.

As already pointed out, many adult and larval nematodes conduct active fermentations when oxygen is absent, and even when it is present. On the other hand, there is no evidence for the occrrence of fermentations in eggs. This is not surprising, as nematode eggs in general are impermeable to organic molecules, whose accumulation as toxic waste products would be unfavourable to survival.

Investigations of nematode fementations have been confined to a few species, each of which has its own peculiarities. The fact that many, and perhaps most, species are able temporarily to survive anoxic conditions strongly suggests that the ability to ferment is widespread. It is, however, a great deal easier to subject an organism to anaerobic conditions and note its ability to survive than it is to extend the experiment to a study of the excreted waste products. As a result, the widely-held view that fermentations are a more or less characteristic feature of nematode metabolism is still not well supported experimentally. As already stated, glycolysis of carbohydrate with formation of pyruvate appears to be a normal anaerobic sequance of reactions in nematodes as in other animals, and it is the subsequent metabolism of pyruvate which determines the peculiarly aerobic or anaerobic features of catabolism. As we have just seen, oxidative metabolism of pyruvate by nematodes may or may not follow the same pathways as in higher animals. It is in the fermentation of pyruvate, however, that their divergence is most notable. This divergence is only relative.

Most higher animals can survive short periods of anoxia, during which time pyruvate is reduced to lactate, most of which accumulates in the tissues until these are irreversibly poisoned, or until it is metabolized upon readmission of oxygen. Nematodes, on the other hand, are often exposed to more prolonged shortages of oxygen. They do not accumulate fermentation products, but excrete them efficiently. This ability to excrete such waste products provides an explanation for the fact that nematodes under anoxic conditions do not incur a pronounced oxygen debt. In higher animals, for example, tissue lactate accumulated during anaerobiosis may be metabolized rapidly upon readmission of air, produceing a temporary and quantitative increase in Qo_2 which is spoken of as the repayment of an oxygen debt. If the lactate were excreted instead, there would, presumably, be no oxygen debt. With some exceptions, the products of nematode fermentation are carbon dioxide and organic acids.

Fermentation acids were first studied in *Ascaris* which, as has been shown since, produces a singularly complex mixture. It is probably fortunate that Ascaris was chosen for Weinland's experiments, for the results presented a challenge to biochemists which has proved to be irresistible, and has led to the investigation of other species, most of which, fortunately, seem to excrete somewhat simpler mixtures. *Ascaris* adults and *Trichinella* larvae excrete a mixture of acids which is considerably more complex than that of species such as *Heterakis*, Litomosoides, or *Dracunculus*. An immediate resemblance to the end-products of bacterial fermentations is apparent, and indeed there is evidence that the mechanism of propinic acid formation by *Heterakis* and in *Ascaris* muscle is similar to that occurring in *Propionobacterium*. Some of the acids, such as the α-methylbutyric acid produced in large amounts by *Ascaris* are obviusly the result of a long sequence of reactions beginning with pyruvate or a similar preursor. Saz, Vidrine and Hubbard (1958) have proposed a mechanism for this reaction, beginning with the condensation of two moles of pyruvate α-acetolactate and CO_2, which was supported in part by experimental evidence.

The *Ascaris* fermentation acids are well tolerated by the parasite itself, as considerable quantities (about 40 mole/litre) are present in the hemolymph. *Heterakis* tissue, however, contain very little acetic and propionic acids, although succinic acid, which is the immediate precursor of propionic acid and carbon dioxide and in this fermentation, occurs in appreciable amounts. Lactic acid, which is the commonest fermentation acid in most animal tissues, is a major excretion product only in the filarial worms *Litomosoides* and *Dracunculus*.

Many microbial fermentation products include non-acid substances, particularly various alcohols. Until recently little attention has been paid to the possibility that nematodes excrete such compounds. However, as Bueding (1951) and Saz, Vidrine and Hubbard (1958) have found that a 4-carbon alcohel, acetylmethycarbinol is a normal excretion product in *Litomosoides* and *Ascaris*, it is possible that non-acid substances occur fairly generally. Carbon dioxide, which is a characteristic product of nematode fermentation, arises from pyruvate or other intermediates in the formation of acid and non-acid fermentation products. It is noteworthy that none of the fermentations which have just been considered are eliminated during aerobic respiration. In air, the *Litomosoides* fermentation is depressed, and the relative proportions of acetic and lactic acids are altered, but in *Ascaris*,

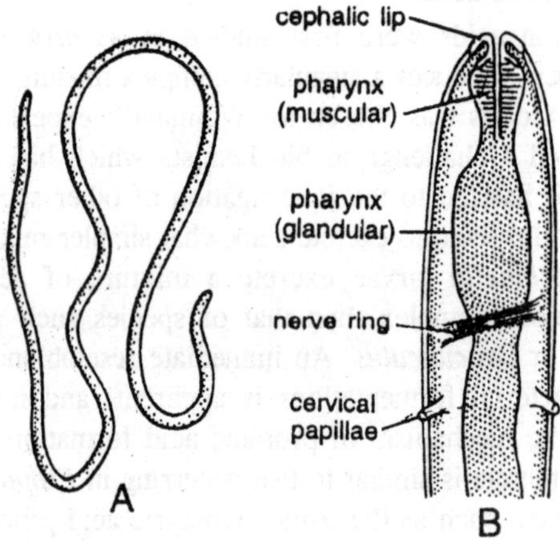

Fig. 3.10. Dracunculus medinensis. A—Entire worm. B—Anterior end of male.

Heterakis, *Dracunculus* and *Trichinella* there is little change in the total acids excreted. Thus, aerobic respiration in these last four species is simply superimposed upon a constant basic fermentation.

Ascaris and *Heterokis* are intestinal parasites which probably suffer a chronic oxygen deficiency; *Litomosoides*, *Dracunculus* and *Trichinella*, on the other hand, live in the pleural cavity, subcutaneous tissue or skeletal muscle of their hosts, in which there is probably no serious lack of oxygen. Moreover, *Trichinella* larvae, owing to their small size and the participation of the cytochrome oxidase system in their aerobic metabolism, make full use of oxygen even at very low partial pressures. Their persistence in maintaining a vigorous anaerobic fermentation is therefore difficult to understand, although it is evidently a fundamental part of their catabolism. Observations such as these illustrate perfectly the need for additional investigation of species occupying widely different habitats. In concluding this brief account of fermentation, it is legitimate to ask why nematodes indulge in such metabolic acrobatics as the formation of five-and six-carbon acids when the simple reduction of pyruvate to lactate is sufficient for most higher animals. One suggestion might be that more of the energy of the glucose molecule is thus made available. This seems doubtful, since the nematode acids such as α-methylbutric are actually more reduced than lactic acid. Another possibility worth considering is based on the chemical properties of the acids themselves.

For the most part, fermentation acids are excreted as acids, and not as salts. Tissue cations are conserved in this way, but even more important may be the fact that the permeability of most cell membranes to ions is low. The efficient excretion of acids, therefore, may depend upon their presence in undissociated form. The fermentation acids which have just been discussed fall into two groups with respect to their ionization constants. First lactic acid, and second, all the others, whose respective constants are similar and about one-tenth that of lactic. If *Ascaris* is chosen as an example, we have already seen that its hemolymph is fairly acid (pH 6.4), and can predict that at this pH the proportion of the weaker acids which are present in undissociated form will be very considerably higher than the proportion of any undissociated actic acid which might be present.

The formation of very weak acids should therefore favour their efficient excretion. In the case of parasitic nematodes, the possible effect on the host should alos be considered, as the formation of relatively strong acids would doubtless be highly damaging. *Litomosoides*, as Bueding (1949b) found, excretes both lactic and acetic acids, and under the semi-aerobic conditions which prevail in the pleural cavity, the excretion of acetic acid may be dominant. Nevertheless, the pH of the pleural cavity of infected rats falls to 6.5, and, as lactic acid is about ten times as effective as acetic in its contribution to the final pH, its relative concentration in the mixture might be an important factor in determining the ability of the host to withstand the infection. It must be understood, however, that this kind of reasoning, to be effective, must be supported by a great deal more factual information. It is perhaps advisable at this point to provide a brief summary of aerobic and anaerobic respiration and biological oxidations. No nematode yet studied has been shown by conclusive evidence to be capable of surviving and reproducing indefinitely under strictly anaerobic conditions, and all species examined have shown some capacity for using oxygen.

The evidence suggests that oxygen may be necessary for muscular contraction, but that it is not used exclusively for this purpose. The properties of the hemolymph and tissue hemolymph are such that it has not been possible to assign to them an oxygen-carrying function. Oxidative metabolism may proceed by way of the Krebs tricarboxylic acid cycle, with cytochrome oxidase as the terminal hydrogen acceptor, but this is not always the case, and the general features of cell oxidations are still not clear. Both anaerobic and aerobic fermentations occur in all of the few species which have been investigated, which are by no means representative. Fermentations are typically complex, with excretion of carbon dioxide and at least two organic acids. Lactic acid has been recognized as a major acid in only two species. Nematode eggs seem to require oxygen for development, and no fermentation has been observed in them. However, this view may require modification as more observations are made.

Non-Protein Nitrogen

Although the extent to which proteins are used by nematodes as a source of energy is debatbale, ther is, as in all animals, a continuing catabolism of protein and other nitrogenous compounds which results in the formation and excretion of products classified collectively as the non-protein nitrogen fraction (NPN). Most of the NPN originates from structural components of tissues which are continually being broken down and replaced, although when the diet contains much protein, this may be responsible for a more direct formation of NPN.

Mature animals are normally in a state of nitrogen balance, where input and output of nitrogen are equal. These well-known principles are mentioned here only because it is important to realize that it has not yet been possible to study NPN in nematodes under conditions which are physiologically satisfying. All investigations have been based on the collection of material excreted during a 24-hour period or longer, which means, particularly in the smaller species, that the nitrogen metabolism is that of a starving animal and is not necessarily the same as that of animals in nitrogen balance. Earlier studies were also complicated in some cases by the presence of a bacterial flora, but this problem can now be solved by judicious use of antibiotics or bactericides. However, it is hardly necessary to stress that a certain amount of caution is desirable in assessing the results which will now be presented. Total nitrogen excreted by several animal parasitic species, expressed as mg per 100 gm body weight per 24 hours is as follows:

Ascaris lumbricoides	39	(Savel, 1955a)
Ascaridia galli	29—41	(Rogers, 1952)
Nematodirus spp.	123—159	(Rogers, 1952)
Trichinella spiralis (larvae)	280	(Haskins & Weinstein, 1957a

Clearly, the rate of nitrogen excretion by the small nematodes *Trichinella* and *Nematodirus* is several times as great as in the much larger *Ascaridia* and *Ascaris*. The need for caution in accepting such results can be illustrated by a simple calculation based on the nitrogen excreted in 24 hours by *Trichinella*. It this nitrogen originates in tissue proteins, and the factor for converting nitrogen to protein is 6.25, then protein metabolized is 1.75 gm. Assuming that total solids in *Trichinella* larvae amount to 20 per cent of the fresh weight, and that these solids are 70 per cent protein, it turns out that surviving *Trichinella* larvae catabolize 12.5 per cent of their body proteins evevery 24 hours. This is a very vigorous metabolism, which may acctually be much more extensive than that occurring under natural conditions in the host muscle. Analysis of the excreted nitrogen in the species listed above has confirmed Weinland's (1904) conclusion that ammonia is a major constituent. As has often been pointed out, this is not surprising, since most nematodes are aquatic invertebrates in which the danger of auto-intoxication by ammonia is probably not great. Its effect on the host in heavy parasitic infections is less certain, although Myuge (1957b) has suggested that it is a major cause of damage in infections of plants by *Ditylenchus destructor*. However, the data summarized in Table shows clearly that the excretions also contain significant amounts of urea, amino acids and peptides.

Rogers (1955a) and Haskins and Weinstein (1957b) have analyzed the last two fractions in some detail, and have considered the possibility that these are not normally waste products of metabolism, but appear in large quantities in the excreta as a result of the somewhat abnormal experimental conditions which have already been discussed. It is of interest in this connection that by simply confining individuals of *Ascaris* in such a way that the ingestion and ejection of water from the intestinal tract was not restricted, although the body itself was not immersed in saline, Cavier and Savel (1954c) were able to show that this species can regulate the relative proportions of ammonia and urea which are excreted.

Under such conditions urea accounted for 52 per cent of the total nitrogen excreted, ammonia for only 2 per cent. As many nematodes live from time to time, or at certain stages of their life cycle, under conditions of wate restriction, a similar regulation of the excretion products may occur. One of the more interesting aspects of nitrogen excretion was recently discovered by Weinstein and Haskins (1955) who obtained evidence for the excretion of small amounts of several aliphatic amines from the excretory pore of *Nippostrongylus muris*. *Trichinella* excretes similar amines. Most of these, such as ethyl amine, propyl amine and cadaverine, can be regarded as decarboxylation products of the corresponding amino acids, and some are known to exert a pharmacological action upon the tissues of higher animals. They comparise 7 per cent of the total nitrogen excreted by *Trichinella*.

Tissue minces of *Ascaris Ascaridia* and *Nematodirus* all liberate ammonia from added 1-amino acids presumably as a result of 1-amino acid oxidase action. The most active enzyme is found in the intestine, which, as stated earlier, also contains active proteinases and peptidases. This then, could be one source of excreted ammonia. However, these species also contain urease, an enzyme which is not common in animals, and which converts urea to ammonia and carbon dioxide. As urea is a common excretion product it might also be an immediate precursor of some of the ammonia formed. In higher animals, of course, ammonia formed by deamination reactions is efficiently converted to the less toxic urea by means of the Krebs-Henseleit ornithine cycle. All the species mentioned above seem to possess some form of this cycle, with the enzymes concerned in it being most active in the intestine. Thus, there appear to be alternative routes for the disposal of the primary product of deamination, which is ammonia. It can be excreted directly or it can be converted to urea; the urea can be excreted, or reconverted to ammonia by the action of urease. The balance between these alternatives is possibly determined by the conditions to which the parasite is exposed from time to time, conditions of water restriction favouring the excretion of urea.

On the basis of enzyme distribution in *Ascaris*, the intestine is the probable site of the reactions involved, and may also be the tissue primarily concerned with nitrogen excretion. Cavier and Savel (1954c) provided direct experimental support for this hypothesis in their studies on *Ascaris*. Certainly the curious and unusual juxtaposition of the Krebs-Henseleit urea-synthesizing cycle and the enzyme urease are deserving of more extended study.

Toxicity of Excretion Products

The long-standing curiosity of parasitologists concerning the toxic properties of the secretions and excretions has never been adequately satisfied. Toxicity is only one of the factors which determines the aggressiveness and pathogenicity of the parasite, and it is impossible here to attempt an assessment of its general role in the establishment of parasitic infections. The following comments, therefore, are made only because it has become clear during the preceding discussion that nematodes do in fact secrete or excrete substances which are apt to be toxic to the host. There seems as yet to be no sound evidence that nematodes secrete toxins comparable to the several well-defined bacterial toxins *i.e.,* substances toxic in very small concentrations which are excreted by the microorganism and which when injected separately from the microorganism produce symptoms that mimic the disease. According to this definition, it will not even be possible to put the criteria for toxin production to an experimental test until at least one pathogenic nematode has been cultured successfully, and the culture fluids examined critically for toxic reactions.

An indirect approach to this problem has been made by Nolte and Kohler (1952) who prepared extracts from leaves of plants infected with *Heterodera schachtii*, *Pratylenchus pratensis* or *Ditylenchus dipsaci* and then exposed leaves of healthy, uninfected plants to the extracts. Curling and wilting similar to that observed in the leaves of infected plants occurred, but, as the authors were aware, such experiments cannot distinguish between nematode toxins and plant toxins produced in response to the infection. Similarly, the grass *Festuca* which has been infected with *Anguina agrostis* becomes toxic to cattle and sheep, which are not harmed by direct injection of *Anguina* into the blood stream. Such observations are interesting, and inconclusive.

Once again it is apparent that the solution to an important problem awaits the development of improved methods for studying the nematodes themselves. The toxicity of excretion products generally, or of substances arising in the host as the result of infection, can be likened to the decomposition products produced by bacteria, and known generically as ptomaines. The doctrine of ptomaines as developed long ago by Brieger has not, in bacteriological science, survived the demonstration that such ptomaines are as a rule only feebly toxic. One should not, however, forget that this kind of restrained toxicity may provide a more satisfactory explanation for the subacute symptoms of many nematode diseases than any other.

In *Ascaris* infections, for example, the adults living free in the intestine excrete considerable amounts of five-and six-carbon fermentation acids whose toxicity to the host does not seem to have been studied. Similarly, nematodes commonly undergo several molts which may result in the introduction of a good deal of protein and protein decomposition products into the host tissues, and it has been suggested that the guanidine liberated from muscle during severe *Trichinella* infections may be responsible for many of the symptoms of trichinosis. Where circulation of host fluids is not highly efficient, much of the local damage due to infection may be the result of the normal living habits of the parasite. Myuge (1957b) has suggested, as the result of his experiments with *Ditylenchus destructor* that infection of potatoes can be described in physiological terms as follows.

The parasite secretes amylas and protease which bring about extracorporeal digestion of starch and protein, and at the same time it excretes ammonia, which causes the pH to rise and initiates necrosis of the host tissue. The plant responds, as most plants do to infection of any kind, by increasing the local respiratory rate. Meanwhile, changes in osmotic pressure caused by hydrolysis of proteins and starch lead to increased turgor pressure within the cells which results eventually in disruption and rotting of the infected portion of the tuber. On the other hand, similar changes in Qo_2 and pH in cucumber roots infected with *Meloidogyne incognita* lead not to necrosis but to gall formation. Clearly, the host response in *Meloidogyne* infections is more dramatic than in *D. destructor* infections, although the initial stimulus provided by the parasite may be similar. It is not necessary to accept the details of such hypotheses as these in order to profit from the concept that toxicity to the host may be the result of the parasite's normal physiological functions of secretion and excretion. On the other hand it is essential to remember that the pathogenicity of nematodes may be due to other factors which are not suitable for inclusion in the present discussion, but which are described elsewhere in this book.

Water Balance and Permeability

All the space between tissues in nematodes is filled with an aqueous solution, the hemolymph. There are no circular muscles in the body wall, so that the longitudinal muscles contract against the incompressible hemolymph, and, as these muscles are divided into dorsal and ventral groups by the lateral chords, effective motion can only occur in the dorsoventral direction. Considerable muscular tone is normally exerted, with the result that the hemolymph always exerts a hydrostatic, or turgor, presure which is easily demonstrated by cutting the body wall and observing the sudden gush of hemolymph. Fielding (1951) reported the persistence of turgor pressure in individuals of *Anguina tritici* and *Ditylenchus dipsaci* which had been dormant for more than twenty years.

The magnitude of the turgor pressure of *Ascaris* surviving in isotonic solution has been measured by Harris and Crofton (1957) who found that it varied in a rhythmical way, occasionally approaching one-half atmosphere, and averaging about 70 mm. of mercury. Harris and Crofton suggested that the peculiarities of musculature and turgor pressure, coupled with the limited distribution of nerve cells which are confined entirely to the head and tail regions, must create problems in neuro-muscular physiology quite different from the reflex mechanism occurring in other animals. Not much has yet been learned about neuro-muscular physiology in nematodes, although the subject is one of great interest from the point of view of drug action. It has been established that acetylcholine stimulates isolated muscle strips of *Ascaris*, and that its action is prolonged, probably owing to the extremely low concentrations of cholinesterase in the muscle.

The lips and tails are extremely sensitive and the head region, including the nerve ring, contains about sixteen times more acetylcholine than the somatic muscles. Thus, although acetylcholine appears to play a part in sensory and nerve-muscle physiology in *Ascaris* its precise role, and the possible role of other humoral mediators, has not been accurately defined. The constant turgor pressure exerted by the hemolymph means that the nematode intestine, which consists of a single layer of cells devoid of muscle fibres, is completely collapsed when empty, and can be filled only by exerting a back pressure of its own. This is made possible by the muscles and valves of the pharynx or esophagus and the anus, which constitute a closed system until a sufficient supply of fluids have been ingested and the hydrostatic pressure has increased to a critical point. The anus then opens and, as anyone who has worked with *Ascaris* knows, the intestinal contents are discharged with considerable violence. Filling of the intestine entails an increase in the total volume of the worm which is possible because of the niceties of the cuticular structure.

The fibre layers are so arranged that relaxation of the somatic musculature permits the extension of the fibre "cage" and the elongation of the nematode. The volume also increase provided that the original value of the angle Θ is greater than 55° As in *Ascaris* this angle is usually about 70° the worm is theoretically capable of very considerable length and volume increase at constant turgor pressure. Presumably other nematodes also make use of this favourable geometry in the fibre layers, for Stephenson (1942) observed that in hypertonic solutions loss of water entailed a 30 per cent *decrease* in the length of *Rhabditis terrestris*, while Vargas (1952) reported that ramoval of the microfilaria of *Onchocerca vovulus* from an isotonic glucose solution into water *increased* the length by 55 per cent. These observations on *Rhabditis* and *Onchoccrca* quite clearly indicate a lack of osmoregulation, and raise the question whether nematodes in general can control the water content of their bodies.

Consideration of the results of several investigations show that many species can, and others cannot. *Ascaris* and *Parascaris*, for example, rapidly absorb water from hypotonic colutions, and lose it to hypertonic solutions. Water is apparently lost or gained freely through the cuticle, for ligatering the worm at both ends does not prevent the changes from taking place. *Ascaris* larvae are also sensitive to osmotic changes in the medium. Many other species, however are able to survive aparently unharmed in solutions which are definitely hypotonic, although they are apt to be more sensitive to hypertonic, solutions. *Ostertagia*, *Haemonthus* larvae, and *Eustrongylides* larvae all survive well in hypotonic saline while adults of *Meloidogyne javanica* live for many days in water alone. Osche (1952) has shown that many saprophytic species readily adapt themselves to the osmotic changes occurring during the progressive decomposition of organic matter, and that a number of free-living species which inhabit environments of low osmotic pressure adapt themselves without difficulty to the higher osmotic pressures normally encountered by parasitic species.

In short, nematodes as a class appear to be well-endowed with the ability to regulate their water content in widely differing environments, although a number of parasitic species can no longr do so. In *Ascaris*, as already mentined, water can be absorbed through the cuticle, which is similar morphologically to the cuticles of other nematodes. It possible, therefore, that the cuticle is generally used for this purpose. In the many species which control their water content, however, it is less certain how water is discharged from the body. A useful study was made by Weinstein (1952), who found that the rate of pulsation of the excretory ampulla in larvae of *Nippostrongylur muris* and *Ancylostoma caninum* is inversely proportional to the degree of hypotonicity of the medium, and that water is actually discharged from the excretory pore of these two species.

Additional observations on other species are now needed to test the general validity of the hypothesis that the excretory system plays a major role in water regulation. Granting that many nematodes are capable of regulating their water content in solutions hypotonic to their own tissues, there remains the problem of salt regulation, concerning which there is still very little reliable information. Despite iitts inability to regulate water, *Ascaris* maintains fairly close control over the concentration of magnesium, calcium and potassium in its hemolymph when placed in solutions of varying ionic composition, and althogh the chloride concentration can be made to fluctuate widely, it tends to be significantly lower than the concentration in the medium. In this parasite, then, there is a fairly efficient regulation of ions, even though there is essentially no regulation of water.

No similar experiment seems to have been made using other species. Variations in the ionic composition of the medium are known to affect survival and respiration, as mentioned earlier, but it is not known whether such effect can be correlated with alternations in the internal ionic environment. Investigation of the ionic composition of soil and plant parasitic nematodes, the degree of control exercised by particular species, and comparison with the composition of the normal environment, might be expected to provide basic bio ogical knowledge of potential practical usefulness. The permeability of *Ascaris* cuticle to respiratory gases, water, chloride and possible to small ions such as the ammonium ion is perhaps to be construed as a general property of the nematode cuticle. On the other hand, there is no evidence for permeability to nonvolatile organic compounds and the cuticle is rightly regarded as providing an efficient barrier between the nematode and damaging substances in its environment. Digestive enzymes, whether those of the host in parasitic species or of microorganisms in saprophytic species, do not normally attack the cuticle. the resistant properties are primarily those of the external cortex, for it has been relized for many years that the other layers are readity digested by enzymes such a pepsin or trypsin.

As previously mentioned there is no general agreement concerning the chemical nature of the external cortex, which is perhaps a somewhat unusual collagen subjected to some degree of tanning. Such proteins are commonly resistant to enzyme or chemical action, and when tanning is heavy, tend also to be impremeable even to water. This change was observed by Ellenby (1946) in cysts of *Heterodera rostochiensis*, which when young are white arer freely permeable to water, but when old and yellow (tanned) are highly resistant to water loss. However, in addition to the external cortex there is present in *Ascaris* an outermost layer which is very thin, osmiophilic and presumably a lipid. This membrane, about which very litle is known, may provide the initial environmental barrier set up the cuticle. Trim (1944) found that antihelminthics such as hexylresorcinol enter *Ascaris* by penetrating the cuticular barrier. Penetration is facilitated by surface active agents which are not themselves absorbed, and worms pre-treated with detergent lose phosphate from their bodies.

These observations suggest that a lipid barrier is of great importance in imaintaining the normal cuticular function. Whether the barrier consists solely of a coating over the external cortex, such as Chitwood (1938) claimed to have isolated from *Ditylenchus dipsaci*, and Bird and Deutsch (1957) found in *Ascaris* by electron microscopy, or whether lipids distributed more generally in the cuticle also play their part remains to be seen. If there is still much uncertainty regarding the intimate cuticular distribution of lipids, it is nevertheless clear that the action of a great many antihelminthics, and most of the useful soil nematocides, is closely relted to their ability to penetrate lipid barriers.

The soil nematocides, for example, are relatively non-polar organic molecules with only slight water solubility but pronounced lipid solubility, and which have an appreciable vapour pressure at physiological temperatures. Many organic compounds possessing these properties apparently, have access to nematode tissue by permeation through the cuticle. Their success as nematocides, however, may depend upon a more specific effect on the tissue. This field, which is that of drug action, is still largely unexplord in nematodes.

It is worth considering however, in the light of Trim's (1949) observations on the leakage of phosphate through *Ascaris* cuticle, that nematocides of the type discussed above may act in part by altering the cuticular structure insuch as way that unfavourable disturbances in water and salt regulation take place. During periods in which nematode larvae or adults are more or less dormant, it might be predicted that their permeability to water andother substances would be much reduced. Infective filariform strongyle larvae, for example, are sheathed in the remains of the second stage cuticle, and are generally considered to be much more resistant to adverse environmental conditions than younger larvae, or those which have entered a host animal. However, the physiological basis of this sesistance does not seem to have been systematically explored. similarly, the prolonged survival of some plant parasitic nematodes in soil suggests that important adaptations, among which might be greatly reduced permeability, are made. If this is so, then susceptibility to nematocides may also be altered. Thus, there is obviously a need for much broader physiological studies than have yet been made.

Permeability of Eggs, and Hatching

Nematode eggs present special problems to the investigator for in these it is increasingly apparent that during development and subsequently up to the time a suitable stimulus is received, the eggs have only a very limited permeability, which suddenly increases markedly in response to the stimulus, and as part of the hatching process. The egg-shell of a great many nematodes consists, as has been pointed out already, of an other layer which may contain chitin or quinone-tanned protein, and an inner vitelline membrane containing a preponderance of lipid-soluble substances.

In *Ascaris* and *Parascaris* this material consists of ascarosides, and reasons for believing that similar glycosides may be present in the vitelline membranes of other eggs have already been presented. Recent work suggests that all shells do not conform to this general structure, but since those which do not, apart from the three species of lungworms studied by Monne (*Dictyocaulus viviparous*, *D. filaria*, and *Metastrongylus elongantus*) are still not well-known, they will not be considered here. There can be little double that the semi-permeability of *Ascaris* eggs is primarily a function of the vitelline membrane, for if the chitonous shell is dissolved in hypochlorite solution the eggs retain their semi-permeable properties. Moreover, heating the eggs to about 75°C melts this membrane, and it is then found that solutes within the eggs diffuse freely through the hard shell. Both the hard shell and the vitelline membrane are also extremely resistant to the action of many corrosive chemicals. The vitelline membrane, apart from its remarkable chemical intertness, seems also to be truly semi-permeable, *i.e.*, only gases are able to pass back and forth through it, and water leaks from it verys lowly. Permeability to respiratory gases is of course self-evident, as the egg metabolism is aerobic.

A striking demonstration of permeability to gases but not to related ions is afforded by the high sensitivity to poisoning by hydrogen cyanide or hydrogen azide (10^{-5}M) and the low sensitivity to cyanide or azide ions (10^{-2}M) Passey and Fair-brain, (1955). The severely restricted permeability to water can be shown by the failure of the embryo to plasmolyze even in concentrated salt solutions. Although other nematode eggs have not been examined in similar detail, the consensus seems to the that they too are chemically resistant and highly impermeable. Wilson (1958) has found, for example, that the relatively thin-shelled eggs of *Trichostrongylus retortaeformis* are, like those of *Ascaris*, incapable of being plasmolyzed in strong salt solutions. As the development of most eggs occurs outside the body of the host, the protection offered by the shell provides an obvious advantages.

At the same time, it presents practical problems of great importance to those who are concerned in public health or agriculture. Fortunately, the lipoid nature of the vitelline membrane tends to produce a sensitivity to organic pesticides similar to that conferred by the cuticle on larvae and adults, with the result that a single agent is sometimes effective against all three stages of development. Most nematode eggs, including those of a great many parasitic species, hatch soon after the embryo is completely developed. This appears to be spontaneous process requiring no alteration of the external environment, and must therefore be due to a physiological stimulus provided by the embryo. Wilson (1958), for example, found that the best hatching of *Trichostrongylus retortaeformis* eggs occurred in distilled water, and that hatching was delayed, rather than accelerated, by salts or sucrose. Just before hatching, the vitelline membrane suddenly became permeable to water, while changes in the hard outer shell were also noted but not detailed.

In contrast to those eggs which hatch spontaneously, there are a number which are not self-stimulating, but require a more or less specific alteration in their external environment which is normally provided by the host. Until such time as these suitable conditions for hatching are provided, these eggs survive and retain their infectivity. Two such eggs are those of *Ascaris* and *Heterodera*, about which useful information has recently been acquired. The first change which occurs in *Ascaris* eggs as the result of a hatching stimulus is, an *Trichostrongylus* eggs, the development of permeability in the vitelline membrane. This is followed, as shown by Rogers (1958), by the secretion or activation of a chitinase which permeates the vitelline membrane and dissolves the hard shell.

The digestion is usually more extensive in one area than in any other, so that the embryo actually leaves the egg through a hole in the shell. These are chemical events which may also take place in *Trichostrongylus* eggs. *Ascaris* differs chiefly in that hatching normally occurs only in response to a stimulus supplied by the host. The stimulus appears to have at least three components which are (1) a pH near 7.2, (2) a partial pressure of carbondioxide approximating 40 mm of mercury, (3) a low redox potential. As these are conditions to be expected in the small intestine of many species during feeding, they explain adequately the well-known fact that host specificity for *Ascaris* is not determined by conditions for hatching, but by conditions for subsequent development of the larvae.

Eggs of the cat and fowl ascarids hatch in response to the same stimulus as *Ascaris* eggs (Rogers, private communication). At the present time it is not possible to state that hatching in *Heterodera* follows the pattern of stimulus and response characteristic of *Trichostrongylus* and *Ascaris* eggs. It is clear, as the result of many investigations, that *Heterodera rostochiensis* eggs hatch when exposed to a hatching factor secreted from the roots of the potato or tomato, and that this factor has chemical and physiological properties similar to those of the cardiac glycosides.

Although recent work strongly supports the view that cardiac glycosides have a profound effect on the permeability of animal membranes there seems to be no clear evidence that hatching factor causes of corresponding increase in the permeability of *Heterodera* eggs. It it does, then it action on the vitelline membrane is direct, and dies not necessarily involve the embryo at all. This does not mean that the embryo can make to contribution to the hatching process, such as digestion of the chitinous shell, but only that the development of permeability in the vitelline membrane is a direct result of the presence of hatching factor in the external environment, rather than an indirect response of the embryo itself to a suitable stimulus. The stimulus-response phenomenon is also responsible for the exsheathment of infective trichostrongyle larvae in the gastro-intestinal tract of the sheep, in which larvae of *Haemonchus contortus*, *Trichostrongylus axei* and other species produce an exsheathing fluid in response to a host stimulus which has not been completely identified, but which is similar to the stimulus affecting ascarid eggs. However, the sheath differs from the eggs in being permeable to small molecules and in having no structure comparable to the vitelline membrane. As a result, exsheathing fluid secreted by the stimulated larva has immediate access to the inner surface of the sheath, which is digested directly without any necessity for the preliminary development of permeability.

Conclusion

While the foregoing account of nematode physiology and biochemistry makes no pretence at completeness, a dispassionate view can lead only to the conclusion that present knowledge is centered about a few animal parasites with occasional excursions into the realm of plant-parasitic and free-living species. In discussing "the extraodinary similarity in from and organization between the very large number of species, genera, orders and families which constitute this class of the very diversiform phylum Aschelminthes," Harris and Crofton (1957) comment that "the elementary student may be forgiven at times for thinking that there is only one nematode but that the model comes in different size and with a great variety of life histories." It is unfortunately not yet possible to formulate criteria for judging the applicability of this view to nematode physiology, although its validity is certainly questionable.

In considering respiration, for example, a reasonable amount of evidence assembled in the preceding pages supports the view that animal parasites, including those dwelling in essentially anaerobic habitats such as the mammlian intestine, are facultative aerobes which continue to ferment actively even while respiring. On the other hand, anaerobic and aerobic fermentations in free-living or plant-parasitic species have not been investigated, although an active respiration has been demonstrated. The questions then arise, whether oxidative metabolism in free-living nematodes is of the cytochrome-catalyzed type, with little ability to ferment and a well-defined Pasteur effect, and are the observed deviations from this pattern in parasites a reflection of adaptations to parasitism? These are obvious questions, representative of many others which might be asked, and which are at the moment unanswerable. In so far as the physiology and biochemistry of parasitism by nematodes is concerned, it is clear that the solution to many such problems depends upon the acquisition of broad comparative knowledge, much of which is now within technical reach of the patient investigator.

4

PLANARIA

Planaria or *Dugesia* belongs to class Turbellaria. The Turbellarians are mostly free-living, unsegmented flatworms that are clothed with a cellular epidermis, which is usually ciliated, sometimes only in parts. Turbellarians are primarily aquatic, and the great majority are marine. Although these are a few pelagic species, most of them are bottom dwellers than live in mud or sand, under stones and shells, or on sea weeds, *Degusia* is among them. The following description mostly belongs to the common fresh water genus, the *Degusia* formarly called *Planaria*. It is a well-known representative of the Turbellaria.

SYSTEMATIC POSITION

Phylum	—	Platyhelminthes
Class	—	Turbellaria
Order	—	Tricladida
Family	—	Planariidae
Genus	—	*Dugesia*

Habits and Habitat

Dugesia is the commonest free-living platyhelminthes. They are usually found on the vegetation or under the stones in ponds, lakes, marshes, rivers and small streams. They generally inhabit pure, cool and clear water. Some species remain in currents, while others like to remain in quiet water. When active, they glide over the surface but when at rest they crawl down to the bottom or under other smooth objects, where light is not bright. *Dugesia* is found under the stone generally attached in great numbers, if pieces of meat are dropped in water near the bank. It starved they become smaller, but if well-fed they lay cocoons containing eggs which hatch to give young. They are carnivorous and extremely cannibalistic.

Morphology

Shape, Size and Colour

These are small, thin, dorsoventrally flattened, bilaterally symmetrical animals measuring about 12 mm long. Colouration is due to the dissolved pigments present beneath the epidermis. The colour may be dark yellow or olive, brown or blackish brown and may have stripes of white, orange or yellow colour. The colouration is light or whitish on ventral side.

Structure

There are two ends in the body. Anterior end is blunt, while the posterior end is tapering. Anterior end is almost square or triangular with its apex pointing forward. This portion of the body is known as head, because with it animal moves forward. There are two lateral projections, called '*acuricles or ears*' at the base of the head. But these ears are not sensory organs. On the dorsal side of the head, near the anterior end, there are two black *eye-spots*. Behind the auricles the body is somewhat constructed to form a so-called *neck*. On the

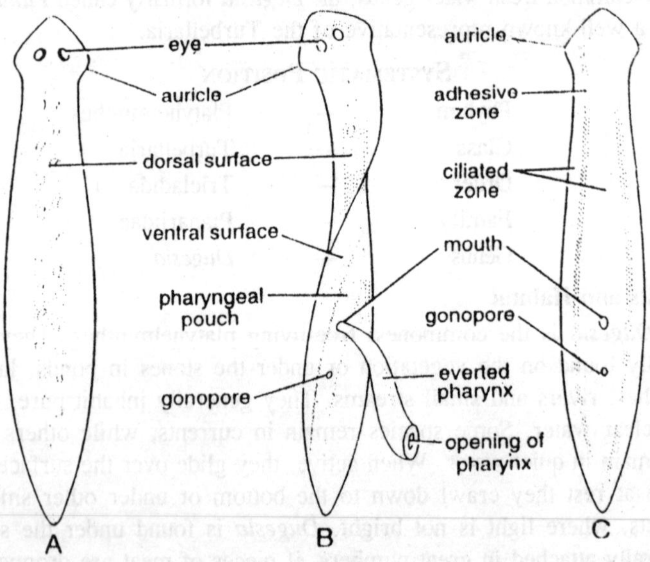

Fig. 4.1. Dugesia. External features. A—Dorsal surface; B—Body twisted to show a part of ventral surface; C—Ventral surface.

ventral side of body, near its middle, is found the mouth through which *proboscis* or *pharynx* protrudes. Thus, in Dugesia mouth is not situated on the head. The muscular walls of the pharynx serve for capturing the food. Behind the mouth in mature worms is situated a small *genital pore*. There are also excretory openings, the *nephridiopores*, on the dorsal surface situated laterally, but they are very minute and are, therefore, invisible. The ventral surface of the body is covered with cilia, which help in locomotion.

Body-wall

The body of a planarian is covered by a ciliated epidermis. The epidermal cell on its free surface also show a number of microvilli. Cilia are more prominent on the ventral surface. The epidermal cells rest on a basement membrane. Among the epidermal cells the following other types of cells are also present.

(i) *Gland cell* secreting adhesive substances.

(ii) *Rhabditogen cell* which secretes and contains rhabdites.

(iii) *Sensory cell.*

Rhabdites are arranged at right angles to the body surface. They are hyaline elliptical structures secreted by rhabditogen cells. When they are expelled out they swell on contact with water forming a sticky froth around the body. This may be protective, help in gliding or even in capture of food. Some workers like *Reisinger* and *Kelbetz* (1964) suggest that the rhabdites and nematocysts are evolutionarily related. Below the epidermis are granules and rods of pigments. The gland cells are unicellular, some occur in the epidermis but most of them are in the mesenchyme. They have long necks opening on the

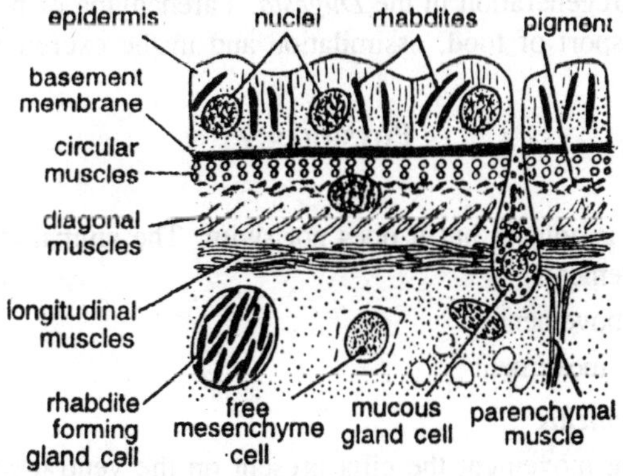

Fig. 4.2. V.L.S. dorsal body wall.

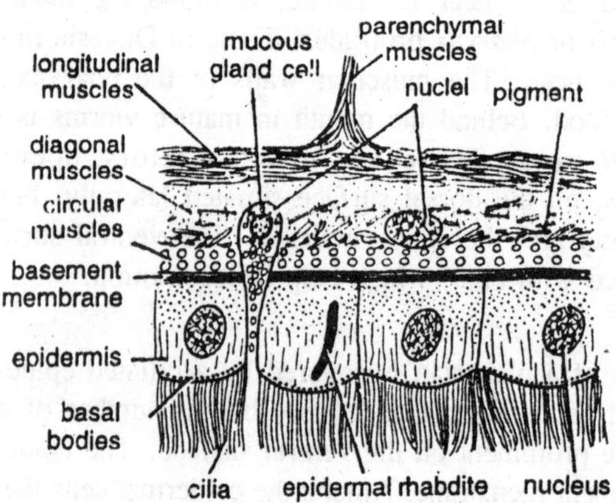

Fig. 4.3. V.L.S. ventral body wall.

surface and secrete mucous. Below the epidermis is a thin elastic basement membrane at which the epidermal cells rest. It maintains the general form of the body and provides surface for the attachment of underlying muscles.

The muscle layer consists of outermost this sheet of *circular muscle fibres*, middle broad *diagonal muscles* and innermost is the longitudinal muscles. On the ventral body wall the *longitudinal muscles* are more developed as this surface is important in the locomotion of the animal. Some muscle fibres run dorsoventrally in mesenchyme-*Mesenchyme*. The bulk of body-wall and the interior is filled with mesenchyme, which consists of connective tissue (net-like syncytium containing nuclei), free wandering cells and fluid filled spaces. The wandering cells multiply by mitosis and differentiated into other types of cells. These cells help in regeneration in the *Dugesia*. Parenchyma or mesenchyma helps in transport of food, assimilation and in the excretion of waste products.

Physiology

Locomotion

Dugesia is aquatic but it does not swim. The animal exhibits two types of movements:

1. Gliding movement
2. Crawling movement

Gliding movement

In gliding movement the cilia present on the ventral surface beat against a track of mucus secreted by the mucous glands. As a result

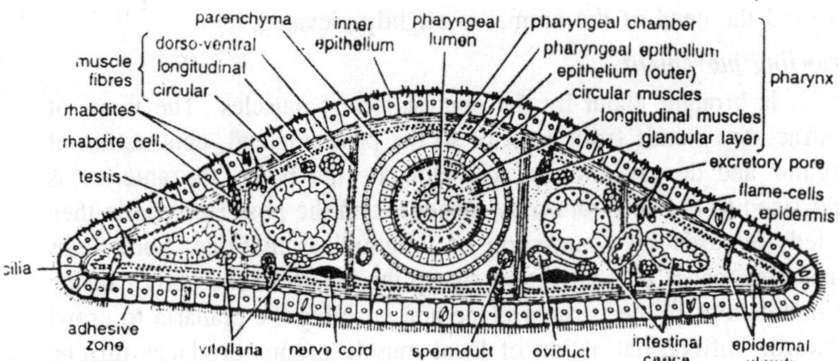

Fig. 4.4. T.S. through pharyngeal region (semi-diagrammatic).

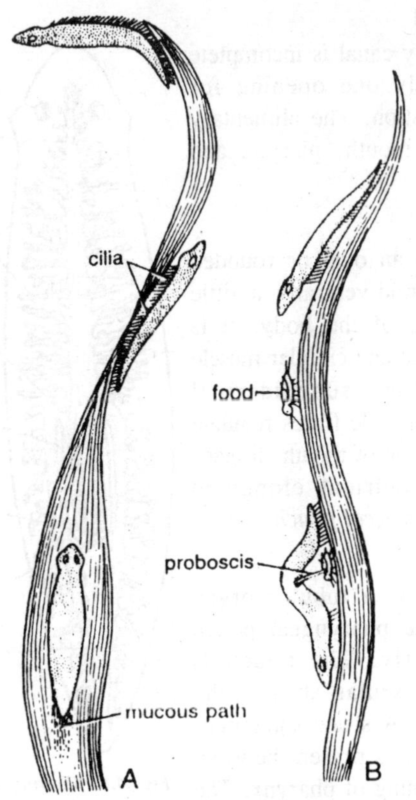

Fig. 4.5. Dugesia. A—Locomotion; B—Feeding.

of ciliary action, the animal is propelled forward in a gliding manner. The wave of ciliary action proceeds from the anterior to the posterior end and the head of the animal is slightly elevated.

Crawling movement

It is brought about by the contraction of muscles. The wave of contraction extends from anterior to the posterior end, contraction of circular and dorsoventral muscle elongate the body, the anterior is held fixed to the substratum by mucous, and the posterior end is then pulled forwards by contraction of longitudinal muscles. Sometimes transverse waves of contraction sweep the length of the body causing its undulations in the vertical plane and enabling the planaria to crawl forward. Differential action of local muscle groups produces turning or twisting movements.

DIGESTIVE SYSTEM

Alimentary Canal

The alimentary canal is incomplete *i.e.,* there is only one opening for ingestion and egestion. The alimentary canal comprises: mouth, plarynx and intestine.

Mouth

The mouth is an oval or rounded aperture situated mid-ventrally a little behind the middle of the body. It is surrounded by radial and circular muscle fibres from the subepidermal musculature. The muscle fibres regulate the opening or closing of mouth. It leads into a large cylindrical elongated chamber. The *pharyngeal pouch.*

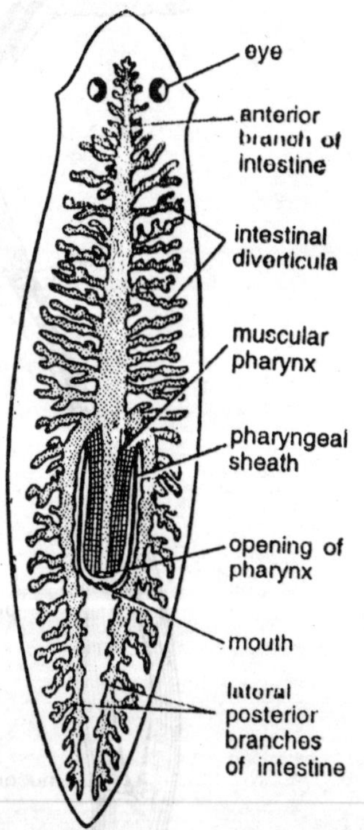

Fig. 4.6. Showing the alimentary canal.

Pharynx

The pharynx is a tubular organ situated within the pharyngeal pouch projects posteriorly. The pouch is bounded by a muscular sheath, the *pharyngeal sheath*. A space sometimes called buccal cavity is present between mouth and the opening of pharynx. The free end of the pharynx projected out of

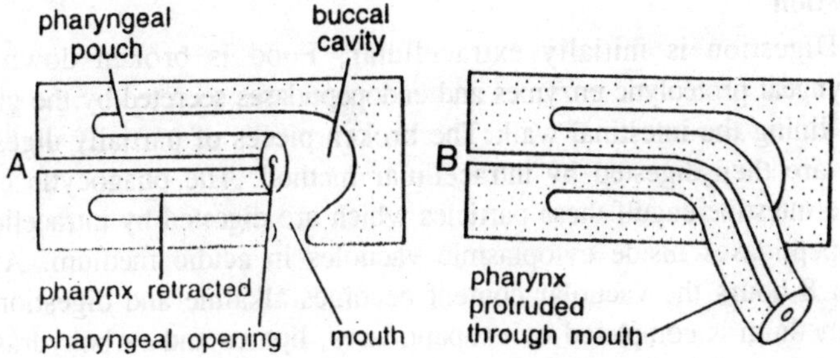

Fig. 4.7. Diagram to show: A—Retracted pharynx; B—Everted pharynx.

the mouth during feeding. The projecting pharynx is called *proboscis*. This is brought about by the action of muscular wall. Pharynx opens into the intestine by way of a short *oesophagus*.

Intestine

The intestine immediately divides into three branches. The median branch extends forwards upto the head while the other two lateral branches extend backwards upto the posterior end, one on either side of pharyngeal cavity. All the three branches give off numerous branching and blind diverticulal which end into the mesenchyme. The much branching intestine is a means of increasing the surface area for digestion, absorption and distribution of food.

Histology of alimentary canal

The pharyngeal wall shows a complicated histological structure. It consists of nine layers from the surface to the lumen: epithelial cells, longitudinal muscle layer, circular muscle layer, outer gland cells, nerve plexuses, inner gland cells, longitudinal muscle layer, circular muscle layer and an endodermal epithelial lining. The intestine is simple and thin consists of a single-layered epithelium, the gastrodermis. The gastrodermal cells are of two types phagocytic cells and granular cells.

Food and Feeding

Normally *Dugesia* is *holozoic* and canivorous subsisting for the most part on crustaceans, insect larvae and small worms. During feeding, it glides through the action of cilia and the adhesive glands secrete adhesive secretion. When prey comes in contact with the secretion it becomes firmly adherent. *Dugesia* then folds the anterior end of its body over the prey and thus immobilizes it. The pharynx is then produced out through the mouth and through its peristaltic action sucks up small particles of the prey into its intestine.

Digestion

Digestion is initially extracellular. Food is broken down by pharyngeal proteolytic enzymes and endopeptidases secreted by the gland cells lining the intestinal wall. The broken pieces of partially digested food are then digested by intracellular method. The phagocytic cells of the intestine engulf these particles which are digested by intracellular endopeptidases inside cytoplasmic vacuoles in acidic medium. After about 8 hours the vacuolar content becomes alkaline and digestion in this medium is completed by exopeptidases, lipases and carbohydrases. The excess of food is stored as fat and proteins in the intestinal cells and mesenchyma. Freshwater planarians are able to withstand prolonged experimental starvation. In extreme cases, they utilize part of enteron and all the mesenchyma and reproductive system. In fact, the body volume may be reduced to as little as 1/300 of the original.

Egestion

Since there is no anus, the egestion of undigested food material takes place through mouth. It is brought about by the contraction of the muscles of body-wall.

Respiration

Mode of respiration is aerobic but there are no special organs for this. Oxygen from the surrounding water is directly into the body and carbon dioxide is diffused out through the entire body surface. Flattened body is an admirable adaptation for this type of respiration.

Excretory System

The excretory system or *protonephridial system* of *Dugesia* consists of a pair of *longitudinal excretory canals* and numerous tiny enlargements known as flame cells.

Excretory Canals

A pair of longitudinal excretory canals running on each side of the body throughout the length. They are highly inter-coiled and open to the dorsal surface by several minute poses called nephridiopores. Anteriorly two canals are united by a *transverse vessel*, in front of eyes. Each canal divides into a number of branches which further divides into very fine blind *capillaries* which bear flagella at the inner end (*Wilson* and *Webster*, 1974). The terminal structure is composed of flame cells or cryptocytes.

Flame Cells

Each flame cell represents the unit of excretory system. It is a large, tubular, protoplasmic body. It is produced into numerous

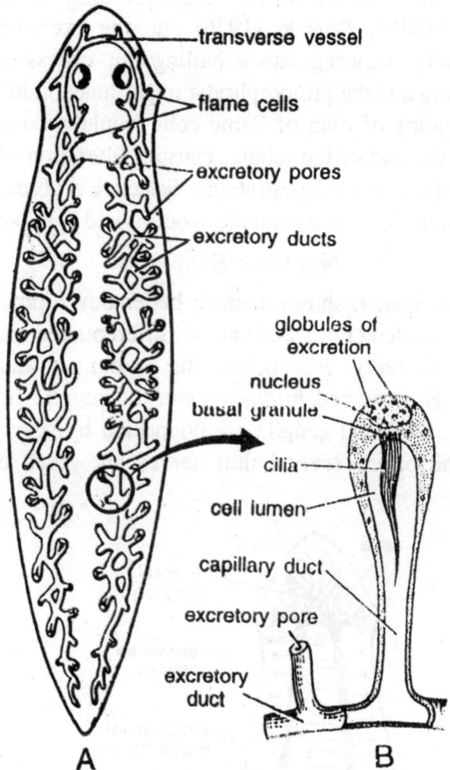

Fig. 4.8. A—Excretory system. B—A single flame cell in section, connected with an excretory duct.

protoplasmic processes reaching the mesenchyme. Its terminal part is swollen into a bulb and contains a conspicuous nucleus. The flame cell has an intracellular space which is continued into capillaries. The space encloses a bunch of cilia which vibrate giving the appearance of flickering candle flame. Besides the flame cells, certain glandular cells called the ***arthrocytes*** or ***paranephroytes*** occur in close contact with the excretory canals and also help removal of excretory products.

Physiology of Excretion

Excretory substance is collected from the mesenchyme and is transferred into the cavities of flame cells. The beating of cilia of flame cells causes hydrostatic pressure by which the fluid waste passes into the longitudinal trunks and goes out of nephridiopores. According to Storer, Wolcott, Guyer, Hickman etc. the flame cells simply help in excretion of nitrogenous wastes. But according to some workers like Mckanna, (1968); *Prusch*, (1976) etc. the excretory system in *Dugesia* is largely osmoregulatory bailing out excess of water. As osmoregulatory organs, the protonephridia might function in the following manner. The beating of cilia of flame cells would produce a negative pressure within the end of the tubule, causing filteration of fluid across the membrane covering the fenestrations. Selective reabsorption of ions, especially $K+$ and $Cl-$, by the tubule would yield a hyposmotic fluid.

Nervous System

The nervous system shows a much better development than what we found in the coelenterates and can be differentiated into the *central* and *peripheral systems*. Just below the epidermis and behind the eyespots, there is a bilobed *brain* or *cerebral ganglia* in the form of an inverted 'V'. Both the ganglia are connected by several transverse fibres. From the brain, several thin nerves are given off anteriorly

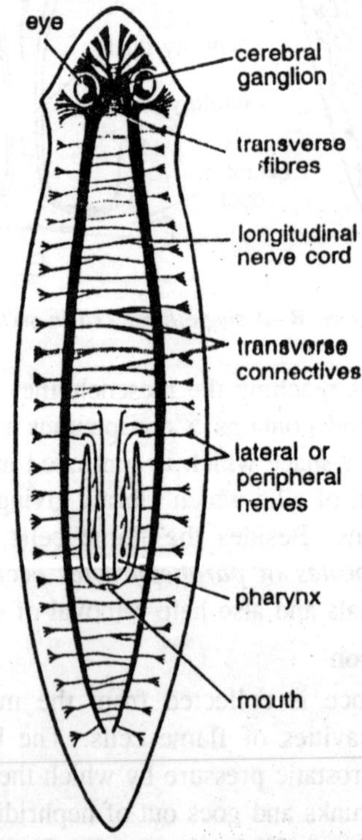

Fig. 4.9. Nervous system.

and to the eyes. From the brain two *lateral longitudinal nerve cords* extend posteriorly along the ventral side. The two nerve cords are connected together by *transverse connectives*. The peripheral system is present in the form of a *sub-epidermal plexus* (just below the epidermis) and a *sub-muscular plexus* below the muscle layers in the mesenchyme. The nerve-cells or neurons may be unipolar, bipolar or multipolar. All the neurons possessing neurites are similar. These neurites cannot be differentiated into axons and dendrons.

SENSE ORGANS

Dugesia possesses special sensory organs, called receptons. These are photoreceptors or eyes, auricular organs, tangoreceptors and rheoreceptors.

Photoreceptors or Eyes

The eyes are two dark spots on the dorsal surface of the anterior end, one on either side of the median line. Each eye has the form of a cup whose opening faces anterio-posteriorly. Structure of each *eye* is very simple. It is made up of highly pigmented *retina*. Inside the cup are situated numerous nerve-cells, the *fibrillae* of which are in contact with the retina and their opposite ends are continued into an optic nerve, which is joined with *brain*. The nerve cells enter the cup through the opening. Ectoderm over the eye is not pigmented. *Lens* is absent in eye. Eye are capable of a crude discrimination of the direction of light. The pigment cup serves as a shield and light can enter only

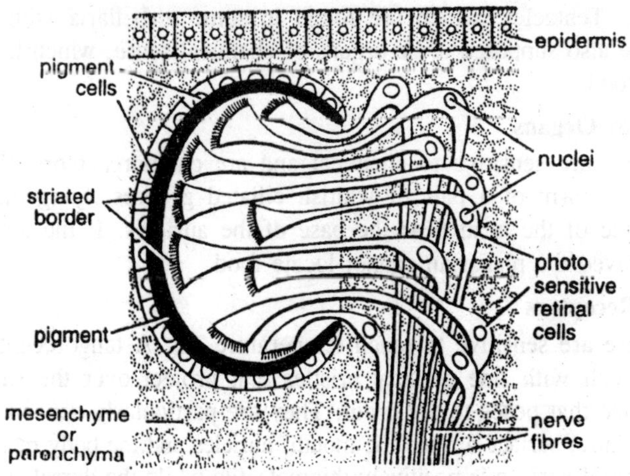

Fig. 4.10. V.S. planarian eye.

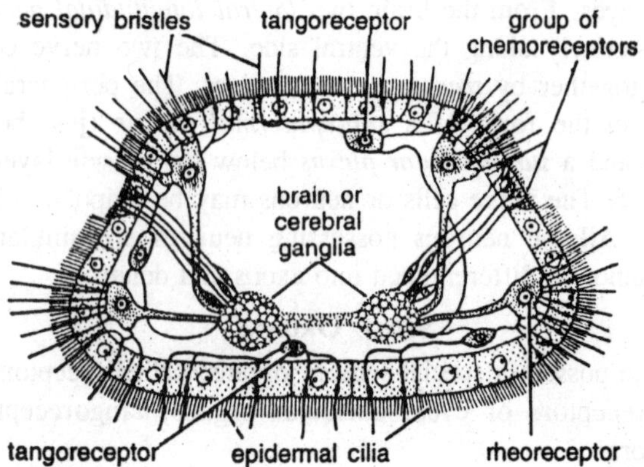

Fig. 4.11. A diagrammatic T.S. of the anterior end of a planarian showing the various receptors.

through its opening to stimulate the photosensitive expanded ends of retinal cells, thus the animal can detect the direction light. The animal is negatively phototactic and is most active at night. It the eyes are removed it can still react to light. However the reaction is much slower. This shows that it possesses some light sensitive cells over the general body surface.

Ciliated Pits

Besides eyes, there are *ciliated pits* which are also sense-organs. They are situated on either side of the head, having special sensory cells with long cilia. These ciliated pits are connected with nervous network. Tentacles, which are found in some turbellaria with long cilia, are also supposed to be organs of chemical sense, which help in finding food.

Auricular Organs

These are sensitive to chemical and are chemoreceptors. These are in the form of a pair of whitish ciliated groores, lying one on either side of the head near the base of the auricles. If the auricles are removed the planarian cannot locate food.

Tango Receptors

These are sensitive to touch temperature etc. A tangoreceptor is a small cell with fine sensory bristles that project over the surface except one that poses to the brain. They are particularly concentrated on the ventral surface around the mouth, auricles and the body margins. Thus ventral surface is positively thigmotactile while the dorsal surface is negatively thigmotactic.

Rheoreceptor

These are sensitive to water current. They are of general occurrence and their sensory processes or bristles project much beyond the level of cilia.

REPRODUCTIVE SYSTEM

Dugesia reproduces with asexually as well as sexually.

Asexual Reproduction

Asexual reproduction occur by the transverse binary fission. The plane of fission usually forms posterior to the pharynx. The posterior part attaches to substratum by mucus, the anterior part stretches forward until the animal snap into two. Each half regenerates the wanting parts resulting in the formation of two complete planarians. Some planarians fragment into two or more pieces each growing into a complete individual. Asexual reproduction is restricted to some species only. Some species reproduce asexually in one season, and may develop sex organs in the other, but these are some species that reproduce exclusively asexually and rarely develop sex organs.

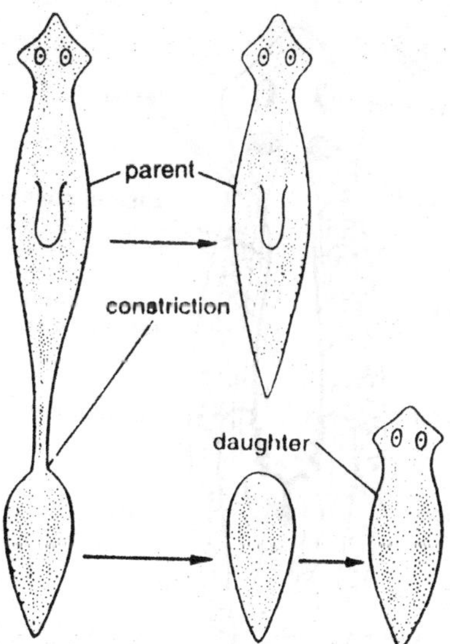

Fig. 4.12. Asexual reproduction by fission.

Sexual Reproduction

The reproduction organs are more complicated in these animals. The gonads *i.e.,* testes and or aries are derived from the meroderm and there is a system of tubules and chambers in which fertilisation occurs. The animal reproduces in early summer and the gonads are developed temporarily in the breeding season. The *Dugesia* is bisexual or hermaphrodite but self-fertilisation does not take place and the cross fertilisation is the rule. After breeding season the gonads degenerate and disappear and the worm reproduces asexually.

Male Reproductive System

Male reproductive system consists of a large number of small spherical testes, which are situated on both the sides of the body. Each testis is connected with minute tubes, called *vasa efferntia*. There form two large ducts called *vasa deferentia*. The two vasa deferentia run parallel to one another towards the posterior end, where these bend mesially towards the penis. Each vas deferens dilated to form the seminal or spermlducal vesicle with thick muscular wall. The sperms are stored here. The penis shows two regions, the anterior large spherical *penis bulb* and the posterior narrow conical *penis papilla* with muscular wall. The lumen of penis is very narrow and called

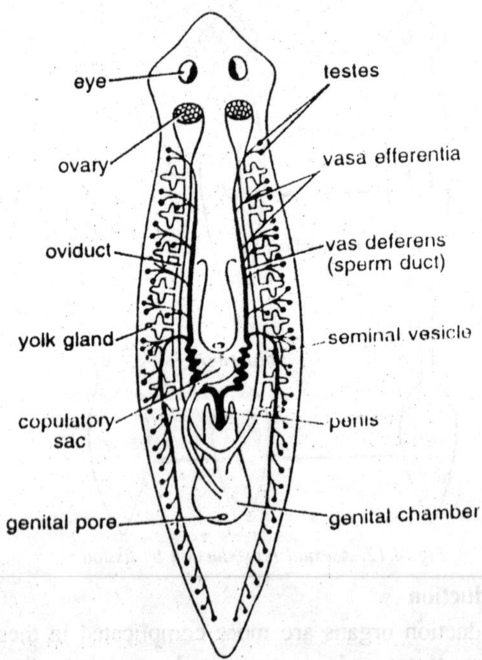

Fig. 4.13. Reproductive system.

ejaculatory duct. The penis papilla opens into the genital atrium which terminates in the common gonopore, situated mid-ventrally behind the opening of month. The lumen of penis bulb is formed by *bulbar cavities*. The bulb contains numerous *unicellular glands*, the prostate glands which open into it lumen.

Female Reproductive Organs

The female reproductive organs consist of a pair *ovaries*, which are small apherical bodies situated laterally just behind the head. From each ovary arises an oviduct which extends posteriorly through the parenchyma. Immediately after its emergence each oviduct presents a dilation, which is known as *seminal receptacle*. The two oviducts unite to form a common vagina which opens into the *genital artium*. The oviducts along their whole length are surrounded by numerous vitelline glands which open into oviducts by short *vitelline ducts*. Because of the close association of oviduct and vitelline glands, these are sometimes collectively referred to as the ovovitelline ducts. Opening into the genital chamber is found a muscular sac-like *copulatory sac* or *uterus* or *bursal canal*. The bursal canal opens into a large sac the *bursa conplatrix* which receives penis and temporarily stores the sperms received during copulation. It opens below into female genital atrium.

Copulation

Although *Dugesia* is hermaphrodite, yet *cross-fertilization* is a rule. *Copulation* takes place between the two worms in which their posterior

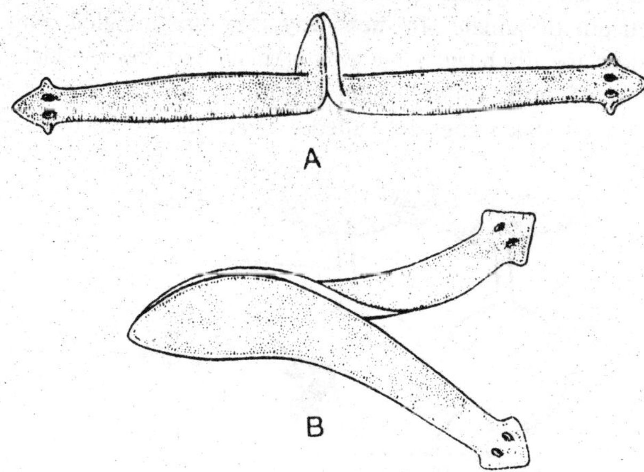

Fig. 4.14. Copulation. A—Head ends facing in opposite directions; B—Head ends facing in the same direction.

and ventral surface come together and the penis of each is inserted into the *genital atrium* of the other. In this way sperms are deposited in the *seminal receptacles* of the mating partner. The sperms are stored inside the copulatory sac for some time and then travel along the oviducts to reach the seminal receptacles. The eggs are deposited in the seminal receptacle and are fertilized. Thus the fertilization is internal. During breeding season, each worm may copulate many.

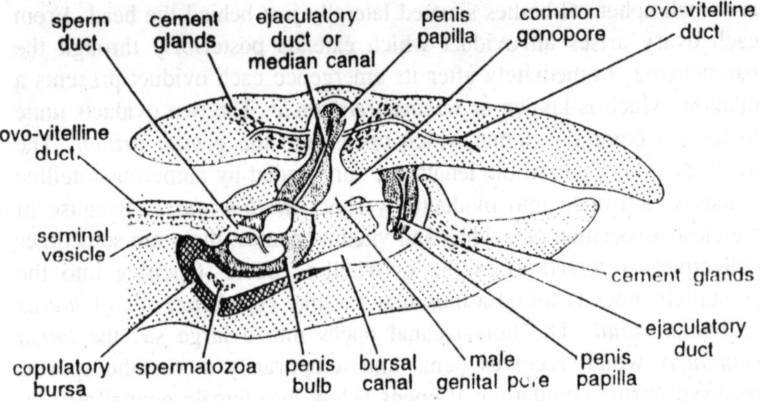

Fig. 4.15. A copulating pair in section.

Cocoon Formation

The zygote thus formed, pass backwards through the ovovitelline ducts and reach the genital atrium of ovovitelline ducts and reach the genital atrium of female. In their way they are mingled with yolk cells which are discharged by yolk glands into the oviducts. The deposition of yolk outside the zygote is unique condition in platyhelminthes. Such eggs are called *ectolecithal*. When the eggs are

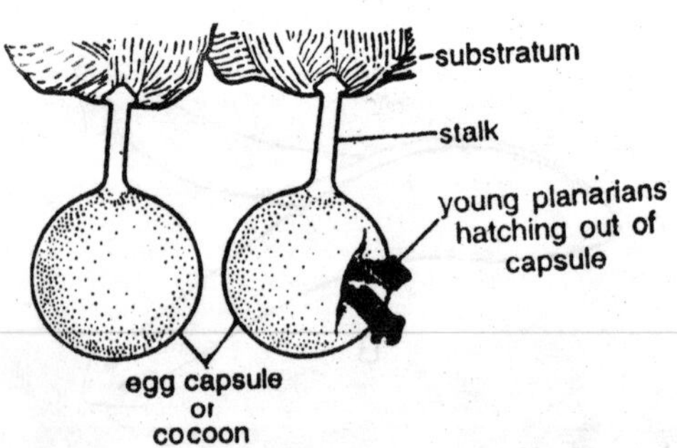

Fig. 4.16. Young planarians hatching from an egg capsule.

reached in genital atrium 5-10 eggs along with yolk cells are enclosed in a protective, proteineous shell forming a cocoon or an egg-capsule. The capsule is coated by a sticky substance secreated by the cement glands that surround and open into the common genital atrium. The capsule is now extruded through the genital pore. As it is passing out, sticky coating is drawn out into a sticky stalk that fastens it to some object chiefly the under surface of stones. The capsules are laid in succession at intervals of a few days.

Development

Development starts soon after the cocoon is laid and the hatching occurs in 2 to 3 weeks depending upon the temperature of water. The cleavage is *spiral* and *determinate* type. No larval stage is formed hence the development is *direct*. A young planaria or *juvenile* glides out of the capsule from each egg.

Regeneration and Transplantation

Those Turbellaria that reproduce asexually also possess high powers of regeneration and consequently have served as material for innumerable studies on regeneration. In other Turbellaria regeneration of a head is limited to pieces from anterior regions or to pieces containing the cerebral ganglia; or else the animals are incapable of

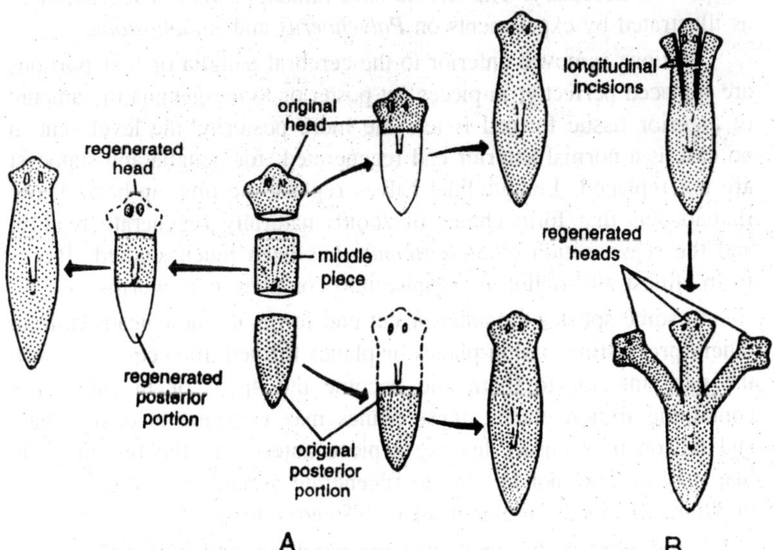

Fig. 4.17. Regeneration in planarians. A—Three individuals regenerate from an individual cut into three parts. B—Formation of a heteromorph with three heads.

regenerative processes beyond wound healing. Many forms can replace posterior regions, including the pharynx and the copulatory apparatus, when the cuts are made posterior to the cerebral ganglia. The region anterior to the ganglia is incapable of regeneration in any case. Polarity is strongly retained in pieces, *i.e.*, the anterior cut surface regenerates a head, the posterior cut surface, a tail; but alternations of polarity occur or may be induced or increased by experimental means. The capacity to regenerate, the rate of the process, the amount of new tissue produced, and often the type of anterior regenerate are markedly related to level, being greatest or most normal anteriorly and declining posteriorly. In short, the flatworm body like the coelenterate is a polarized system; the polarisation consists in some sort of physiological gradation along the anteroposterior axis. The characteristics of and evidence for the existence of such an axial gradation have recently been summarized by Child (1941).

Regeneration involves two processes: regeneration proper, or *epimorphosis*, in which new tissue grows out from the cut surface in partial replacement of the part removed; and *morphallaxis*, or the working over of the old tissue and organs to fit into the new animal. The relative roles of the two processes vary with different species and with size and level of the piece; but in any event reorganisation of the old parts is necessary. The Acoela have limited powers of regeneration, as illustrated by experiments on *Polychoerus* and *Aphanostoma*.

Portions removed anterior to the cerebral ganglia or rear portions are replaced perfectly; in pieces cut posterior to the ganglia the amount of anterior tissue formed is less the more posterior the level, but in no case is a normal anterior end regenerated, for ganglia and statocyst are not replaced. Longitudinal halves replace the missing half. Those rhabdocoels that form chains of zooids naturally regenerate readily, and the regeneration of *Stenostomum* has been much studied. Pieces from almost any region reorganise into complete new worms.

In some species the anterior cut end forms the new head while in others preexisting fission planes or planes formed after cutting absorb the part anterior to them and become the head; or in cut pieces containing fission planes, these planes may continue to differentiate and fission may ensue, leaving a piece anterior to the fission plane that may or may not be able to regenerate a head. The regeneration of *Microstomum* is similar to that of *Stenostomum*.

Some other rhabdocoels may regenerate the tail tip but in general rhabdocoels that lack asexual reproduction and alloeocoels are incapable of any regeneration, although wounds are healed by epidermal migration.

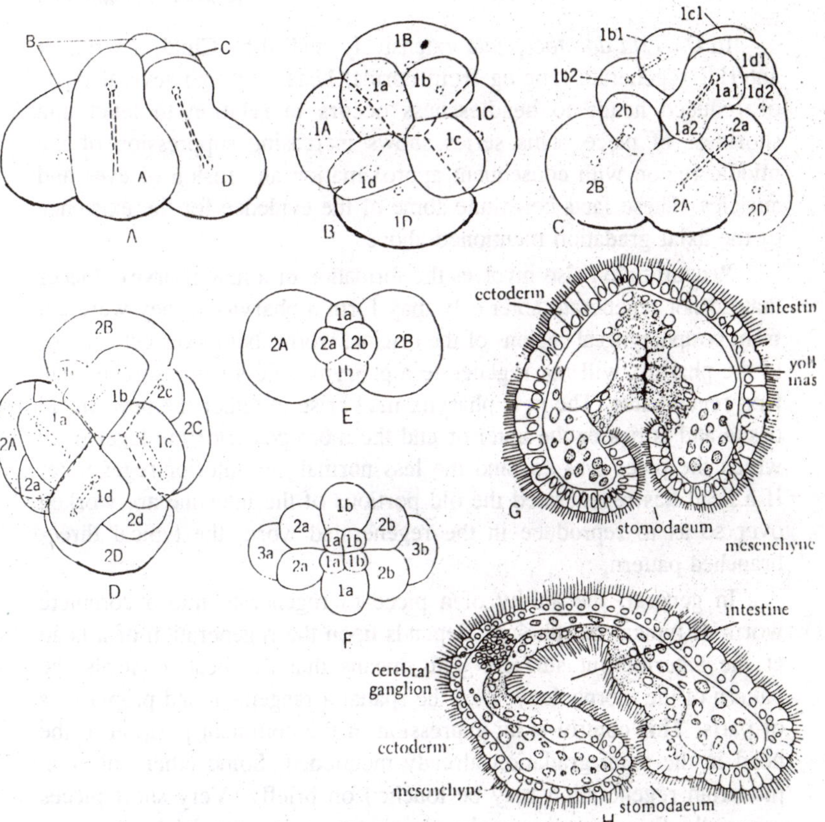

Fig. 4.18. A-D—Spiral determinate cleavage in Hoploplana, with cell lineage traced by numbering. E-F—Spiral cleavage in an acoel, Polychoerus, resulting from the transverse cleavages of the two-celled rather than the four-celled stages. G-H—Later stages in the development of Hoploplana, showing the shift in the axis of the embryo.

Fresh-water planarians of the genera *Dugesia*, *Phagocata*, *Polycelis*, etc., display remarkable powers of regeneration and have been the object of numerous researches. In general, pieces of moderate size from any level form a complete new worm. With reduction in size of piece, complete worms may regenerate from even very short pieces in some species, while in others short pieces tend to regenerate reduced or imperfect heads or may remain headless, and this tendency is greatest the more posterior the level. The size and normality of the regenerated head the rate of anterior regeneration, and the length of tail produced are generally greater the more anterior the level from which the piece was cut and decline in posterior levels.

In *Dugesia dorotocephala* extensively studied by Child, a series of anterior regenerate ranging from normal heads through several types of reduced heads to headlessness occurs in relation to level and shortness of piece. This series shows increasing suppression of the middle region with consequent approximation and fusion of eyes and auricles. These facts constitute some of the evidence for the existence of the axial gradation mentioned above.

Reconstitution also involves the formation of a new pharynx. Pieces that do not regenerate anteriorly may form a pharynx if they were cut from postpharyngeal regions of the original worm; but pieces cut anterior to the pharynx will not regenerate a pharynx unless some sort of head first regenerates. The new pharynx may arise in either the new or old tissue and is nearer the anterior end the more posterior the level from which the piece was cut and the less normal the anterior regenerate. If a head has regenerated the old portions of the intestine are worked over so as to reproduce in the regenerated worm the typical three-branched pattern.

In general, the ability of a piece to regenerate into a complete worm of typical morphology depends upon the regeneration of a head at the anterior cut surface. This means that the head controls the pattern of the morphology, *i.e.*, the spatial arrangement and proportions of parts. This control is an expression of the dominant position of the head in the axial gradation already mentioned. Some other points in planarian regeneration may be touched on briefly. Very short pieces especially from anterior regions may regenerate biaxial heads, *i.e.*, form a head at each cut surface. Isolated heads cut off just behind the eyes also often regenerate a reversed head at the cut surface. In some species short posterior pieces produce biaxial tails. After oblique cuts, a head usually forms at the most anterior point, a tail at the most posterior point, of the cut surfaces.

Lateral longitudinal pieces usually regenerate with reference to the original polarity; but short lateral strips, or those cut lateral to the ventral nerve cords, or small triangular lateral pieces tend to regenerate the new head at the cut surface, *i.e.*, at right angles to the original polarity (Morgan, 1900b; Olmsted, 1918; Beyer and Child, 1930). By means of appropriate splits, double or multiple heads or tails may be induced at will; heads will grow out from the body sides behind forwardly directed gashes, tails behind posteriorly directed ones one or more "crotch" heads may arise in the angle of posterior longitudinal splits if these are carried far enough forward (Beissenshirtz,

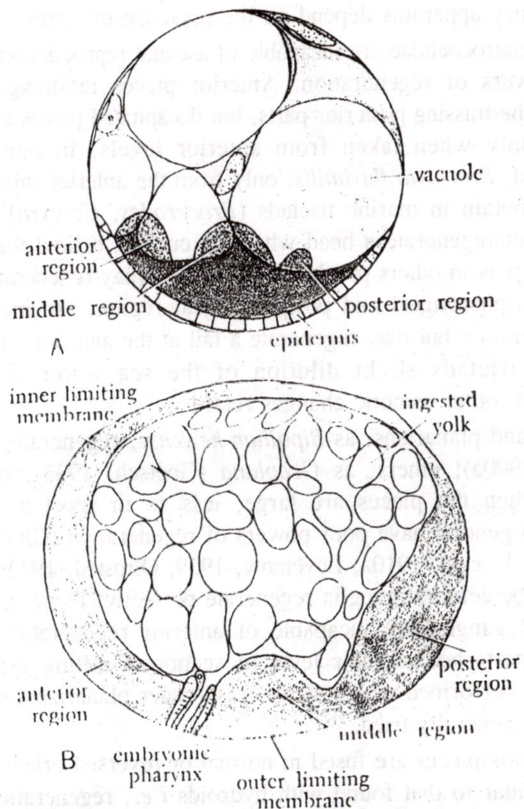

Fig. 4.19. A—A stage in the development of a rhabdocoel with ectolecithal ovum, showing the three embryonic regions. B—A stage of triclad development, with yolk material already ingested through the embryonic pharynx, and with the three embryonic regions visible.

1928; Silber and Hamburger, 1939; Hull, 1940); heads will form along the edges of windows through the body and in fact an endless variety of bizarre forms can be produced by appropriate cuts.

Regeneration is also affected by external factors such as temperature, osmotic pressure, and chemical composition of the medium; many experiments have been performed on the effect of various salts, anesthetics, radiations, etc., on regeneration. If a sexually mature planarian is cut in two between the pharynx and the copulatory apparatus, the latter degenerates and the piece regenerates into an asexual worm. The large anterior piece regenerates a new copulatory apparatus if active testes are present. The presence of ovaries is not pertinent. This result indicates that the formation and maintenance of the copulatory apparatus depend on the presence of testes.

The Dendrocoelidae are incapable of asexual reproduction and have limited powers of regeneration. Anterior pieces retaining the head regenerate the missing posterior parts, but decapitated pieces can reform the head only when taken from anterior levels, in our common dendrocoelid, *Procotyla fluviatilis*, only from the anterior third. Similar conditions obtain in marine triclads (*Procerodes*, *Cercyra*), some of which cannot regenerate a head when the cut lies behind the cerebral ganglia whereas in others prepharyngeal pieces may regenerate a head. Pieces from pharyngeal and postpharyngeal regions are incapable of head regeneration but may regenerate a tail at the anterior cut surface. In marine triclads slight dilution of the sea water accelerates regeneration; other osmotic changes retard.

Some land planarians, as *Bipalium kewense*, regenerate very well (Morgan, 1900a); others, as *Geoplana* (Goetsch, 1933), regenerate perfectly when the pieces are large, less so in short pieces. The polyclads in general have poor powers of regeneration (Child. 1904a, b, c, 1905a, b, c, d, 1910a; Levetzow, 1939; Olmsted, 1922b). Pieces containing the cerebral ganglia regenerate perfectly. Pieces cut behind the cerebral ganglia are incapable of anterior regeneration but can regenerate parts behind their level of section. Grafting experiments may also be performed with suitable fresh-water planarians; species of *Dugesia* are generally used.

When crosspieces are fused in normal or reversed orientation, the result is similar to that found with hydroids *i.e.*, regeneration occurs according to the original polarity of each piece except that the polarity of very short pieces fused to larger pieces may be altered. Triangular or rectangular pieces may be excised and grafted into windows made into other parts of worms (Miller, 1939, Okado and Sugino, 1937; Santos, 1931) in normal or altered orientation. Such grafts from the cephalic region containing much nervous tissue are more effective in inducing a head-like outgrowth and causing reorganization of adjacent parts of the host with production of one or more pharynges, the more posteriorly they are grafted into the host. Similar grafts from tail regions implanted anteriorly are influenced by the host to develop into a tail regardless of their original polarity. Heads regenerated from grafts or at a cut influence adjacent regions for a certain distance, reorganising them to a greater or less degree into postcephalic regions (with pharynx and intestinal branches), and prevent the formation of another head at a cut surface or otherwise within a certain distance.

The histology of the regenerative process has been repeatedly studied in planarians with not entirely concordant results; the differences are probably real, depending on the species employed. The wound is considerably closed by muscle contraction (this contraction is omitted if the cutting occurs in an isotonic medium) and the adjacent epidermal cells flatten, become syncytial, and creep over the cut surface. If the exposed surface is sufficiently large, regeneration cells may participate in the formation of this covering membrane, which later differentiates into the typical epidermis. A mass of cells, variously known as regeneration cells, formative cells, etc., accumulates beneath the covering membrane and differentiates into the various missing parts. In the case of anterior regeneration strands of these cells attach to the cut nerve cords and differentiate into brain and eyes. They also adhere to the cut intestinal branches and build on appropriate terminations, and differentiate into muscle cells and gland cells in appropriate locations. It appears, however, that in some species or with regard to some systems, the new parts may be formed by proliferation of the adjacent cells of the old parts, i.e., Bandier (1936) states that in the regeneration of land planarians both cut nerve ends and intestinal branches grow from their own cells. The source of these regeneration cells is still under dispute.

There is little doubt that they comprise preexisting cells that migrate to the cut surface either with or without multiplying by mitosis. According to some authors, they consist entirely of the free cells of the mesenchyme; others state that, after cutting, various body cells, such as gastrodermis, yolk glands, gland cells, etc., undergo dedifferentiation into amoeboid cells and join the free mesenchymal cells to form the mass of regeneration cells. An extreme view is to the effect that the regeneration cells are persistent embryonic cells held as a reserve to function in regeneration and formation of the reproductive system. It appears to the author that there is no evidence in support of such a view and that some of the evidence cited above oppose it.

Curtis and Schulze have found that there are conspicuously fewer free mesenchymal cells in planarians with poor powers of regeneration, as *Procotyla fluviatilis*, than in those with high powers, as species of *Dugesia*; but their tabulation fails to show any marked decline in number of free mesenchymal cells in those regions of *Procotyla* that cannot regenerate a head as compared with regions that can. Further regions unable to regenerate a head can regenerate a tail and also can produce (from free mesenchymal cells) the entire complicated copulatory apparatus. Attempts to shed some light on these problems have been made by studying the effect of X rays and radium on regeneration. (Bardeen and Baetjer, 1904; Gianferrari, 1929; Meserve and Kenney, 1934; Schaper. 1904; Strandskov, 1937; Weiand, 1930). The subject has been reviewed by Curtis (1936) who has researched extensively on the matter.

In general, exposure to sufficient dosage of X rays and radium inhibits regeneration altogether in most species, although *Dugesia dorotocephala* will regenerate after any exposures that do not kill. The failure to regenerate is caused, according to some of the above workers, by the killing off of the free mesenchyme cells, according to others by the inability of these cells after radiation to migrate or to differentiate. The result therefore only proves what was already known, namely, that cells of the type of free mesenchymal cells are responsible for regeneration. It does not settle the questions whether these cells are in fact embryonic rests and whether they get contributions from dedifferentiated tissues. The difference in susceptibility to radiation between the mesenchyme cells and other body cells evidently varies with different species.

5

GROWTH OF PARASITES

A parasites must not only be successful in reaching the next host in its particular life-cycle, but must also establish itself, grow and mature if reproductive success is to be achieved. Sexual maturity will normally occur only in the final host. In the intermediate host (or hosts, should the life -cycle be indirect), the parasite may grow in size and may increase its numbers dramatically by asexual multiplication. Alternatively it may develop into a quiescent stage, such as the metacercaria of the Digenea, awaiting transmission to the next host, either by ingestion or death of the intermediate host. In this chapter we shall consider the array of factors that permit parasite establishment in a potential host animal.

A simplified sequence of events, depicting the barriers to invasion and establishment that confront a parasite during its infection of a single host. This cycle will clearly be repeated twice, or even three times, in parasites that infect two or more hosts during a more complex life-cycle. The first barrier to invasion is primarily an ecological one and is outside the scope of this book. The second barrier represents the physical limitations to invasion. This physical barrier applies both to parasites that penetrate the host surface and to those that enter either the digestive system or hypodermically, via a blood feeding vector, and pertains to the conditions immediately encountered by the parasite once contact with the host has been made. The third barrier is a complex of physico-chemical and biotic factors that the parasite must be able to overcome or tolerate over a more extended time scale once access to the host body has been gained. These include the immune defence mechanisms of the host. Here we shall consider the physico-chemical and interactive biotic factors that may allow or prevent parasite establishment, growth and maturation. Our attention will be focussed primarily on parasite development within the final host, since this has, in general, attracted the most attention from parasitologists.

Hatching Mechanisms

Parasites that invade the host a cystic or egg stages enter the host passively during feeding. Such parasites excyst, or hatch, into larvae or pre-adults in the alimentary canal of the host, where they may either remain and develop to maturity, of from which they may migrate to other tissues on the host's body. The variety of stages that enter via the mouth are listed below. Protozoans usually produce infective cystic stages that occur free in the environment, e.g. the oocysts of the Coccidia, and these are acquired when the host accidentally ingests them.

Many digeneans from infective metacercariae, which are usually encysted and located within the tissue of an intermediate host; a few metacercariae inhabit the external environment (e.g. *Fasciola hepatica*). The cestodes form a variety of infective stages; these include eggs, that are liberated into the external environment; cysticercoids, cysticerci and hydatid cysts, that are encysted in the tissues of intermediate hosts; and the unencysted parasitic larvae, the procercoids and plerocercoids. Acanthocephalans are infective to the final host as the cystacanth larva, which is encysted in an invertebrate intermediate host.

Nematodes are infective either through an embryonated second-stage larva within the egg or an ensheathed third- stage larva, the former occurring free in the external environment, the later encapsulated within an intermediate host or free on herbage (trichostrongyles). Almost without exception, the infective stage of those parasites that are acquired by being eaten by the final host have a reduced level of metabolism, i.e. they are quiescent. Possible exceptions to this general rule include the progenetic larvae, such as pseudophyllidean plerocercoids (Cestoda) and strigeid metacercariae (Digenea), which grow actively and, as larvae, are very close to the adult stage in their development, lacing only ripe gonads.

On ingestion by the hosts, the infective parasite must be able to withstand the rigours of life in the upper alimentary tract, particularly the stomach, but at the same time, it must be capable of utilizing a variety of stimuli to free itself from any protective covering and activate its hitherto quiescent metabolism. Precocious activation could prove lethal to the parasite, so a very delicate balance of factors must operate to ensure that hatching and activation take place in that region of the gut where the prevailing conditions can be tolerated by the liberated, immature parasite. This location is very often the duodenum in mammals and birds, but there are exceptions to this.

Protozoa

Excystation and activation of infective protozoan cysts has been studied on only a small number of genera, including *Eimeria*, *Isospora*, *Toxoplasma*, *Sarcocystis* and *Entamoeba*. Most of these investigations have involved *in vitro* systems and may, therefore, shed little light on the events that occur naturally. Excystation and activation can usually be accomplished, *in vitro*, by raising the ambient temperature to 37°C for mammalian parasites and to 43°C for avian forms, using a culture medium of neutral pH, containing reducing agents and with a low O_2 and a high pCO_2. Bile salts may be an additional requirement for optimal excystation and activation.

There is evidence that excystation and activation of protozoan cysts is a biphasic process, therefore, with *Eimeria* and *Entamoeba hystolytica*, initial activation of the encysted parasite is brought about by the elevation of temperature, the presence of reducing agents and a high pCO_2. Subsequent excystation is due to the activity of proteolytic enzymes, produced either by the parasite itself of by the host gut. Excystation in *Eimeria bovis* requires pretreatment with CO_2 and reducing agents, followed by trypsin and bile treatment.

Mechanical disruption of the cyst walls may also be of importance in the initial processes, allowing the permeation of host enzymes. Rather unusually, the infective stages of *H. Histomonas meleagridis* are transmitted unencysted, within the eggs of parasitic nematodes (*Heterakis*) and in this way they are able to survive the potentially inimical conditions of the bird stomach. In the Coccidia, the infective stages are sporozoites that are contained within sporocysts, themselves within the oocyst. Rupture of the oocyst micropyle is influenced by treatment with carbon dioxide in mammalian parasites, while in avian forms mechanical damage is thought to be more important. Excystation occurs when sporocysts are released from the oocyst and are exposed to bile and trypsin or chymotrypsin in the duodenum, thereby freeing the sporozoites into the gut.

In many species of *Eimeria* and some species of *Isospora*, the sporocyst contains a structure, termed the *Stieda body*, which breaks down under the influence of host bile and trypsin to release the sporozoites. In other species of *Isospora* (*I. canis*, *I. arctopitheci*, *I. endocallimici* and *I. begemina*) the sporocyst lacks a Stieda body and excystation occurs following the general breakdown of the sporocyst wall on exposure to proteolytic enzymes. In general, the process of excystation of coccidian sporozoites begins in the stomach or rumen and is completed in the upper small intestine. Coccidian inhabitants of the large intestine may, however, delay excystment until they reach a more posterior region of the gut, e.g. *Eimeria tenella*.

Digenea

The infective stage for the final host in many digeneans is the metacercaria, a quiescent larva that may be encysted either within an intermediate host or attached to vegetation, or alternatively, unencysted within an intermediate host, as is the case with some strigeid metacercariae, e.g. *Diplostomum spathaceum* and *D. gasterostei* within the eyes of many species of fish, and *D. phoxini* within the brain of minnows. The majority of metacercariae are enclosed within a cyst in the tissues of an intermediate host. The structure of the cyst varies widely with species, and thus the requirements for optimum activation and excystation may be very different. In those species that encyst on vegetation (Fasciolidae, Paramphistomidae and Notocotylidae), the

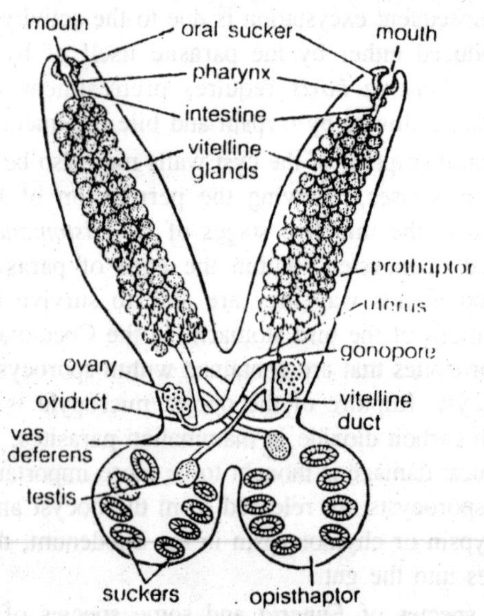

Fig. 5.1. Diplozoon.

metacercarial cyst wall is a much-thickened structure, capable of withstanding dessication and abrasion. In such cases excystation may be a complex process, requiring elevated temperature, bile salts and high pCO_2 to activate the larva, possibly stimulating it to secrete endogenous lytic enzymes.

At the same time host proteases, such as pepsin in the stomach and trypsin in the duodenum, digest the cyst wall from the outside. In the case of *Fasciola*, the larva emerges from the cyst through a digested mucopolysaccharide plug in the cyst wall. Triggers for excystation are quite specific, since many metacercariae can pass through the alimentary canal of an unsuitable host and fail to excyst, yet retain their infectiveness. The chemical structure of the walls of metacercarial cysts has not been adequately determined. The number of distinct layers that occur varies from one to four, depending upon species. *Parvatrema timondavidi* metacercariae are surrounded by a single layer of unidentified viscous material, and the metacercariae of *Posthodiplostomum* and Xiphidiocercariae and surrounded by two-layered cysts, composed of a layer of glycoprotein external to a layer of protein. Three-layered cysts walls are found in *Holostephanus luhei* and *Cyathocotyle bushiensis* (mucopolysacharide, protein and host tissue),

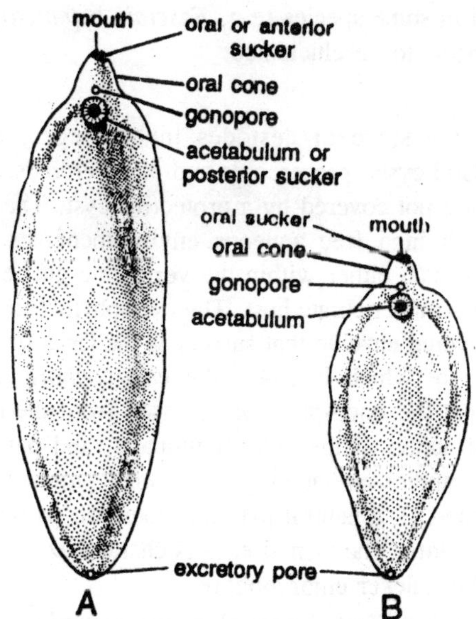

Fig. 5.2. Liver flukes. A—Fasciola gigantica. B—Fasciola hepatica.

Parorchis acanthus (protein, glyco-protein and mucopolysaccharide) and in *Notocotylus attenuatus* (acid mucopolysaccharide and two protein layers). Four-layered cysts walls are a feature of the metacercariae of *Fasciola hepatica* (tanned protein, tow layers of mucoprotein plus mucopolysaccharide, and an innermost keratin layer) and *Psilotrema oligoon* (mucopolysaccharide, protein, mucopolysaccharide and an innermost protein layer).

There is a large gap in our knowledge concerning the structure of the metacercarial cyst and the relationship between its structure and the optimal conditions required for activation and excystation of the enclosed larva. With this wide variation in cyst structure, one would expect that the process of excystation in the vertebrate gut is equally varied in its biochemistry. Limited studies carried out *in vitro* suggest that this is, indeed, the case. In some species, such as *Parvatrema*, elevation of the ambient temperature is sufficient to cause excystation, while other species require pepsin pretreatment, followed by trypsin-bicarbonate or pancreatin treatment. It is not clear what contribution is made by the larval digenean itself in the process of excystation, but it may aid its liberation from the cysts by active physical movements or by secreting lytic enzymes. Host bile salt seem to be prerequisites for excystation in some species (e.g. *Fasciola hepatica*) but their role in general remains to be elucidated.

Cestoda

The infective stage of cestodes include eggs, cysticercoids, cysticerci, hydatid cysts, procercoids and plerocercoids, of which only the latter two are not covered by a protective cyst. The eggs of many tapeworms hatch in a free aqueous environment, but those of the Cyclophyllidea hatch either within the vertebrate gut (e.g. Taeniidae) or in the gut of an invertebrate host. The cyclophyllidean eggs possesses a thick, multilayered envelop that surrounds the larva (the oncosphere), the outer layer of which is called the *embryophore*. In non-taeniid cyclophyllideans, the oncosphere is released from the egg, partly by the mechanical action of the host mouthparts, and partly by enzyme action. *In vitro* studies on hatching in hymenolepidid eggs suggest that bicarbonate, free CO_2 a neutral pH and proteases, particularly trypsin, are important factors. The taeniid eggs is characterized by the presence of a considerably thicker embryophore.

Hatching in these cestodes involves the enzymic disruption of the embryophore (by pancreatin in *Taenia psiformis*, and by pepsin in *T. saginata*) and activation of the oncosphere, which is then released from

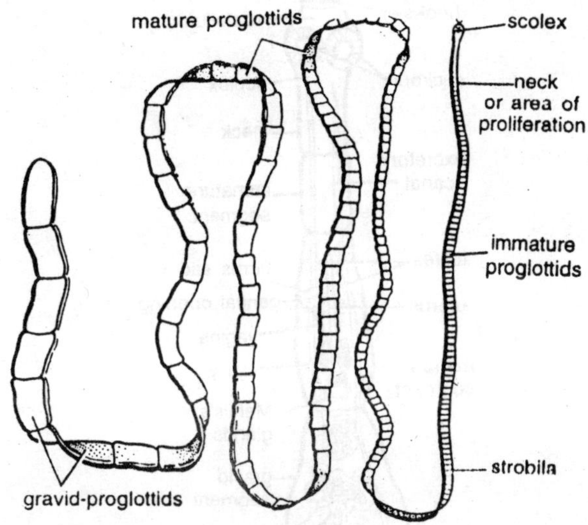

Fig. 5.3. Taenia solium.

the enclosing oncospheral membrane. Bile salts are obligatory prerequisites for activation in these two cestodes.

Excystation of cysticercoids, cysticerci and hydatid cysts takes place in the vertebrate intestine following their ingestion along with the tissues of the intermediate host. Elevated temperature, a variety of host enzymes and bile salts have all been implicated in the processes of excystation and evagination of the larval scolex. In some, e.g. *Echinococcus granulosus*, pepsin alone is sufficient to cause excystation, but more usually trypsin plus bile or pancreatin plus bile are the minimal effective requirements *in vitro*.

It may also be assumed that the prevailing physicochemical conditions are important contributory factors for both excystation and scolex evagination, but the roles of pO_2, pCO_2, pO_2, pH, temperature and redox balance have not been examined thoroughly. The plerocercoid larvae of the Pseudophyllidea are not encysted and are released from the tissues of the intermediate host during proteolysis and digestion in the upper intestine of the final host. These larvae are progenetic, resembling closely the adult worm in terms of size and somatic morphology. Sexual maturation in these larvae is probably induced by the elevated temperature in the bird or mammal gut in which they develop, along with the physico-chemical conditions that prevail. Because of their progenetic nature, the plerocercoid larvae of *Diphyllobothrium*,

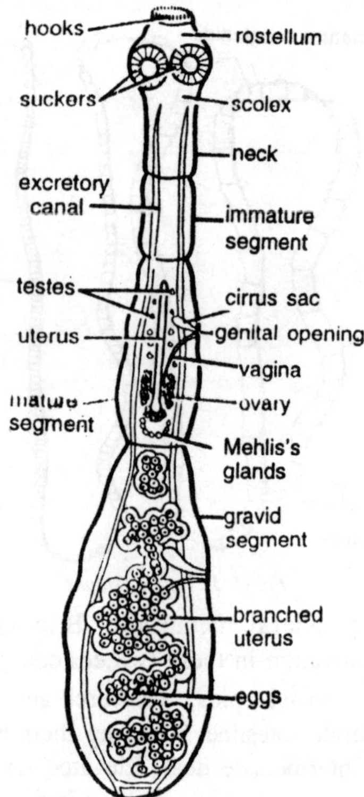

Fig. 5.4. Echinococcus granulosus.

Ligula and *Schistocephalus* have prove relatively easy to cultivate to maturity *in vitro*.

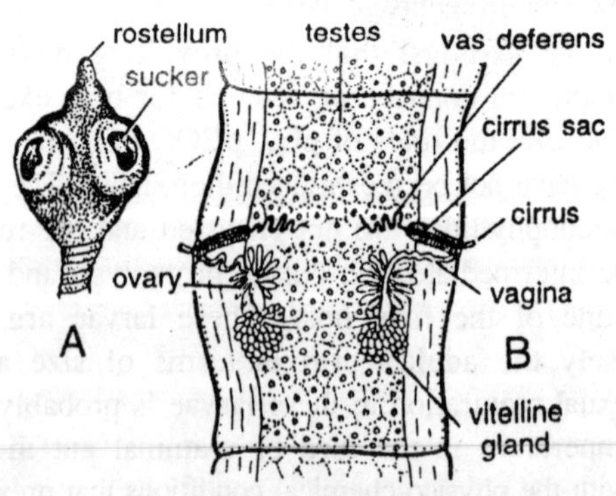

Fig. 5.5. Diphyllobothrium.

Acanthocephala

Excystation and activation of acanthocephalan cystacanths has been examined in only three species: *Moniliformis dubius*, *Polymorphus minutus* and *Echinorhynchus truttae*. Activation of the larva occurs after mechanical or enzymic disruption of the thin envelop that forms the cysts. Evagination of the proboscis occurs only after the cystacanth has left the stomach of the host and it is thought to be stimulated specifically by bile salt and a high pCO_2. The exact role of bile is not fully understood—the cystacanths of *P. minutus* can be activated *in vitro* in the absence of bile salts, but the addition of bile salts to an activation medium increase both the percentage activation and the rate of activation of the systacanths.

Nematoda

The eggs of some nematodes will hatch only after ingestion by a suitable host animal. In the case of the ascaridoid nematodes the process of hatching is initiated by the conditions in the mammalian or bird but, including the secretion of various enzymes by the enclosed larva. These enzymes, include chitinases, esterases and proteases, and it is their action that allows egress of the larva though a local, damaged region of the egg shell. Elevated temperature, neutral pH, low pO_2 and high pCO_2 are responsible for triggering the activation of the larva within the egg. Many animal-parasitic nematodes are transmitted to the final host, not as the egg stage but as third-stage larvae, ensheathed in the moulted (but not shed) cuticle of the second-stage.

Prior to any development inside the host, exsheathment must take place in the host gut. *Ancylostoma*, *Haemonchus* and *Trichostrongylus* all exsheath before development occurs and require specific stimuli to initiate this event. The major stimuli for exsheathment are high pCO_2, elevated temperature, the appropriate pH and the presence of reducing agents. Digestive enzymes of host origin do not appear to be involved in exsheathment in *Haemonchus contortus* and *Trichostrongylus axei*, whereas other species may require host proteases and bile salts for effective exsheathment to occur. Initiation of the exsheathment process is accompanied by the production of exsheanthment fluid by the larval nematode. There is some controversy over the exact nature of exsheathing fluid, but it probably contains proteolytic enzymes, e.g. leucine aminopeptidase has been identified in the exsheathing fluid of *T. Colubriformis*. The release of this fluid appears to cause the digestion of a localized, anterior region of the sheath, allowing the larva to escape into the gut.

Biochemical Considerations of Early Establishment

Role of Bile Salts

The role of bile salts as biochemical determinants of host specificity in gut parasites has been the subject of considerable speculation by parasitologists. Unfortunately there is, at present, insufficient evidence to lend support to this attractive and plausible hypothesis. Bile salts can be demonstrated to play important roles in the establishment of many of the parasites that enter the host through the alimentary canal. At least five distinct functions for bile are known.

1. Effects on membrane permeability.
2. Initiation of encapsulated larval activity.
3. Lytic effects on parasite surfaces.
4. Synergystic action with host digestive enzymes.
5. Metabolic effects upon both establishing and established parasites.

Because of the amphipathic nature of bile salts, i.e. each molecule possesses a hydrophilic and a hydrophobic end, they act as powerful surface-active agents, or *surfactants*. As such, bile salts can exert a profound effect on biological membranes, affecting both the lipid and protein moieties. The action of bile salts upon the membranes of parasite eggs and cysts increases their permeability, allowing the entry of water and digestive enzymes, though what was formerly an impermeable covering. In this way bile salts are thought to assist in hatching, excystation and exsheathment of many encapsulated parasites in the vertebrates gut.

The story is more complex that this, since synthetic surfactants, such as Tween 80, do not necessarily act in the same way as naturally - occurring bile salts, and some may even inhibit the release of the parasite. Bile salts stimulating movement of encysted larvae has been recorded in a number of species of parasite, e.g. *Eimeria* (Protozoa), *Cyathocotyle bushiensis*, *Echinoparyphium serratum*, *Fasciola hepatica*, *Holostephanus luhei* (Digenea), *Cysticercus pisiformis* (Cestoda), *Moniliformis budius* and *Polymorphus minutus* (Acanthocephala), but neither the extent of this effect, nor the way in which it is effected is understood. One of the mechanisms by which bile salts might influence the host specificity of either a permanent or a temporary parasitic resident of the vertebrate gut is suggested by the lytic effect of bile salts on certain parasites. This phenomenon has been demonstrated *in vitro*, with the protoscoleces of the hydatid organism, *Echinococcus granulosus*.

Bile that is rich in deoxycholic acid, such as that of hare, rabbit and shepp, readily lyses the surfaces of the protoscoleces and thereby suggests a possible biochemical explanation for the unsuitability of these mammals as hosts for the tapeworm. Bile from suitable hosts, such as dogs, is typically low in deoxycholic acid. Although the extent of the lytic effects of bile have not been investigate, the limited data at present available do support the working hypothesis that bile may be one of the determinants of host specificity. Many parasite cysts rupture, releasing their larvae, under the direct influence of bile, in a region of the small intestine that is proximal to the opening of the bile duct. In the presence of unsuitable bile salts, therefore, parasite larvae may either fail to hatch or hatch but fail to survive due to the lytic effects of the bile. This hypothesis could readily be tested for a wide range of parasites by employing a relatively straightforward *in vitro* assay.

Bile may act synergistically with host digestive enzymes and this may assist in enzymic breakdown of parasite cyst walls. Mammalian trypsin and lipase activities are both increased in the presence of bile salt *in vitro*. Bile can additionally exert metabolic effects upon established as well as establishing parasites, e.g. experimental cannulation of the rat bile duct will both prevent the establishment of the tapeworm *Hymenollepis diminuta* and reduce the size and fecundity of already established infections. The effects of such experimental manipulations must be interpreted with caution, however, and addition evidence is required before any generalisations can be made on the metabolic effects of bile on parasites.

Metabolic Changes during Establishment

In many parasites, there is a typically complex life-cycle, involving free-living larval stages as well as parasitic larvae and adults. It is not unexpected, therefore, to find that metabolic differences exist in various parts of the life-cycle, though this has been investigated in only a small number of cases. Consequently, there is a marked lack of information on the stimuli, or triggers, that are responsible for bringing about the dramatic changes that must take place when, for example, a free-living stage of a parasite invades a mammalian host. From a biochemical standpoint there is one obvious and important difference between free-living, infective forms of parasites and fully parasitic adults; energy metabolism of the infective larvae tends to be aerobic, while the adult relies heavily, if not exclusively, upon anaerobic carbohydrate metabolism.

Parasitic larval stages, such as sporocysts and rediae in the Digenea, resemble the adult worm and tend towards facultative anaerobiosis. Therefore, we note that the infective larvae of several species of nematode, the miracidia and cercariae of digeneans and, perhaps, the free-swimming coracidia of cestodes are all typically aerobic organisms. They possess a functional glyocolytic pathway, Krebs cycle and β-oxidation of stored lipid, and their electron transfer system terminates in cytochrome oxidase. Parasitic larvae, such as sporocysts, rediae, procercoids, plerocercoids, and acanthocephalan cystacanths, more closely resemble their adult forms in their metabolic pathways of energy metabolism, and are fundamentally anaerobic, lacking a complete Krebs cycle, relying upon glycolysis, and possessing branched electron transfer chains. It is not at all apparent just what physiological and biochemical events control the dramatic switch from aerobiosis to anaerobiosis.

During the process of invasion the parasite is seen to lose certain biochemical pathways and modify others. Isoenzymes, i.e. different proteins that catalyze the same chemical conversion, must be important during invasion, since there are often marked alterations in pH and temperature confronting a parasite on its journey from a free-living environment to the internal tissues of a host animal. It is not known whether the enzymes required for the adult stage are present, but inoperative, in the juvenile parasite, or whether they are synthesized *de novo* during and immediately after invasion. Temperature and a high pCO_2 appear to be the most likely candidates for orchestrating the metabolic with common to many parasites during invasion. Other factors, such as pO_2, pH, rH and osmotic pressure may also be involved.

There is evidence that some larval digeneans (e.g. *Zoogonus rubellus* and *Himasthla quissitensis*) may be preadapted for the higher temperatures encountered in the body of the final host, since there is a direct relationship between Q_{10} of the larva and the temperature of the final host. A number of parasites undergo metabolic changes as the ambient temperature is raised above 30°C, e.g. *Nippostrongylus brasilensis*, *Necator americanus*, *Strongyloides ratti* and *Schistocephalus solidus*. Conversely, parasites whose final host is a poikilotherm ("cold-blooded") must respond to factors other than temperature to initiate metabolic switching, but this has not been investigated in any detail.

Parasites are probably genetically rather complex, since the adult genome carries information for metabolic pathways that operate at other stages of the life-cycle. It would be interesting to speculate that

Table 5.1 Metabolic differences between adult and larval parasites.

Parasite species	Summary of larval *metabolism*	Summary of adult *metabolism*
Ascaris lumbricoides	glycolysis-high pyruvate kinase activity; functional TCA cycle; β-oxidation of lipid; functional glyoxylate cycle; terminal oxidase is cytochrome oxidase	glycolysis to PEP- low pyruvate kinase activity; carbon dioxide fixation forming oxalacetate; mitochondrial reduction of fumarate to succinate; no TCA cycle; β-oxidation enzymes present but inoperative; no glyoxylate cycle; terminal oxidases are cytochromes o and a_3
Haemonchus contortus	functional TCA cycle	fixation of carbon dioxide
Schistosoma mansoni	*miracidia and cercariae* functional glycolysis and TCA cycles; cytochrome oxidase present; Pasteur effect demonstrable; probable β-oxidation of stored lipid *sporocysts.* utilize oxygen; cytochrome oxidase present; lactate excreted	typically anaerobic
Fasciola hepatica	*miracidia and cercariae* functional glycolysis and TCA cycles; large lipid stores *sporocysts* functional glycolysis; no cytochrome oxidase (present in rediae)	functional glycolysis; fixation of carbon dioxide; partial reverse TCA cycle; no lipid catabolism; branched cytochrome chain

the more complex the life-cycle of the parasite, the greater the quantity of functional DNA and RNA required. This hypothesis has not been tested, but evidence, from a single study, suggests that the genome of some cestodes and nematodes is more complex than that of equivalent free-living animals by a factor of between two and three.

Migration and Site Selection

Patterns of Migration

Three readily discernible patterns of migration may occur once a parasite has invaded the host's body. The first migratory activity serves to remove the parasite from its site of entry to the chosen target organ, where it will reside as a larva or as the adult stage. Such migrations are frequently accompanied, particularly in the helminths, by growth and development of the parasite, and conclude in the maturation of the worm. These are called *ontogenetic migrations*, since relocation and parasite development are closely related. Although we know very little about the events that take place during ontogenetic migrations of parasites, it is evident that a strick sequence of triggers exists, presumably emanating from the host and recognized by the parasite, that stimulate the sequence of normal development. This may be exemplified by the aberrant migrations that can occur when certain parasites invade a host in which complete development cannot occur, yet they may survive for some considerable time. This is not uncommon in skin penetrating forms, such as larval hookworms and larval digeneans.

In the hookworms, penetration of an unsuitable host (e.g. dog worms entering man) may lead to the production of what are termed cutaneous or visceral *larva migrans*. These are migratory forms that fail to develop, lacking the necessary signals from the host, and they wander though the peripheral or deeper body tissues, often for extensive periods, causing pathological responses in the host during their migration and often becoming encapsulated in organs and tissues. In the correct host these larvae would migrate via a strict route through the body, ultimately reaching the alimentary tract, where they become sexually mature and produce eggs. Another example is that of avian schistosome cercariae, which will penetrate the skin of man but never develop to the adult worm; a mild pathological condition, known as schistosome dermatitis, is produced at the site of entry.

In man the schistosomules of avian schistosomes begin to migrate but this is brought to a half in the dermal layers of the skin. Ontogenetic migrations by parasites may result in either the adult stage reaching is preferred target organ or in the production of a quiescent larval stage that will await transmission to the next host digenean metacercariae are examples of the latter. It should also be noted that a great many parasites undergo ontogenetic migrations in their invertebrate intermediate hosts, e.g. trypanosomes and malarias in insects, digeneans in molluscs, cestodes in aquatic invertebrates and nematodes in insects. In these instances, successful transmission to the next host can only be accomplished after the completion of the ontogenetic migration, when the parasite is at the correct developmental stage and in a suitable location within its invertebrate host, e.g. tapeworm cysticercoids will be infective to vertebrates only after they penetrate the insect gut wall and complete their development in the haemocoel. A third type of migration has been characterized in the tapeworm, *Hymenolepis diminuta* and is thought to be related to the feeding patterns of the rat host. In worms that are eight days old, or older, there is a posterior migration in the intestine of the rat during the day and an anterior migration by night. This takes place in both single and multiple worm infections when the rats are feeding by night and fasting by day, which is their normal routine.

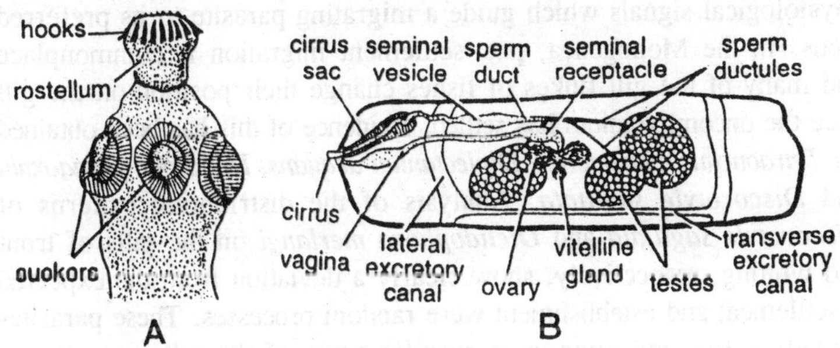

Fig. 5.6. Hymenolepis. A—Scolex. B—Mature proglottid.

The pattern of migration can be reversed, experimentally, if rats are fed by day and denied access to food by night. This migration represents a real and considerable movement of tapeworms in the host's intestine, but the factors that initiate this migration have not been isolated. It has been suggested that either the passage of food or the gradually diminishing levels of free glucose in the intestine are responsible for stimulating the posterior phase of the migration, while the subsequent anterior migration might be along a gradient of glucose or some other substance, such as bile. Alternatively, the posterior movement could simply be passive, the worms being swept backwards, with the food, by peristalsis; once in the lower small intestine it may be that the worms recognize the unsuitability of their surroundings and actively migrate anteriorly to the duodenum. Certainly, they grow less well in the posterior intestine than in the more anterior regions. It is not known whether this type of diurnal feeding migration is common occurrence in gut-dwelling parasites (it would be most likely to occur in parasites that were not firmly attached to the mucosa of the host intestine).

Site Selection and Recognition

The great majority of invading parasites do not settle at the site of entry of the host but migrate to their preferred locus. Additionally, most parasites appear to be rather specific in their site requirements and rarely does one encounter ectopic parasitisms in the normal host. These facts strongly suggest that parasite migration from the site of entry to the final location is a directed event, involving the active participation of the parasite itself. The ubiquity of migrations associated with site selection should, ideally, allow us considerable insight into the mechanisms involved in the processes of site recognition and location. Unfortunately, this is not the case, and we known very little about the physiological signals which guide a migrating parasite to its preferred locus. In the Monogenea, post-settlement migration is commonplace and many of the gill flukes of fishes change their position on the gill once the oncomiracidium has settled. Evidence of this has been obtained for *Tetraonchus monenteron*, *Diplectanum aequans*, *Diplozoon paradoxum* and *Discocotyle sagittata*. Analysis of the distribution patterns of *Discocotyle sagittata* and *Diclidophora merlangi* on the gills of trout and whiting, respectively, show clearly a deviation from the expected if settlement and establishment were random processes. These parasites actively select and migrate to specific areas of the gills, usually on the most anterior gill.

A rather special case exists for the amphibian parasite *Polystoma integerrimum*. As a juvenile worm, the parasite inhabits the internal gills of the tadpole and, during the metamorphosis of the host, it migrates along the ventral, external surface of the host to the cloaca, from which it enters the urinary bladder, where it resides as an adult worm. Sexual maturation of *Polystoma* coincides with gonadotrophin release by its host, ensuring that parasite eggs are released into the water at a time when larval frogs are abundant. Fertile, neotenic larvae of *Polystoma* are formed it the external gills of the tadpole are invaded, and a secondary cycle of development takes place. The stimulus that controls the migration of juvenile *Polystoma* from the gills to the

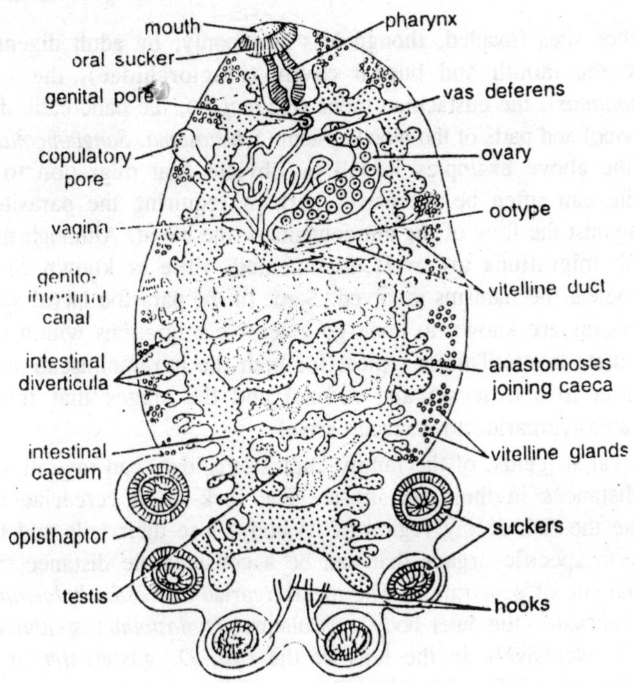

Fig. 5.7. Polystoma.

urinary bladder is probably endogenous, since larvae placed on metamorphosing tadpoles migrate first to the gills and then to the cloaca. The majority of adult digeneans parasitize the alimentary tract, and its associated structures, of vertebrates.

Little is known about their ability to select sites, but it may be assumed that this facet of their physiology is highly developed, as judged by their patterns of distribution within the host. Ingested digeneans will hatch or excyst in the upper small intestine where there is an abundance of proteases and bile salts, and from there they may migrate to other regions of the gut. Preferred sites in the Digenea include the stomach (*Otodistomum* in elasmobranchs, *Hemiurus* and *Derogenes* in teleosts), the duodenum (*Centrovitus*, *Glypthelmins* and *Loxogenes* in amphibians, *Mesocoelium* in reptiles, *Acanthoparyphium* in birds, *Nudacotyle* and *Alaria* in mammals), the posterior small intestine (*Lasiotocus* in teleosts, *Plagiorchis* and *Echinostoma* in birds, *Echinostoma* in mammals), the rectum (*Stephanostomum*, *Zoogonides* and *Podocotyle* in teleosts, *Diplodiscus* in amphibians), the intestinal caeca (*Catatropis*, *Postharmostomum* and *Notocotylus* in birds).

Other sites invaded, though less commonly, by adult digeneans include the mouth and buccal cavity (Plagiorchidae), the lungs (*Paragonimus*), the eustachian tubes (*Halipegus*), the pancreatic ducts (*Eurytrema*) and parts of the blood system (*Schistosoma*, *Sanguinicolidae*). From the above examples it will be obvious that migration to the final site can often be extensive, perhaps requiring the parasite to move against the flow of the gut contents or the blood. Although many of these migrations are well documented, little is known of the physiological mechanisms involved. Some of the parasitic larval stages of digeneans are known to undergo extensive migrations within their intermediate hosts. This is typical for both the intramolluscan stages (miracidia to sporocysts and rediae) and the stages that inhabit vertebrates (cercariae to metacercariae).

Larval strigeids, of the family Diplostomatidae, can migrate over large distances in their fish hosts. The fork-tailed cercariae may penetrate the fish at any region of its body, lose their tails and then migrate to specific organs that can be a considerable distance from the initial site of penetration. The metacercariae of *Posthodiplostomum cuticola* encyst in the outer body musculature, *Diplostomum spathaceum* localized exclusively in the lens of the eye, *D. gasterostei* in the retinal tissue and *D. phoxini* in the ventricles of the brain of their respective hosts. Studies on *Cercaria X*, which is probably *D. spathaceum*, demonstrate that migration can occur via the blood system, the lymphatic system and through the tissues; larvae that do not reach their target organ within approximately 24 hours of penetration are thought not to survive. Adult cestodes are all inhabitants of the vertebrate gut and their site selectivity within the alimentary tract is known for only a few species.

The majority of tapeworms are located in the small intestine, or the equivalent region that lies immediately posterior to the opening of the stomach. Occasionally tapeworms occur in other regions, e.g. *Hymenolepis microstoma* in the bile duct of mice, and *H. fausti* in the caeca of ducks. Within the intestine, a tapeworm prefers a particular region for attachment, through this apparent preference can alter on a diurnal basis, as we have seen for *H. diminuta*. In the chicken gut, *Raillietina cesticillus* excysts close to the opening of the bile duct in the duodenum and, during growth and development of the worm over the next seventeen days, its position of attachment gradually shifts in a posterior direction, so that the fully mature worm inhabits a region 25 cm posterior to the opening of the bile duct. Transplanted *H.*

diminuta will migrate to a preferred region of the small intestine regardless of the site of inoculation, but in this species, the notion of a preferred site is complicated by the occurrence of a diurnal feeding migration. The fish tapeworm, *Proteocephalus filicollis*, inhabits the rectum of the three spined stickleback while immature and the intestine when mature.

All of the above three cestodes, then, exhibit ontogenetic migrations and a pronounced degree of site selection. In some other cestodes the site of attachment is determined by the morphology of the scolex in relation to the morphology of the host gut mucosa. This is illustrated by two tapeworms that cohabit the intestine of rays (*Raia radiata*): *Pseudanthobothrium*, which has a deep cup-shaped scolex, is restricted to the anterior three tiers of the intestinal spiral valve, where the gut villi are long: *Phyllobothrium* possesses a much flattened and foliaceous scolex and attaches only to the posterior tiers of the spiral valve, which have shorter villi. Larval cestodes may be somewhat less specific in their choice of site, as they often inhabit the body cavity or musculature of the intermediate host. In the case of *Echinococcus granulosus*, the hydatid cysts develop in the first suitable site encountered after penetration through the gut wall, though this may be partly determined according to whether the blood or lymphatic system is used as a migratory route.

Adult acanthocephalans all inhabit the vertebrate gut, in which they are attached to the mucosa and submucosa by the armed proboscis. The proboscis itself is freely retractable so that the potential for movement and change of location within the gut exists. This has not been observed, however, except by implication in concurrent infections of *Moniliformis* with *Hymenolepis*, and *Neoechinorhynchus* with *Proteocephalus*. From observations on the location of acanthocephalans in natural infections it seems probable that each species has a preferred locus in the host's gut. Some species are found in the stomach, particularly in fishes (*Acanthocephalus lucii* in perch, *Acanthorhynchus borealis* in *Lota* and *Silurus*), but a large majority inhabit the upper intestine (*Pomphorhynchus*, *Neoechinorhynchus*, *Acanthocephalus ranae*, *Moniliformis dubius* and *Polymorphus minutus*), or the more posterior regions including the large intestine, the caeca in birds and the rectum in fishes.

Larval acanthocephalans inhabit the body cavity of their invertebrate intermediate hosts, but it is not known to what extent they select specific sites within the haemocoel. Nematodes parasitize a great many organ systems of invertebrates and vertebrates and each species appears to possess its own specific requirements in terms of sites selected. The much quoted study by Schad (1963) clearly demonstrates that several closely related species of pinworms, of the genus *Tachygonetria*, can coexist within the colon of turtles by virtue of their differences in site specificity, along with divergence in their nutrient requirements and feeding methods. Few other studies have examined site specificity in nematodes in such detail.

It is evident, from the predilection of many nematodes for complicated patterns of migration around the host's body, that they must be able to recognize and respond to a variety of sequential stimuli concerned with their development, and that they are capable of actively choosing the site in which to reside. The general subject of site selection in parasites has been largely overlooked by parasitologists, even to the extent that simple descriptions of the sites actually occupied by naturally-occurring parasites are often vague and imprecise. There is a whole field of parasite physiology lying fallow at present, which would be likely to reveal some very interesting aspects of host-parasite interaction. The most fundamental unanswered questions are:

1. What factors permit or prevent the establishment of a parasite at a particular locus in the host's body?
2. How do parasites recognize suitable environments for their continued development?

Invasion of Host Tissues

It is possible to recognize two separate processes of invasion in the life history of many parasites. The first invasion involves the entry of the parasite into the host's body, while the second invasion occurs then the parasite invades the particular tissues of the host is which it resides. Biphasic invasion, therefore, is typical of tissue parasites, but does not necessarily apply to those gut parasites that inhabit the lumen. Obvious examples of parasites with distinct invasion processes include the malarias and babesias that invade red blood cells, coccidians, nematodes and cestodes that invade the gut mucosa, schistosomes that invade the blood system and trypanosomes that invade nervous tissue. Recent attention has focussed on the invasion of mammalian and avian red blood cells by malarial parasites, and of intestinal epithelia by coccidians.

The accumulating evidence points to a high degree of specificity in this secondary invasion, and this feature may be important in determining the host specificity of certain parasites. Malarial parasites

(*Plasmodium*) invade the vertebrate blood stream as sporozoites, during the feeding of an infected mosquito on the host blood. The sporozoites do not remain in the perpheral blood of the mammal for long, but invade liver tissue where they multiply asexually, by schizogony, resulting in the production of merozoites. These forms are released into the blood stream and invade host red cells (erythrocytes), where a second schizogonic cycle takes place. The mechanisms by which the merozoites locate and then penetrate the erythrocytes are complex and far from completely understood. The merozoite is surrounded by a thick surface coat, which is lost during invasion of the red cell. The surface coat, carries a net negative charge, and is covered with fine filaments, that are involved in the initial attachment of the parasite to the host cell. It can be removed experimentally by treatment with proteases. At the anterior end of the merozoite is an apical protruberance which contains a complex of organelles (rhoptries, micronemes, polar rings and microbodies). These organelles characteristically disappear shortly after entry into the red cell is accomplished.

The actual process of invasion is initiated when the merozoite attached to the surface of the red cell by way of the surface filaments, and the apex of the merozoite is appended directly to the host cell surface. This latter event causes local bending of the red cell membrane. Since the surfaces of both the parasite and the host cell are negatively charged, the initial attachment must be regarded as a "captive" process that overrides the natural electrostatic repulsion of the two cells. As entry proceeds, the red cell membrane invaginates, and this is thought to be due to the release of material from the rhoptries and micronemes. There is, therefore, no penetration as such, of the red cell membrane and the engulfed parasite comes to lie within what is termed, a *parasitophorous vacuole*, bounded by the red cell membrane. The entire act of invasion may be very rapid, e.g. occupying as little as one minute in *Plasmodium knowlesi*.

The mechanisms by which the *parasitophorous vacoule* is formed are not known, but a histidinerich protein has been isolated from various stages of *P. lophurae*, and this induced cupping of red cell membranes at low concentrations, *in vitro*, and membrane lysis at higher concentrations. Whether such a protein is secreted by the apical organelles of the merozoite to facilitate invasion remains to be clarified. The sloughed-off parasite surface coat appears to plug the opening to the *parasitophorous vacuole* and may also aid the movement of the parasite into the red cell. Once invasion is completed, the merozoite transforms into the trophozoite which remains within the parasitophorous vacuole. There is evidence that invagination of host cells is a common occurrence during tissue invasion by many protozoan parasites, and it may prove to be the typical method of invasion in both sporozoans and coccidians. *Babesia*, *Eimeria* and *Toxoplasma* all invade by the invaginative route. In these genera, the parasitophorous vacuole often break down shortly after invasion, resulting in an intimate contact between the parasite and the host cell cytosol.

It has been alternatively suggested that *Toxoplasma* invades the host mucosal epithelium by inducing phagocytosis and avoiding subsequent digestion. However, inhibitors of phagocytosis, such as colchicine and cytochalasin B, do not interrupt invasion unless the parasite itself is treated directly, implying active invasion by the merozoite. In the same way that the merozoites of *Plasmodium* recognize and attach to red blood cells, so *Eimeria*, among the coccidians, exhibits a marked preference for intestinal epithelial cells. So strong is the apparent degree of site specificity that sporozoites of *Eimeria* will develop only in gut cells, regardless of whether the site of entry to the host's body is via intravenous, intramuscular, intraperitoneal or parenteral injection. Little is known about the invasion of gut epithelia *in vivo*, but there is a growing body of information based on *in vitro* studies. In culture, *Eimeria* sporozoites will invade and develop in cells from at least seventeen different hosts. Entry into these cells is rapid and appears to be an active process. The role of the apical organelles in this process is not clear.

There is controversy concerning whether the sporozoites of *Eimeria* are phagocytosed by the host cell or whether they actively induce phagocytosis. Treatment of cell cultures with anti-phagocytic drugs suggests that the latter event occurs. Recognition of the suitable host cell by a tissue parasite will depend upon surface receptors situated on the host cell membrane. There may even by receptors that are specific for attachment and others specific for invasion, since the merozoites of *Plasmodium* will attach to a variety of cell types, but will only invade erythrocytes.

Invasion of red cells by malarial parasites is unaffected by neuraminidase treatment, but is reduced by treatment of red cells with proteases, concanavalin A and wheat germ agglutinin. Although the chemical nature of these surface receptors remains elusive, it is evident that they are labile entities that are readily lost or neutralized. Tissue invasion by helminth parasites has been inadequately investigated and little is known about the processes of cell recognition or invasion. It is not known whether helminths invade host cells by the invaginative route or by direct penetration.

Factors Inhibiting Parasite Growth and Development

Crowding Effect

Intraspecific crowding of parasites, within host tissues or within the gut lumen, may result in a reduction in their rate of growth, the maximum size attained and their fecundity. Only a small number of examples of the crowding effect have been described, yet it seems likely to be a common phenomenon amongst the helminths. Probably the best described example of crowding is that of *Hymenolepis diminuta*. When this tapeworm is grown in carefully controlled infection densities in rats, there emerges a clear relationship between the intensity of infection and a reduction in growth rate, maximum worm weight and length, and egg production. Any increase in the number of worms in an infection, however, has no effect upon the rate of maturity of *H. diminuta*, and egg production always commences on the sixteenth day after infection.

There is, as yet, no unequivocal explanation for the crowding effect in this cestode. Competition for space, for oxygen and for essential nutrients, such as glucose, have all been proposed as likely candidates. Alternatively, the production of toxins or growth inhibitors by the worms may be responsible, and the immune responses of the host may be directly or indirectly involved. Crowding of *H. microstoma* in the mouse bile duct affects these worms in a similar manner. The crowding effect has been observed in some species of parasitic nematodes, e.g. the length of adult *Skrjabingylus nasicola*, which infections of increased intensity. *Ancylostoma caninum* infections of dogs are also subject to a crowding effect.

Interspecific Interactions

Interactions may take place between cohabiting parasites of different species, and these can result in reduced parasite growth and fecundity, and also be accompanied by alterations in site selection. Concurrent laboratory infections of rats with *Moniliformis dubius* and *Hymenolepis diminuta*, result in a marked restriction in the distribution of each species within the rat's small intestine, as well as a reduction in size of both (the tapeworm shows the greatest size decrease in this relationship). A similar situation can be found in natural infections, e.g. co-occurring *Proteocephalus filicollis* and a species of

Neoechinorhynchus mutually alter the site of each other in the gut of the three-spined stickleback, so that each species is restricted in its range as a result of the interspecific interaction. This is termed *competitive exclusion*, but the factors that bring it about are no known. Other classes of interspecific interactions may exist between the parasite and its host, e.g. variations in the size and fecundity of parasites in relation to the size or sex of their host. Furthermore, a single species of parasite may grow and develop quite differently in a range of host species. Such interactions will involve a variety of physiological factors that include the immune response of the host, its immunological status, as well as the physico-chemical conditions that promote or limit establishment and growth of a parasite.

Labile Growth Patterns

Members of the phylum Platyhelminthes possess a remarkable ability for degrowth and dedifferentiation under adverse nutritions conditions. This is characteristic of the free-living Turbellaria and the Cestoda, but is known whether it occurs in the Monogenea of Digenea. In the tapeworms, degrowth is termed *destrobilation*, whereby the majority of the proglottids are shed, leaving only the scolex and the neck region attached to the mucosal surface. Destrobilation can be induced in experimental animals by depriving the host of carbohydrate, and this has been demonstrated with *Raillietina cesticillus* and *Davainea proglottina*. Natural destrobilation has been observed in *Dibothriocephalus*, a parasite of bears, when the host's diet changes prior to hibernation.

Natural destrobilation may be common in cestodes that inhabit animals which hibernate, where it would serve as an overwintering response by the parasite. Experimental deprivation of carbohydrate may induce a marked reduction in tapeworm size, without the obvious shedding of proglottids. This occurs in *Hymenolepis diminuta*, *H. citelli*, *Lacistorhynchus tenuis* and *Oochoristica symmertrica*, and it reflects the dependence of many cestodes on the availability of carbohydrate. Other tapeworms, e.g. *Schistocephalus solidus* and *H. nana*, do not possess such a high degree of sensitivity of depletion of dietary carbohydrate. Cestodes with a typically high rate of growth are more sensitive to a carbohydrate deficiency than those that grow more slowly. A second feature of labile growth and development is found in the development of the hydatid cysts of *Echinococcus granulosus*.

The protoscoleces within the cyst are capable of development of differentiation in one of two separate directions according to the environmental signals available. Should the cyst burst and release the protoscoleces into the body cavity or the tissues of the intermediate host, each protoscolex can become yet another cyst, in which new protoscoleces will develop; this is termed *secondary hydatidosis*. Protoscoleces that enter the alimentary canal of a suitable host will grow into mature tapeworms. A similar degree of lability in development os seen in the related cestodes *Taenia serialis* and *T. multiceps*. *In vitro* studies on *E. granulosus* suggest that the stimulus for strobilar development to the adult worm is contact with a proteinaceous substrate.

Cultivation of protoscoleces in a liquid medium always results in the formation of secondary hydatid cysts, whereas provision of a punctured protein base induces strobilar development. The ability of protoscoleces to become secondary hydatid cysts represents a particular problem for the surgeon, and he must exercise considerable care during extirpation of cyst to prevent its rupture and the release of vast numbers of protoscoleces. Although it is not fully understood how environmental stimuli control the direction of development in *Echinococcus*, it is evident that a complex interaction between operator and repressor genes will take place in the switch from larva to adult worm, or larva to secondary larva.

Moulting in Nematodes

A detailed consideration of growth and the phenomenon of moulting in the parasitic nematodes is outside the scope of this book. Suffice it to stress here that, because of their cuticular covering, all nematodes grow by distinct stages and moult, or shed the cuticle, between each growth stage. Commonly there are five growth stage, with four moults, between egg and adult, and in the parasitic nematodes the third-stage larva is normally infective to the host. In some species, the cuticle of the second stage is retained after moulting, forming a protective sheath around the larva; exsheathment takes place in the gut of the host.

6

REPRODUCTION

This chapter contains a very brief account of the processes of asexual and sexual reproduction in parasites, from Protozoa to Nematoda. Limitation of space precludes a detailed description of the physiology of reproduction so that subjects such as sexual attraction, mating, spermatogenesis and oogenesis, development of the fertilized eggs, egg shell formation and hatching of the mature egg, can only be treated in the most superficial of ways, or in some cases neglected altogether. The most reproductive biology of most parasites has not been studied in any great depth and, while the details of the morphology or reproductive systems as well known, relatively little physiological information is available.

ASEXUAL REPRODUCTION

Asexual reproduction is achieved by the splitting or budding of a single cell or an entire organism, and therefore does not involve the fusion of two haploid gametes, with a subsequent meiosis. Asexual multiplication is a feature of many parasitic protozoa, all of the Digenea and a small number of the Cestoda. In many of these parasites there is an alternation of sexual and asexual phases during the life-cycle, sometimes occurring in the same host. In the Protozoa there are at least five distinguishable asexual processes the occur.

Binary fission is by far the most common from a asexual multiplication and involves the parent cell dividing into two daughter cells. The division is preceded by a nuclear mitosis; the plane of division is transverse in the Ciliophora and Sporozoa and longitudinal in the Mastigophora. *Multiple fission* the secondary division of daughter cells before they become separated from the parent cell. It is a feature of the gregarines and trypanosomes and gives rise to spheres or chain of daughter cells. A third form a asexual reproduction is *schizogony*, a form of multiple budding, in which the parent cell (the schizont) produces daughter cells (merozoites) first by undergoing one or more nuclear division, and then by dudding off many uninucleate daughters. Schizogony is typical of the Sporozoa, such as the Haemosporina, (e.g. malarial parasites). *Endodyogeny*, in which two daughter cells are budded internally within the parent cell, is only known to occur in some sporozoans, such as *Toxoplasma* and related forms. A fifth type of asexual reproduction, *single budding*, occurs in free-living but not parasitic protozoans.

Asexual multiplication in the helminths occurs exclusively by *internal budding* or *polyembryony*. In the Digenea, the asexual phases are found only within the snail intermediate host, and involve the sporocyst and redia stage. During the intramolluscan phase of development of a digenean there may either be two generations of sporocysts, i.e. mother and daughters, or the initial sporocyst (arising from the penetrated miracidium) may give rise to a redia, within which are developed daughter rediae- the precise pattern depends upon the species involved. Cercariae are developed asexually within daughter sporocysts or daughter rediae. The intramolluscan phase of the digenean life-cycle is highly proliferative, since one miracidium may produce as many as several million cercariae by asexual development involving sporocysts or rediae. The sporocyst is generally a hollow, elongated organism, with an anterior birth pore and well developed excretory system, but lacking a gut.

At maturity, a mother sporocyst (derived directly from a miracidium) will be replete with growing daughter sporocysts or rediae; these emerge through the birth pore and migrate within the snail, usually to the hepatopancreas (digestive gland). Daughter sporocysts are morphologically similar to the parent sporocyst, and produce, by internal budding of germinal tissue, large numbers of cercariae, which escape through the birth pore when "mature". The redial stages are not unlike the sporocyst but possess a mouth and gut. Asexual multiplication in the Digenea involves the division of *germinal masses* within these larvae; these masses contain large numbers of germ cells, which proliferate during development.

In the mother sporocyst germ balls are formed, each of which will become a daughter sporocyst—many hundreds of daughters can be produced by the division of this germinal tissues. Daughter sporocysts and rediae, likewise contain germinal tissue from which embryonic cercariae are produced. Cercarial production by individual sporocysts or rediae may continue, uninterrupted, for several months, or even perhaps years, as has been recorded in one case in a laboratory infection. There is some argument concerning the exact nature of asexual multiplication in the Digenea. Some authors regard this form of multiplication as parthenogenesis (ameiotic or meiotic), while others consider it to be regenerative budding or polyembryony. It has even been suggested that the intramolluscan stages of digeneans reproduce sexually and that sporocysts and rediae are not larval forms at all. These vies have not been substantiated and the debate is really philosophical and adumbrated by semantic problems. Nevertheless, however one regards the germinal lineage of the intramoluscan stages of digeneans, a clear function emerges, that of enormously increasing the numbers of infective individuals.

This may be an adaptation to the great wastage that must occur once cercariae are liberated into the external environment. In the cestodes, larval development involving asexual budding occurs in only a few groups (Diphyllobothriidae, Hymenolepididae and Taeniidae). The majority of tapeworms form only a single larva from each egg; of the many types of larval stage, only a small number are capable of proliferation by asexual budding. External budding occurs in two larval types, the urocystis (an budding cysticercoid) and the urocystidium (a budding strobilocercus).

Internal budding is the more common pattern and is typical of the polycercus larva (*Parictotaenia paradoxa*), the coenurus larva (*Taenia multiceps*) and the hydatid larva (*Echinococcus granulosus*). Larval scoleces are budded from the cyst wall in the polycercus and coenurus, but they develop within brood capsules in the hydatid cysts, budding from the wall of the capsule. Asexual budding in the cestodes is a highly effective way of increasing the number of individuals during the stay in the intermediate host, e.g., hydatid cysts can grow continually, reaching a diameter of several hundred millimetres and containing several million larval scoleces, each a potential adult tapeworm. Coenurus larvae normally possess a less dramatic reproductive potential and contain only a few hundred larval scoleces.

Sexual Reproduction

Protozoa

Sexual reproduction in one form or another occurs in the Sporozoa, the Ciliophora, the Opalinata and the Hypermastigida (insect gut parasites). One problem of sexuality in the Protozoa concerns the differentiation between fusion of gametes and fusion of individual organisms, such as occurs in conjugation; these processes are often difficult to distinguish. Where recognizable gametes are formed, they may be morphologically identical (iso-gametes) or dissimilar (anisogametes), e.g. microgametes ('male") and macrogametes ("female"). In the Sporozoa, an alternation of sexual and asexual phase frequently takes place, with gamogony leading to gamete production and sporogony leading to the asexual production of sporozoites (schizogony is an addition asexual phase that is found in some sporozoans).

The major division of the Sporozoa into Coccidia and Gregarinida is based upon overt differences in sexual reproduction; in the gregarines gamete production occurs in both "sexes", while the coccidians form only microgametes, that resemble metazoan spermatozoa. In the Opalinata, which are typically parasites of the posterior gut of amphibians, sexual reproduction involves the fusion of anisogametes which are released from cysts produced only during the breeding season of the host. Fertilization takes place within the newly infected tadpole. In the flagellates, particularly those that inhabit the gut of insects such as termites, sexual reproduction involves the fusion of flagellated gametes. This process is under hormonal control and related to the moulting cycle of the host. Where two different hosts are involved in the life-cycle (e.g. in the plasmodia) sexual reproduction will take place in the invertebrate host, while asexual sporogony and schizogony take place in the vertebrate. By contrast, asexual multiplication in the trypanosomatids occurs in both invertebrate and vertebrate hosts.

Monogenea

All monogeneans are hermaphrodite and there usually common genital opening for both male and female systems. Cross-fertilization, i.e. between two different individuals, is probably the normal method of reproduction, but self-fertilization may also occur. The male system consists of testes, which vary in number and disposition, seminal duct for sperm transport, various glands and a copulatory organ ("penis" or cirrus). The female system tends to be variable in structure from group to group, but basically contains a single ovary, extensive vitelline glands, various ducts, including a vagina which may have one or several openings to the outside, and, in the Polyopisthocotylea only, a genito-intestinal duct, whose function is unknown. The ootype is a muscular extension of the oviduct, surrounded by Mehlis' gland, a mucous and

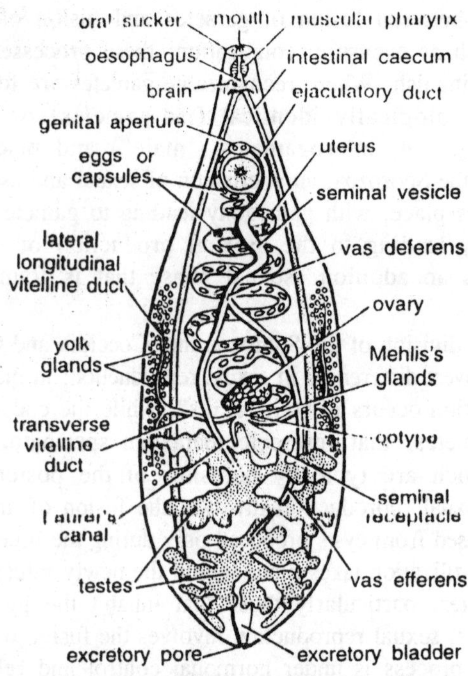

Fig. 6.1. Opisthorchis (=Clonorchis) sinensis.

serous gland. The function of Mehlis' gland is not known. It was originally believed to contribute shell material to the developing egg, but this function is now thought to reside exclusively with the vitelline glands.

Monogeneans are normally oviparous, laying shelled eggs into water, but one family, the Gyrodactylidae, is viviparous, bearing one, or frequently several developing embryos, one within the other. Monogenean eggs are operculate, with filaments at one or both ends. The eggs shells are composed to tanned sclerotin, formed in the vitellaria and hardened by the quinone tanning system. The majority of eggs are liberated into water where they are presumed to sink to hatch, though the filaments may allow them to attach to a new host and hatch there; the latter is thought to be a rare occurrence.

The period of incubation of the egg varies with species, e.g. it may be as short as three days, as in *Dactylogyrus* at 22-23°C, or it may take many weeks, as in *Dictyocotyle* at 10°C. The process of hatching of the monogenean egg has not received a great deal of attention and the roles of temperature, light, pH pCO_2, pO_2 and salinity have not been determined. Light is an important factor for activating the developing larva within the egg and this may facilitate hatching, whereby the free-swimming oncomiracidium emerges through the opening of the operculum.

Digenea

The majority of digeneans are hermaphrodite and probably reproduce sexually by cross-fertilization. Only in two families, the Schistosomatidae and the Didymozoidae, are separate male and females found. The male system consists of paired testes, vasa efferentia, a single vas deferens, a seminal vesicle, an ejaculatory duct, and a cirrus enclosed within a sac. Spermatogenesis is typical of the phylum Platyhelminthes and sperm are stored within the seminal vesicle until copulation takes place. The female reproductive system consists of a single ovary and oviduct, a seminal (sperm) receptacle, paired and extensive vitellaria, an ootype surrounded by Mehlis' gland, Laurer's canal, which is homologous with the monogenean vagina, and an

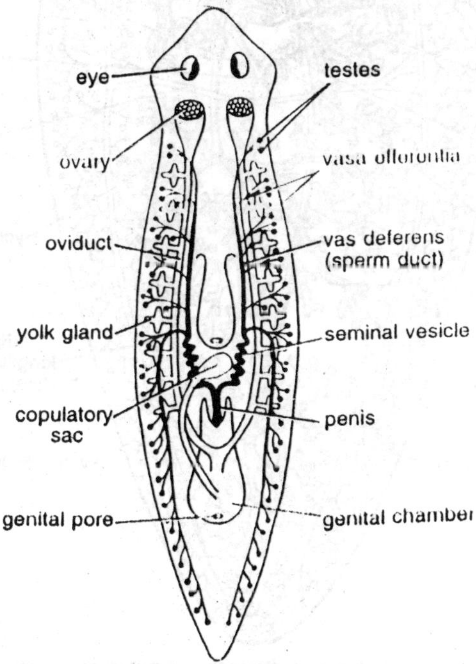

Fig. 6.2. Dugesia. Reproductive system.

extensive uterus. There is a common genital opening for male and female systems, usually situated in the anterior half of the body.

The male system normally develops first (protandry). At copulation the cirrus of one partner is inserted either into genital pore of the other partner or into the opening of Laurer's canal. The development of the digenean egg has been extensively studied and the process of

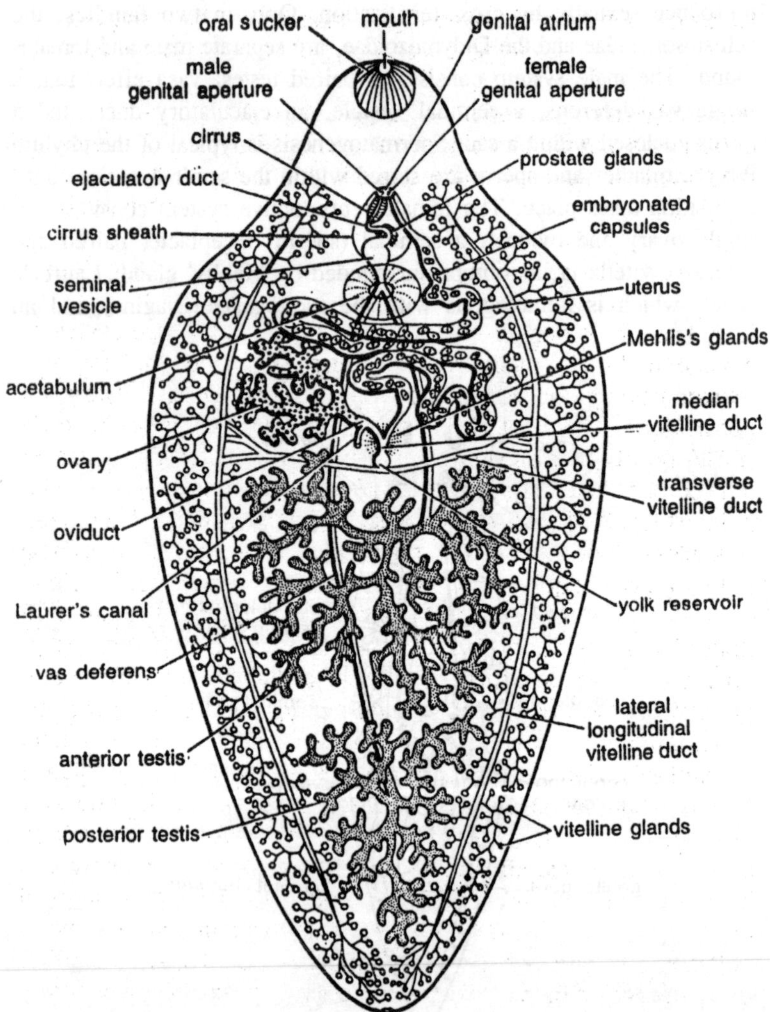

Fig. 6.3. Fasciola hepatica. Reproductive system in ventral view.

egg shell formation is well understood. Ova are produced in the ovary and are liberated periodically into the oviduct. Simultaneously, vitelline cells-from the vitellaria—and spermatozoa—from the seminal receptacle—are released. Fertilization of the egg takes place within the ootype and then the vitelline cells release their stored, shell precursors to form a soft shell around the fertilized egg. The role of the secretions of Mehlis' gland is unknown, but they may have a nutritional function. Shell formation proceeds as the fertilized egg is transported along the uterus.

In a mature digenean the uterus is typically filled with eggs whose shells are opaque and brown due to the process of tanning. Quinone tanning is regarded as the most common method of providing the egg with a resistant covering, but this form of tanning may not be universal, e.g. *Fasciola* egg contain proteins that are cross-linked by disulphide and dityrosine bridges rather than by quinones. Despite this, *Fasciola* does posses phenolases the enzymes which convert o-phenol to o-quinone during the process of quinone tanning. Thus the position remains somewhat enigmatic. The tanned digenean egg is released into the external environment in host faeces or excreta. In most species the egg remain unembryonated until it is outside the host, but some do release eggs that contain well developed embryonic miracidia.

The eggs are normally viable and if they are liberated into an aqueous environment since, despite the tanned shell, they lack the ability to resits dessication. In suitable conditions the egg develops to produce a single miracidium which is release from the egg during hatching-this depends upon the occurrence of appropriate stimuli such as light, temperature, gaseous conditions and salinity. Light appears to be a major hatching stimulus, stimulating the enclosed miracidium to become active or, perhaps, acting on light sensitive substances within the egg that facilitate the loosening of the operculum.

Cestoda

All cestodes, with the single exception of *Dioecocestus*, are hermaphrodites. Male and female systems are present in each proglottid, and the reproductive apparatus is therefore greatly replicated in an individual tapeworm. The Caryophyllaeidea, being unsegmented cestodes, contain only a single reproductive system. In the segmented worms the most mature proglottids are those that are furthest removed from the scolex and protandry marks the normal course of sexual development. The male reproductive system is morphologically similar to its digenean counterpart, consisting of either discrete or diffuse testes, associated ducts, and a well developed cirrus within a pouch. The female reproductive system is also of the basic digenean pattern; the vagina is homologous with Laurer's canal and the morphology of the uterus is somewhat variable. Both self-fertilization and cross-fertilization occur, but little is known of the details of these processes for many species.

The cestode egg varies in its morphology according to life-cycle. In the Pseudophyllidea, the egg is operculate with a thick surrounding capsule and matures outside the tapeworm, normally in water. The cyclophyllidean egg, by contrast is, surrounded by a thin non-operculate capsule, matures within the uterus of the gravid tapeworm proglottid, and hatches in the gut of the host (either an arthropod or a vertebrate, depending upon species). The fully developed embryo within the egg is

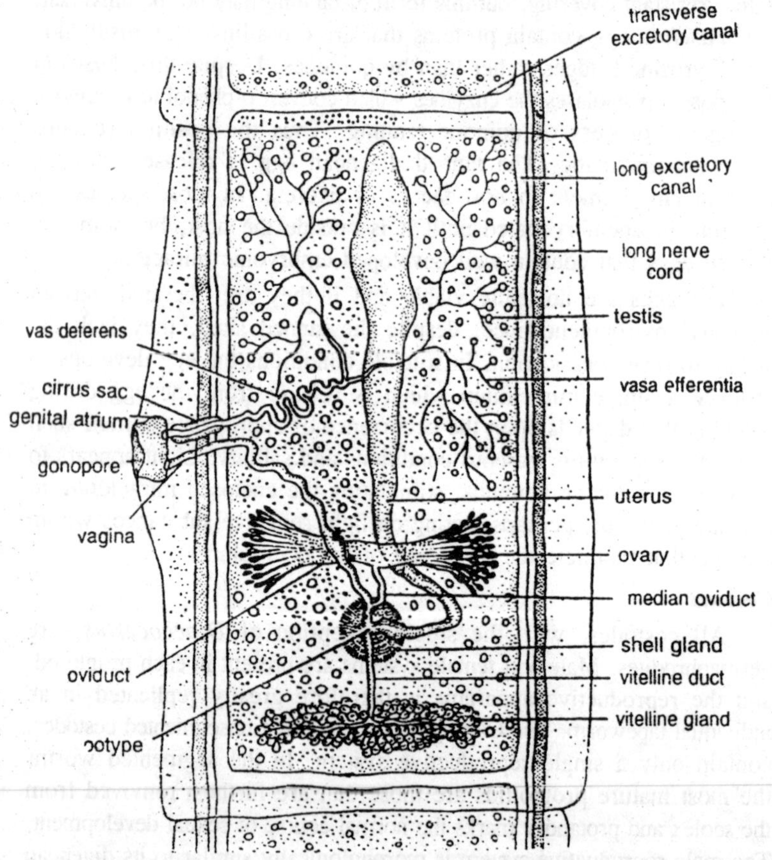

Fig. 6.4. Taenia solium. Mature proglottid showing reproductive system.

termed the *hexacanth*, since it possesses six hooks. In the Pseudophyllidea the hexacanth develops into a coracidium larva upon hatching in water. The precise stimuli associated with hatching of the cestode egg will depend upon whether the egg hatches in the external environment or within the gut of the next host in the life-cycle. Pseudophyllidean eggs may require light to stimulate the opening of the operculum, though hatching of diphyllobothriid eggs has been recorded in the absence of light.

Other factors that may be involved include temperature, pO_2 and salinity, but little precise information is available. There is rather more information on the factors that contribute to hatching of the cyclophyllidean egg, a two part process involving firstly, the breakdown of the embryonic membranes, and secondly, the activation of the hexacanth larva. The hatching mechanisms of taeniid eggs are regarded as somewhat different from those of other cyclophyllideans, due perhaps to the thin embryophore of the latter in contrast with the thicker embryophore of the taeniids. *In vitro* studies on hatching of taeniid eggs show that digestive enzymes, such as pancreatin and pepsin, and bile salt bring about the breakdown of the egg membranes, while an increased ambient temperature and bile salts appear to be responsible for larva activation. The eggs of other cyclophyllideans will hatch in saline solutions and do not require enzymes or bile salts—usually mechanical disruption of the egg membranes is an important factor for hatching.

Acanthocephala

Separate sexes and considerable sexual dimorphism are features of the Acanthocephala; the female worm tends to be rather larger than the male, and is thought to be longer lived. Though the latter is a matter of some controversy. The male reproductive system is completely enclosed within a ligament, which attaches anteriorly to the proboscis sac. The system contains two testes, a sperm duct, cement glands and a copulatory bursa. In the female worm, the reproductive system is in contact with the ligament, but is not enclosed by it. The female system consists of the ovary and ovarian balls, the uterine bell, uterus and vagina. At copulation spermatozoa gain access to the pseudocoelom, which contains the eggs, by way of the vagina, uterus and uterine bell.

Copulation involves the insertion of the male bursa into the vagina of the female and after copulation the male plugs the vaginal opening with secretions that emanate from the cement glands. It seems improbable that female acanthocephalans can mate more than once. In the pseudocoelom, the spermatozoa fertilize the eggs, which are contained in the ovarian balls. The fertilized eggs are then liberated from the balls and continue their development within the pseudocoelom, entering the uterus when mature. The uterine bell acts as a sorting device allowing passage of "mature" eggs, but keeping "immature" eggs in the pseudocoelom to complete their development. Ripe eggs are released from acanthocephalans at varying rates, e.g. *Polymorphus minutus* can liberate 2000 eggs per day, while *Macracanthorhynchus hirudinaceus* released up to 260000 eggs per day; the prepatent period is from three to nine weeks depending upon species.

The eggs, released into the external environment along with the faeces of the host, are surrounded by a protective covering consisting of three to four layers, which contain keratin, chitin and proteins. The shelled embryo, or acanthor, can withstand a range of environmental conditions and will only hatch after being ingested by the arthropod intermediate host. The egg, therefore, acts as a resting stage in the life-cycle which may remain quiescent for extended periods. Hatching in the arthropod gut is dependent upon pH and pCO_2, as well as on digestive enzymes, such as chitinases, to disrupt the shell membranes.

Nematoda

The great majority of nematodes parasites have separate sexes, and exhibit varying degrees of sexual dimorphism, but hermaphroditic and parthenogenetic species also occur. The male reproductive system consists of a single testis, a seminal vesicle, a vas deferens and a cloaca, but it is though that the system was originally double. Spicules, which are hardened, extrusible, cuticular structures used to open the vulva of the female during copulation, are present in the males of most species, as either single or paired structures. They are located at the posterior extremity of the male.

The female reproductive system contains paired ovaries, oviducts and uteri, which connect with a single vagina, opening via the vulva. Frequently, part of the vulva is modified to form a muscular ovijector, responsible for extruding the eggs. In the male system, spermatogonia develop either at the distal end of the testis, or, lets commonly, along the entire length of the testis. In the female, the oogonia often develop in association with a cytoplasmic rachis, whose function is unknown. Subsequent growth and maturation of the oocytes takes place following their detachment from this rachis. Fertilization of the egg occurs within the seminal receptacle of the female. After fertilization the egg becomes

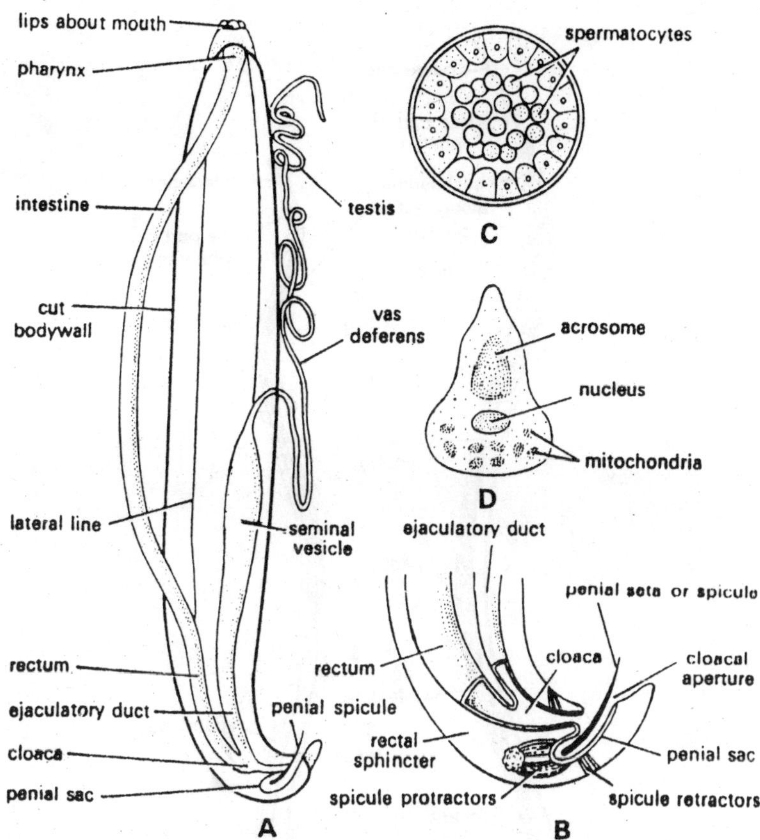

Fig. 6.5. Ascaris lumbricoides. A—Male reproductive system. B—Posterior end of male Ascaris in lateral view showing cloaca and spicule. C—T.S. vas deferens. D—A sperm.

surrounded by three shell layers; an inner ascaroside (lipid) layer, a thick chitinous layer and a thin outer layer. These layers derive from the egg itself, while the uterus may contribute a fourth layer, giving the egg a roughened and often sticky surface. In general, nematode eggs are remarkably resistant to chemical damage and dessication, due, in part, to the ascaroside layer which is composed of esterified glycosides.

The eggs of ovoviviparous nematodes, such as *Trichinella* and the filarial worms, are, not surprisingly, less complex in their structure since the larvae hatch *in utero*. Hatching of the eggs of the parasitic nematodes takes place either in the external environment, to release a

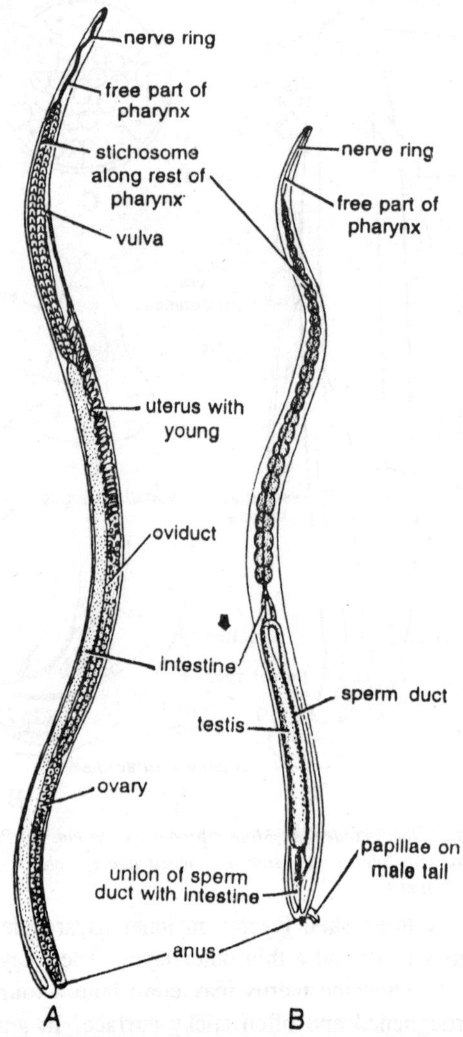

Fig. 6.6. Trichinella spiralis. A—Female. B—Male.

free-living larva, or within the gut of a host animal, either intermediate of definitive host depending upon the life-cycle pattern. In *Trichostrongylus* and *Meloidogyne*, both of whose eggs hatch in the external environment, hatching is brought about by the egg shell becoming permeable to water and rupturing due to increased turgor pressure. The larvae of *Heterodera* actively cut their way to of the shell. Environmental factors such as light, temperature, pO_2 and humidity may all play a role in providing suitable conditions for hatching. Eggs that hatch within a host gut respond to the action of digestive enzymes and the prevailing physico-chemical conditions especially temperature and pCO_2. The factors that render one host suitable and another less suitable for a particular species of nematode, by affecting the hatching of ingested eggs, have not been well investigated.

Synchronization of Parasite Reproduction with Host Cycles

There are a small number of cases known in which the reproductive activities of a parasite are coordinated or synchronized to the reproductive or other cycles of its host. Such synchrony often serves to bring about the co-occurrence of infective parasites and juvenile animals of the host species. We have already noted that, amongst the Protozoa, the Opalinata, which parasitize amphibians, release their gametes only during the breeding season of the frog or toad, and fertilization occurs within the recently infected tadpole larva. A related ciliate genus, *Nyctotherus*, which also parasitizes the posterior gut of amphibians, switches from asexual to sexual reproduction during the breeding season of its host. Experiments with various hormone treatments, using both *Opalina* and *Nyctotherus*, suggest that the levels of host sex hormones directly affect the reproductive activities of these parasites.

In the hypermastigid flagellates inhabiting the arthropod gut, sexual reproduction is initiated by the moulting hormones of the host, e.g. in termites, the hypermastigid gut fauna is lost with each moult, so that sexual reproduction of the parasite is synchronized to the moult cycle; this is one method of reinfection where no cystic stage is produced. Reinfection is essential for the host since the flagellates produce the cellulases that digest the cellulose diet. Only one species of coccidian, *Coelotropha durchoni* in the polychaete worm *Nereis diversicolor*, shows reproductive synchrony with its host, though other sporozoans, including *Plasmodium*, *Leucocytozoon* and *Haemoproteus*, all synchronize to some extent with their hosts' breeding cycles. Amongst the helminths, examples of reproductive synchrony are extremely rare and are limited to the monogenean *Polystoma integerrimum* in amphibians, and to synchrony of mitotic rate in the cestode *Diphyllobothrium dendriticum* with the circadian rhythms of the host. Many other cases will undoubtedly occurs but these are yet to be described.

7

TRANSMISSION OF PARASITES

To the student of parasitology, the life cycle of parasites can often present a bewildering array of detail, with no obvious pattern appearing at first glance. Nevertheless there are certain events that are common in the life-cycles of all parasites, one of which is transmission to the next host. Transmission may occur more than once during a single life-cycle, should the parasite develop in one or more intermediate hosts. The adult of mature parasite completes its development and reproduces in the final or definitive host, releasing eggs, larvae or other infective stages that must be transmitted to another host, whether it be of the same species (as in a direct life-cycle) or to hosts that are of different species (if the life-cycle is indirect). The transmission of a parasite between consecutive hosts in the life-cycle can take place in one of three ways.

First, transmission can occur when the potential host feeds on the egg, larval stage or on the intermediate host harbouring the larval parasite. Secondly, an intermediate host, or vector, may acquire the parasite and subsequently reinfect the definitive host, during feeding on host blood or tissues. Thirdly, transmission can often involve the active penetration of the host by free-living larval stages of the parasite. In an individual species of parasite, more than one of these modes of transmission may be employed at different stages of the life-cycle. A parasite may inhabit widely differing environments at various stages in its life-cycle, and it will be adapted for life in each of these environments, however different they may be.

Rather surprisingly, little attention has been paid by parasitologists to this ability of parasites to accommodate, both morphologically and physiologically to several quite dissimilar environments, differing perhaps in ambient temperature, oxygen availability, salt and water content, pH and food content. The DNA content a parasite will not alter during a single life-cycle, yet the parasite can perhaps adopt many forms and develop physiological attributes each unique to a particular phase in its life-cycle. Parasitic animals demonstrate this phenomenon to a much greater extent than do free-living animals. Little is known of the factors that bring about the "switching on" of some parts of DNA while other parts are "turned off". We presume that environmental stimuli are responsible for initiating such switching mechanisms via the action of operator genes which can activate the appropriate areas of DNA and inactivate the inappropriate areas.

When we consider the factors that control the establishment and development of parasites. Suffice it to say here that these putative environmental triggers, which operate throughout the life-cycle of a parasite, will play a particularly important role during the transmission process, because a parasite must be able to establish in a new host once that host has been located, i.e. its DNA must be switched to the appropriate regions once successful transmission has been accomplished. Parasites typically large numbers of offspring in the form of either eggs or larvae, and this high degree of fecundity is usually regarded as an adaptation to the peculiar hazards of the life-cycle, involving time spent both inside and outside the host or hosts. Transmission between hosts probably represents a period of considerable loss in the life of most parasites, being a time when a great many individuals either fail to locate and enter a a suitable host, or fail to establish successfully within the right host.

Mechanisms for Locating the Host

A variety of different stages are involved in the act of host location. These include oncomiracidia and adult worms in the Monogenea, miracidia and cercariae in the Dignea, coracidia in the Cestoda and various larval stages (usually the third larval stage) in the Nematoda. All of these individuals are active, free-living organisms that usually penetrate the host at its surface, or, in the case of the Monogenea, simply attach to the host surface. Research into the mechanisms employed both by parasite larvae and adults for host location involves behavioural studies, attempts to identify chemo-attractants and various biochemical approaches.

Monogenea

The mechanisms by which monogeneans transfer from host to host have fascinated parasitologists for a very long time, but rather few species have been studied critically. Probably the best known example is *Entobdella soleae*, parasitic on the body surface of the flatfish *Solea solea* (the common European sole), through the work of G.C. Kearn almost a decade ago. Adult *Entobdella soleae* attach to the surface of the sole by broad, cup-shaped *opisthohaptor*; the worm is quite mobile and capable of limited excursions, such as might be required for mating to be effected. Eggs are laid into the surrounding seawater and these hatch after 3-4 weeks, releasing the free-swimming larva, the oncomiracidium. This larva is small, about 0.25 mm long, and is ciliated in three distinct regions of its body surface.

There are four pigmented eyes at the anterior end, each covered by a lens. Information on the host specificity of *E. soleae* strongly suggests that, under natural conditions, the parasite is quite definite in

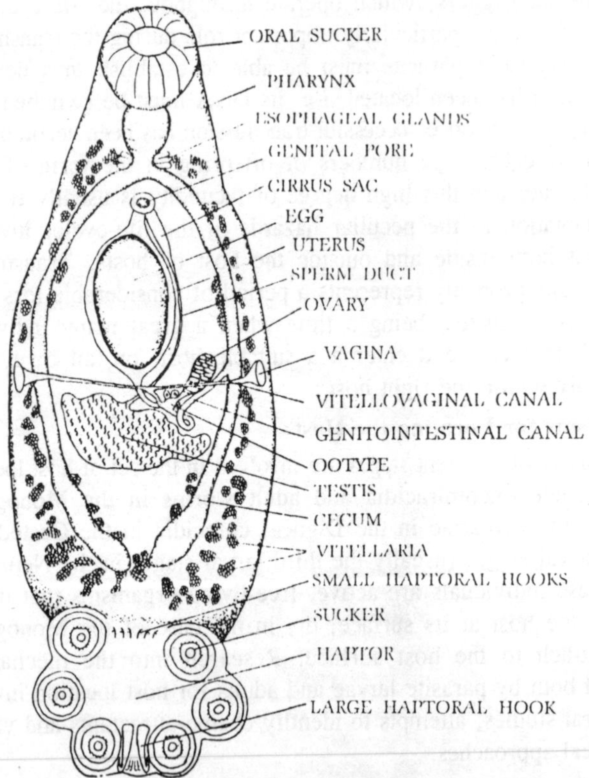

Fig. 7.1. Polystomoidella oblongum, a monogenetic trematode from the urinary bladder of turtles.

its preference for the sole and rarely occurs on other fishes. It this brought about by active selection on the part of the parasite or do the oncomiracidia attach, but fail to develop in a wide range of fish species? These questions have been answered by the ingenious laboratory studies of Kearn, using isolated fish skin and scales. When the oncomiracidia of *E. soleae* are offered a choice of isolated scales of its normal host in the presence of scales of related flatfish, such as *Microhirus* (*Solea*) *variegatum* (the thick-backed sole), *Byglossidium luteum* (the solenette), *Pleuronectes platessa* (the plaice) and *Limanda limanda* (the dab) (all of which co-occur around the British Isles) *Solea solea* scales are selected almost exclusively.

It was noted that the parasite larvae settled, almost in every case, on the remnant of growing skin attached to the excised scale, rather than on the bony element of the scale itself, suggesting that fish skin might possess a specific attractant for oncomiracidia. Attachment to isolated scales occur in complete darkness, ruling out visual perception as a means of host location. The transfer of *E. soleae* from heavily infected sole, placed in a crowded aquarium tank with other species of flatfish, was limited primarily to recruitment of parasites by other sole and not by the other species of flatfish. Additional experiments showed that oncomiracidia will actively select and settle upon agar circles that have been impregnated with extracts of the skin of sole and this finding clearly implicates chemoreception as, perhaps, the major means by which *E. soleae* larvae detect their host.

It is evident that the mucoid secretions, from glands in the fish skin, are the principal attractants, since oncomiracidia are not attracted to sole skin lacing mucus glands, e.g. the cornea. It has been suggested that host-finding by chemotaxis in the Monogenea may be related to their evolution from a carnivorous, free-living ancestor, that found its prey by chemical identification. Many other species of monogeneans are profoundly host-specific, but it is not known whether this phenomenon is brought about by the recognition of specific secretions of a suitable host or by other means.

Monogeneans will also transfer from host to host as adult worms when there is physical contact between host individuals and this is probably a common means of transmission in parasites of gregarious fishes. In fishes that do not shoal, parasite transmission many occur when the hosts make contact during their reproductive activities. These are areas where out knowledge is limited and they deserve further attention.

Digenea

The stages of the life-cycle concerned with transmission of digeneans are the miracidia, which hatch from the egg and penetrate the molluscan intermediate host (usually a gastropod), and the cercariae, which are released from the mollusc after a period of several weeks development and penetrate the next host, which may either be another intermediate host or the definitive host. The cercaria may, alternatively, encyst on vegetation as a metacercarial stage and reach the final host by the ingestion route, e.g. *Fasciola hepatica*. The majority of miracidia actively penetrate the mollusc, but some species must be eaten by the snail for continuation of the life-cycle, e.g. *Haplometra intestinalis* and *Glypthelmins quieta* are eaten by the snail *Physa ayrina*, then they penetrate the snail gut wall and invade the tissues. The miracidium is a ciliated larva with sense organs that consist of eye spots and a variety of lateral papillae, which contain a nervous supply.

Most miracidia respond emphatically to the physical elements of their environment, and they are thought to resemble their mollusc hosts in this regard. They are usually phototactic, thermotactic, thigmotactic and geotactic and will often respond to alteration in pO_2, pCO_2 and pH. Responses such as these probably function to bring the miracidia into close physical contact with molluscs. Once in the vicinity of a suitable snail, chemical attraction may facilitate the final stages of host location, although chemotaxis in miracidia is a controversial issue of long standing. While there is a substantial body of evidence implicating the release, by snails, of chemical attractants that are

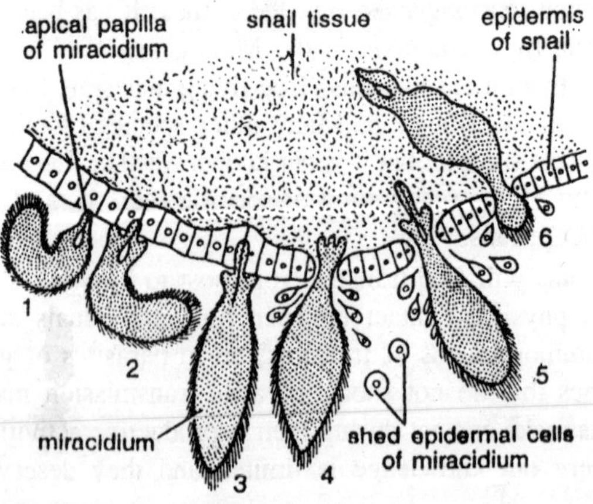

Fig. 7.2. Miracidia of Fasciolopsis buski. Stage of penetration through snail epidermis.

specific for individual species of Digenea and possibly mucoid in orgin, there exist in the literature a great many reports that fail to offer support for the hypothesis of chemical attraction and alternatively suggest that location of snails by miracidia is simply a process of trial and error. Considerably varied experimental approaches have been adopted to test the chemical attraction hypothesis and apparatus employed includes four-armed mazes and Y-shaped choice vessels, collectively called chemotrometers. Other workers have studied miracidial locomotion under the flying spot microscope and yet others have preferred a strictly biochemical approach, using pure chemicals as potential attractants.

Despite this wealth of experimental data, the controversy remains unresolved. Support for chemical attraction hypothesis comes mainly from studies on the miracidia of *Schistosoma mansoni* and *Paragonimus ohirai*. In experimental chambers, the miracidia of *S. mansoni* appear to recognize and swim towards whole snails, crushed snails and snail extracts that have been impregnated in agar blocks. It is not completely certain, however, whether the parasite larvae recognize only those snails in which development can continue of whether unsuitable snails are also detected and approached. Schistosome miracidia detect specific chemicals, such as short-chain fatty acids, certain amino acids and

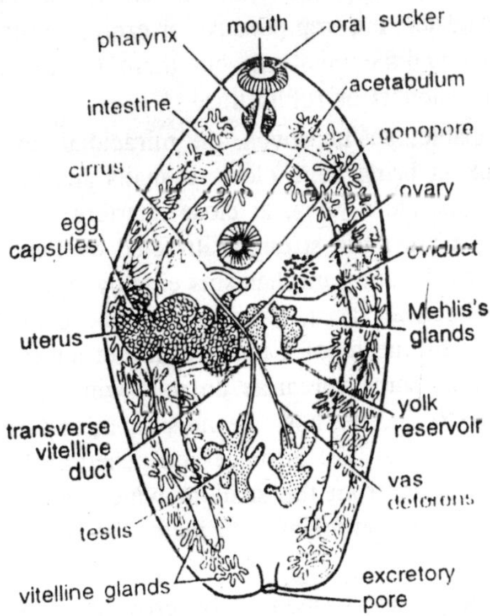

Fig. 7.3. Paragonimus.

sialic acid impregna·ed in agar blocks, and will alter their swimming behaviour according to the nature of the attractant. Unfortunately, this approach to the study of chemotaxis, as pioneered by A.J. McInnis, has not been thoroughly pursued using other species of digeneans, yet it would appear to be a fruitful line to adopt.

Short-chain fatty acids may be recognized by the miracidia of *Fasciola hepatica*. There is, however, no unequivocal evidence that snails release specific substances that will attract only those digneans which can grow in them. The chemical evidence available at present suggests that snail mucus contains a number of distinct components, any which might be involved in attraction. There are many reports in the literature that tend to refute the hypothesis of chemical attraction by snails, including studies on *Fasciola hepatica*, *Fascioloides magna*, *Schistosoma mansoni*, *Austrobilharzia* and *Zoogonoides*.

The age of miracidia can affect their ability to locate the host, but the mechanisms involved are not known, e.g. studies on *Megalodiscus temperatus* have demonstrated that 2-5 hours old larvae are more successful in locating than snail (*Helisoma trivolvis* the either younger or older larvae. On the precise nature of a possible chemical attractant for miracidial host location, snail mucus has been put forward most frequently as the likely candidate. Yet, as we have seen there is no unequivocal evidence to support this notion. It may be, therefore, that other snail exudates, e.g. reproductive or excretory material, faeces or haemolymph, could be involved, but there is little experimental evidence either to support or refute this.

It seems probable, yet unproven, the miracidial responses to the physical environment bring them close to snails and that chemotaxis only operates, if operate it does, at close quarters. What is clear is that most miracidia are more strongly attracted by light than by any other stimulus. Host locating mechanisms employed by cercariae, the other free-living digenean larva, are also surrounded by controversy. These larvae swim by means of a tail and do not appear to use their cilia for locomotion; some cercariae do not swim, at all but crawl along the substrate to find the host. In the swimming forms the tail is lost once contact with the host is made and penetration commenced. Emergence of cercariae from the snail in which they have develop appears to be under the control of such factors as light, temperature, humidity, pO_2 and pH.

In the laboratory, stress (induced by crowding snails) will also stimulate cercarial emergence. Once emerged from the snail, the cercaria exists as a free-living larva for only a short time, up to 36 hours as a rule. The behaviour patterns of swimming cercariae, which represent the most common type, are characterized by alternating periods swimming and resting activity; during the rest period they tend to sink, rising in the water on resuming swimming. This is aided by their positive phototaxis, mediated through pigmented eyespots. Some cercariae do not possess eyespots and are negatively phototactic. These are commonly crawling (rather than swimming) cercaria that infect bottom-dwelling animals.

Superimposed upon the primary response to light are additional responses that assist in the process of host location, e.g. the cercaria of *Posthodiplostomum cuticola*, which penetrate and encyst in a variety of European freshwater fishes, respond to shadows, such s might be produced by a passing fish, and swim spontaneously in a swarm. Spontaneous swimming is also initiated by increased water turbulence and by contact with a moving object. There is no evidence or chemotaxis in *P. cuticola*, however. In general, cercariae are positively phototactic and negatively geotactic, with the noteable exception of crawling forms, but there is doubt as to whether chemotaxis plays any significant role in host location. Early studies on *S. mansoni* and *Clonorchis sinensis* indicated that chemotaxis was unimportant, but more recent work suggests that chemical attraction may play a not insignificant role in host location by cerariae.

Chemotaxis may occur in a small number of species, including *Gorgodera amplicava*, *Schistosoma mansoni*, *Schistosomattium douthitti*, *Trichobilharzia*, *Austrobilharzia*, and *Gigantobilharzia huronesis*. Although it is not clear whether schistosome cercariae actually locate the host skin chemotactically, there is ample evidence that the specific site for cercarial penetration is selected by responses to chemical components in the skin itself, perhaps released by the sebaceous glands. Fatty acids may attract some schistosome cercariae to the host, since it has been demonstrated experimentally that *Schistosomattium douthitti* cercariae fail to penetrate excised mouse ears if the skin is extracted with ether to remove the fatty acids. Penetration resumes if ether-extracted ears are rubbed with fatty acids. The bird schistosome, *Austrobilharzia terrigalensis*, requires contact with sterols, particularly cholesterol, for successful penetration, but these substances do not appear to be attractants for host location by the cercaria.

Cestoda

The only active, free-living larval stage in the cestodes is the coracidium of the pseudophyllidea. This is a ciliated larva, which hatches from the egg after its release into freshwater within the faeces of the bird or mammal definitive host. The coracidium is infective to copepod crustaceans. It differs in two ways from miracidia and cercariae; first, it appears to swim exclusively by random movements, and secondly, it is eaten by its intermediate host, a copepod. As it presents itself as food to its intermediate host, transmission is a passive process. Transmission in all other cestodes also occur exclusively by passive mechanisms involving the ingestion of eggs or larvae by the host.

Nematoda

Many plant-parasitic and some animal-parasitic nematodes, such as the hookworms of vertebrates, posses active, free-living larval stages that are concerned with locating a new host. Chemical attraction may play a, particularly important role in host location by many of the plant parasites. Parasitic nematodes, in general, are will endowed with sense organs, including the antero-lateral amphids and the postero-lateral phasmids, which are thought to be chemoreceptors. Little direct evidence is available to support this hypothesis, however. Host location in plant-parasitic nematodes is accomplished by random movements that operate when the parasite larva is some distance from the target

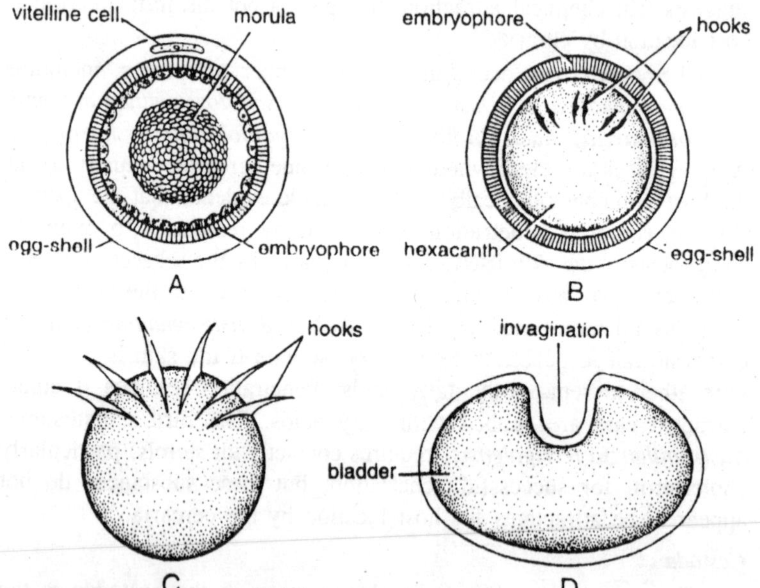

Fig. 7.4. T. solium. Stages in the life cycle. A—Young onchosphere; B—Mature onchosphere; C—Free hexacanth; D—Bladderworm with invagination.

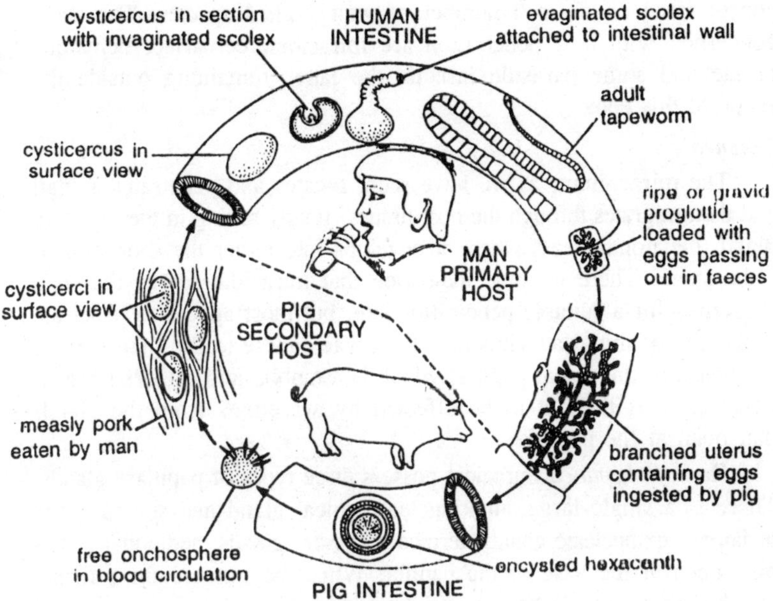

Fig. 7.5. Taenia solium. Life cycle.

plant, and by chemoreception when the plant is near. Mechanical stimulation may also be important for successful host location.

Considerable exploratory behaviour characterized plant nematodes when searching for a suitable entry site once the plant is reached. By contrast with the initial randomness that is a feature of the plant nematodes, the hookworms possess behavioural patterns that are clearly adaptation for locating a warm blooded (homoiothermic) animal in a terrestrial environment. Larval mammalian hookworms, such as *Ancylostoma* and *Necator*, tend to be inactive for much of the time as are only stimulated into activity by he increased temperature due to a passing homoiotherm. *Nippostrongylus* larvae seem to search actively for their rat hosts by characteristic waving movements of the head and once a host is located the larvae spiral rapidly down the hairs following a thermal gradient. Nematode larvae are more sophisticated in their behaviour patterns than the equivalent transmission stages of platyhelminths. They respond to a wide range of stimuli, including light, heat, gravity, various chemicals, mechanical events and electrical fields.

Mechanisms for Penetrating the Host

Information on the processes of active penetration of the host epidermis by larval parasites is rather scanty and tends to be concentrated on a small number of well studied species. The stage concerned with host penetration are miracidia, cercariae, nematode larvae and some parasitic insects, the latter remaining outside the scope of this book.

Digenea

The miracidium, as we have seen, locates and penetrates a snail and then migrates through the host tissues, finally setting in the digestive gland (hepatopancreas) where it develops into either the sporocyst or redia stage. There is some indication that miracidia search the snail epidermis for a suitable penetration site, but most appear to penetrate at the site of initial attachment. The larvae adhere to the snail surface by means of the apical papilla, which is a mobile and suctorial organ. Penetration is thought to be effected by secretions from the glands that open at the papilla.

Fasciola hepatica miracidia possess three types of papillary glands. There is a single large, multinucleate apical gland and several pairs of lateral, uninucleate glands, termed accessory glands; additional glands may open at the base of the papilla. When the miracidium a firmly attached to a suitable host the secretions of these glands are extruded by muscular contractions of the entire body of the larval parasite.

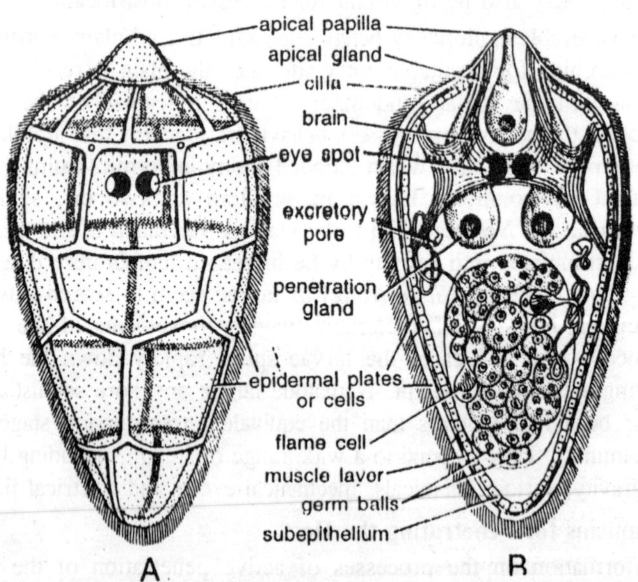

Fig. 7.6. Miracidium larva. A—External structure. B—Internal structure.

Miracidia of *S. mansoni* and *S. mattheei* possess similar papillary glands. It is postulated that the secretions of miracidia contain a mucoid lubricant to aid both attachment and penetration and that lytic enzymes are responsible for breaking down the host epidermis.

At (or perhaps just prior to) penetration, some miracidia, e.g. *Fasciola hepatica* and *Fascioloides magna*, shed their ciliated epidermal plates, whereas other species keep these plates intact during penetration, e.g. *mansoni*. The relevance of these events of obscure at present. Penetration of mammalian skin by the cercariae of digeneans, most notably the schistosomes, has received considerable attention. Mammalian skin is a complicated structure made of three distinct regions: an outer layer of keratinized dead cells (the *stratum corneum*), below which are living layers, the epidermis and the dermis, separated by a basement membrane.

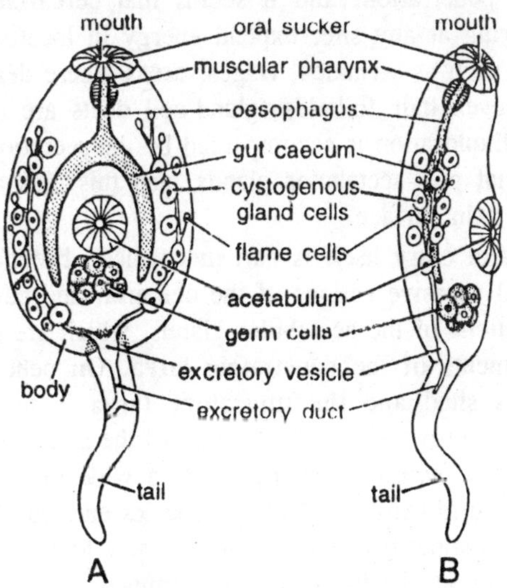

Fig. 7.7. A—Cercaria in ventral view. B—Cercaria in lateral view.

The layers of the skin vary in thickness and composition between different species of mammals, between individuals of the same species and even between different regions of the surface of the same individual. These factors may play an important role in determining both host specificity and the success of parasite establishment in any particular host. There is some evidence that skin lipids are essential for stimulating penetration by schistosome cercariae, e.g. cholesterol is the major stimulant for the avian schistosome, *Sustrobilhrzia terrigalensis*, while mammalian schistosomes require free fatty acids. These lipids are thought to be important stimulants only after the cercariae have actually attached to the host skin, and they are probably not chemical attractants themselves. The penetration of mammalian skin by the cercariae of *Achistosoma mansoni* has been categorized into six stages by Stirewalt (1966), and these may be summarized as follows:

1. Exploration of the skin.
2. Secretion of mucus for attachment.
3. Piercing and penetrating the *stratum corneum* through the muscular action of the cercarial body.
4. Abrasion of host tissue.
5. Mucus secretion to soften the *stratum corneum*.
6. Secretion of (putative) lytic enzymes.

Exploration of the host surface s much more a feature of cercarial than miracidial penetration, and it seems that cercariae, although capable of entering at any site, expend energy in locating the most suitable region, such as wrinkles, ridges, areas where dead cells are flaking off, or even hair follicles (gland-cell ducts are not used as sites of entry). Exploration is accompanied by the secretion of mucus from the cercarial post-acetabular glands, and this aids adhesion of the larva to the skin surface.

The process of entry itself is part mechanical, brought about by the vigorous and abrasive actions of the cercaria, and part enzymic, due to the secretions of the acetabular glands, which are expelled by the body movements of the penetrating larva. On penetration, the cercarial tail is shed and the migratory larva is now called a *schistosomulum*, a term reserved for species of the genus *Schistosoma*. The cercariae of *S. mansoni* can penetrate a wide of mammals but they do not survive to maturity in all the species entered. Thus, while the skin forms a distinct barrier, it is not the sole location in the host body at which host specificity may be determined.

The transformation from free-living cercaria to tissue-dwelling schistosomulum is a physiologically dramatic one, both in terms of temperature acclimatization (from 25°C to 37°C) and adjustment to a medium with a considerably increased ion content. A schistosomulum that has transformed from a cercariae for only a very short time will die if returned to fresh water at 25°C. Transformation in a facet of skin penetration, but it can also be accomplished experimentally, *in vitro*, by using a biphasic culture medium and also by allowing cercariae to penetrate excised mouse skin. The latter is a routine laboratory technique in the culture of schistosomes.

Nematoda

Some members of the order Strongylida, such as the hookworms and related worms (Ancylostomatoidea, Strongyloidea, Trichostrongyloidea and Heligmosomatoidea) are transmitted to the vertebrate host, usually a mammal, by actively penetrating the skin as a free-living stage. The earliest evidence of skin penetration by hookworms was accidental, when in 1897, the famous parasitologist, Looss, became infected with *Ancylostoma duodenale* in error. Fortunately this mishap has no lasting effect on the pioneering studies of Loss in the field of parasitology. A variety of more controlled techniques has been subsequently employed to study the methods of skin penetration by larval nematodes. These include histological studies, histochemistry, tissue lysis studies, penetration though excised skin and behavioural approaches. Early work on skin penetration by larval nematodes was carried out primarily *in vivo* and led, in general, to the conclusion that enzymes were secreted to effect entry across the skin barrier.

More recent work has involved *in vitro* studies, using modifications of the Goodey "floating raft" technique, in which excised skin is stretched across a window in a sheet of cork that is floated in saline. Studies of this type have revealed that there is some variability in the mechanism of penetration adopted by larval nematodes. Initiation of penetration may come through a response to a thermal gradient, as shown by the hookworm, *Ancylostoma tubaeforme*, and entry occurs most frequently between dead cells of the *stratum corneum*. Some larvae exsheath (shed the cuticle of the previous larval stage) before penetration can commence, e.g. *Ancylostoma caninum* and *Necator americanus*, while others penetrate without exsheathment.

The act of exsheathment, when it occurs, is thought to be initiated by skin lipids. Enzymes of parasite origin have been implicated in skin penetration in a number of species (*Necator americanus* and *Strongyloides fulleborni*), but is by no means certain that all species rely on enzymes to effect entry across mammalian skin; equivocal evidence on this subject exists for *Ancylostoma caninum*, *A. ceylanicum* and *A. tubaeforme*. It is apparent that penetration may required a strict sequence of signals for the larva and this is exemplified by the fact that *A. tabaeforme* larvae cannot cross revised skin (i.e. skin with the dermis presented outermost), *in vitro*. Plant-parasitic nematodes posses a sharp anterior stylet which is used to pierce cell walls during

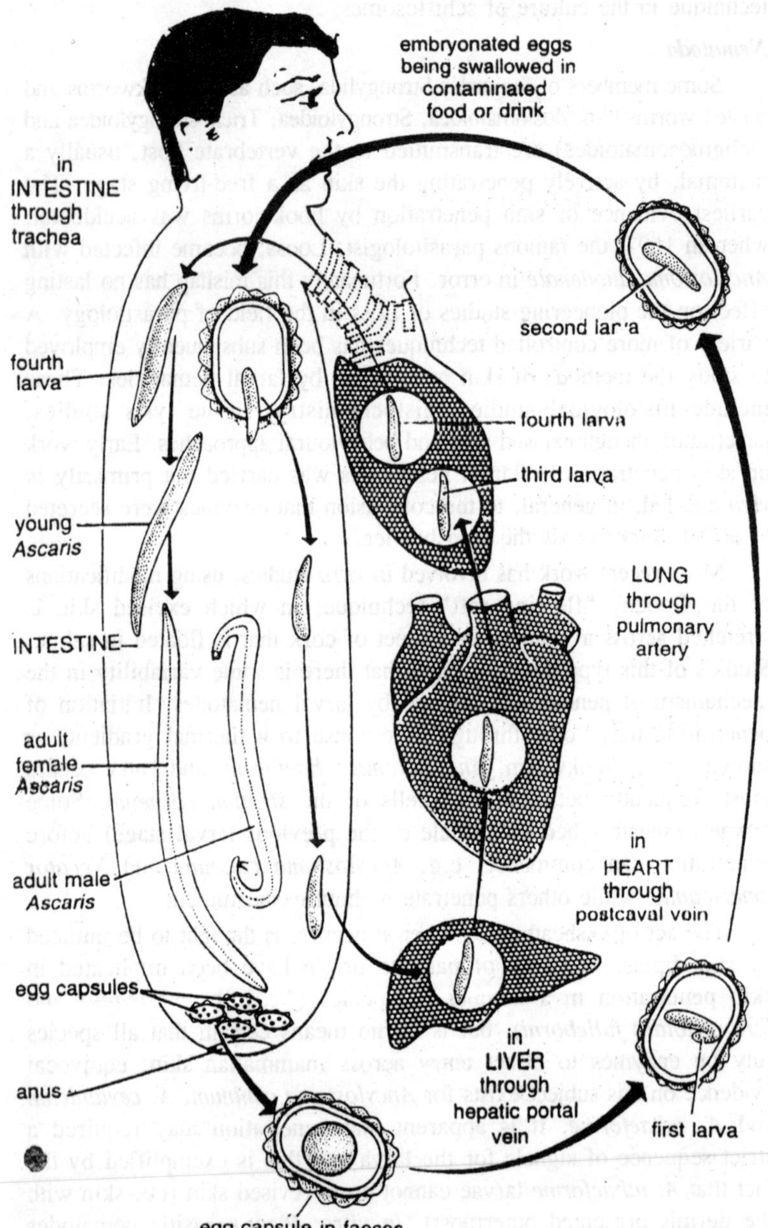

Fig. 7.8. Ascaris lumbricoides. Life cycle.

feeding. The penetration of plant tissues, such as roots, appears to be effected exclusively by the mechanical action of the stylet, although there is some evidence that enzymes might assist with the process.

Circadian Rhythms Associated with Transmission

Circadian rhythms, based on 24-hour cycle of activity, are characteristic features of number of parasites and are frequently associated with transmission either to the final host, if the life-cycle is direct, or to the final or an intermediate host, if the life-cycle is indirect. Hawking (1975) has classified the circadian rhythms of parasites as follows:

1. Rhythms associated with synchronous cell division, e.g. malarial parasites.
2. Rhythms associated with the discharge of infective forms: (a) from a definitive host, e.g. coccidia, pinworms, schistosomes, (b) from an intermediate host, e.g. schistosomes.
3. Rhythms associated with parasite migration, e.g. trypanosomes, malarial parasites, microfilariae (larval filarial nematodes).

Circannual, or yearly, rhythmicity may represent yet another aspect of parasite transmission, particularly when associated with annual cycles of host occurrence. Many species of malarial parasites are strictly periodic in their asexual reproductive cycle in avian or mammalian blood, which cell division, or schizogony, occurring every 24 hours (*Plasmodium knowlesi*), every 48 hours (*P. cynomolgi*) every 72 hours (*P. malariae*). The circadian rhythmicity in malarial schizogony is closely related to the production and viability of the gametocytes, the stage involved in the process of transmission to the mosquito intermediate host, or *vector*. Male gametocytes of several species of *Plasmodium*, including *P. cynomolgi*, *P. knowlesi*, *P. berghei*, *P. chabaudi* and *P. cathemerium*, become ripe (shown by the occurrence of exflagellation) during a limited period of the day, which coincides with the feeding activities of the appropriate mosquito on the final host's blood, e.g. the male gametocytes of *P. knowlesi* are immature between 09.00 and 13.00 hours, mature between 21.00 and 05.00 hours and subsequently become moribund and die.

It has been suggested that both schizogony and the time required for gametocyte maturation are, in each species, related to the biting times of the various mosquitoes involved in their transmission. It is not certain whether the female gametocyte possesses a similar circadian cycle of development. Experimentally-induced hypothermia in monkeys infected with malaria, has led to the conclusion that the synchrony of malarial schizogony is related, or *entrained*, to the daily temperature cycle of the host mammals, since prolonged cooling severely disrupts the parasite's circadian rhythm of development.

In malaria, then, circadian rhythms are associated with the requirement for the infective stage (the gametocyte) to be present in the peripheral blood at time when the transmission to the appropriate biting insect would be most likely to succeed. Other parasites possess similar rhythms which involve the active release of infective stages (eggs or larvae) from the host. The release of infective oocysts of coccidian parasites from the gut of birds is timed to coincide with the optimum period for transmission to another bird, e.g. in the case of *Isospora* in sparrows, the optimum time for transmission would occur in the late afternoon and early evening, since this is the time when roosting takes place, and many birds congregate; at other times of the day sparrows tend to be solitary in their existence.

The life-cycle of *Isospora* is direct and infection is acquired by birds ingesting the liberated oocysts. Mammalian pinworms (*Enterobius vermicularis* in humans and *Syphacia muris* in rats) deposit their eggs

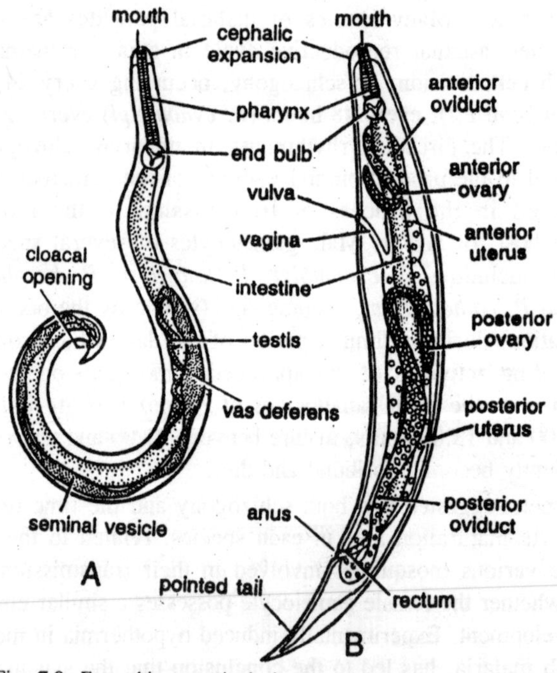

Fig. 7.9. Enterobius vermicularis. A—Adult male. B—Adult female.

in the perianal region of the host by migrating from the rectum, their normal habitat, on a circadian basis. Migration of the female pinworms is associated with the daily lowering of the rectal temperature during sleep. The advantage of laying eggs during this period is probably to avoid contamination with host faeces, an occurrence that might reduce the chances of successful transmission, which takes place via scratching and the accidental ingestion of eggs.

A similar cycle of egg release has been described for human infections of *Schistosoma haematobium*; a peak on the numbers of eggs present in the urine has been found to occur at midday. Hatching of schistosome eggs may also follow a circadian cycle. Schistosome cercariae show strong preferences for liberation from the gastropod host at prescribed times of the day and this appears to be related to the optimum availability of the final host *S. mattheei*, *S. mansoni*, *S. haematobium* and *S. bovis* cercariae tend to emerge during the early part of the day, while *S. japonicum* and *S. rhodaini* show a preference for emergence during the early evening and night.

The emergence of *S. japonicum* cercariae at night suggests that this schistosome may have evolved as a parasite of nocturnal mammals and that man's involvement is perhaps accidental. Many parasites commonly migrate within the tissues of their hosts and in certain cases these migrations are associated with transmission to an intermediate host; their function is to locate the parasite in the most suitable position within the host body for successful transmission to occur. Circadian periodicity in the distribution of microfilariae (larval filaria nematodes) in the mammalian blood system is a well known example of just such a migration. Hawking (1975) has examined the various microfilariae and recognizes four categories of microfilariae, according to their disposition within the hosts:

1. Numerous in the peripheral blood by night, rare or absent by day, e.g. *Wuchereria bancrofti* and *Brugia malayi*.
2. Numerous in the peripheral blood by day, rare or absent by night, e.g. *Loa loa*.
3. More numerous in the peripheral blood in the evening, e.g. *Dirofilaria immitis*.
4. Present in the peripheral blood over the entire 24 hour period, but more numerous in the afternoon, e.g. *W. bancrofti* in the Pacific region.

Microfilariae are transmitted by various species of mosquito and the feeding habits of these insects coincide with the time of appearance

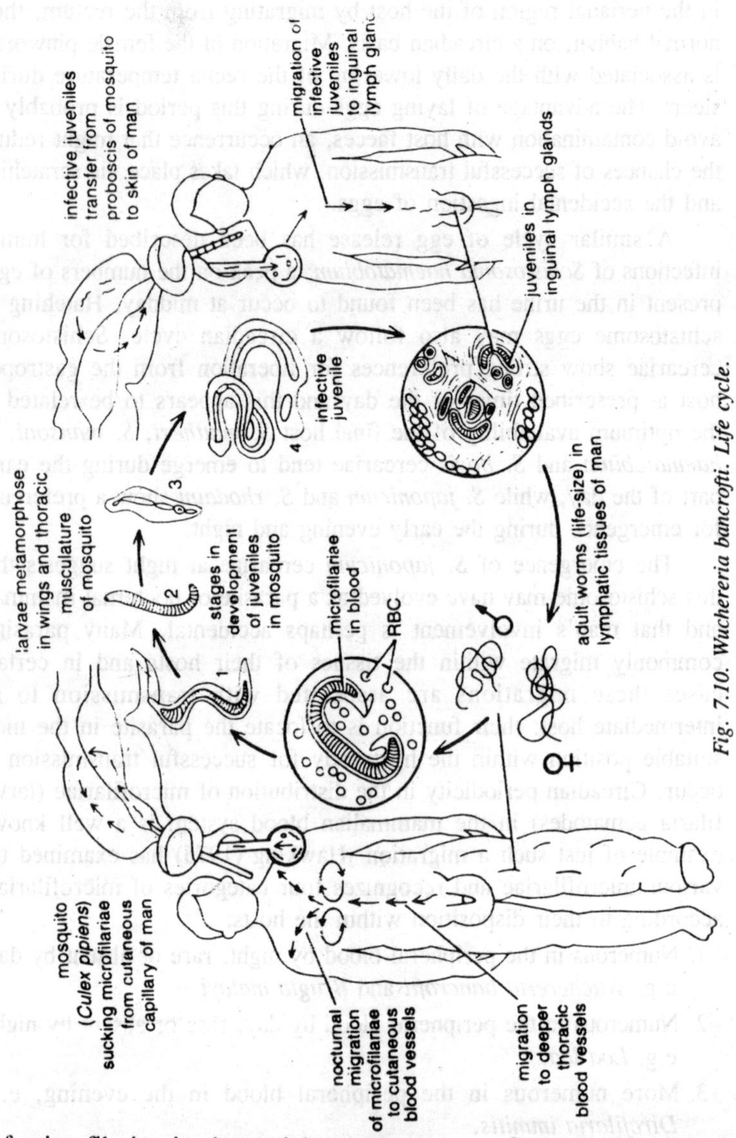

Fig. 7.10. Wuchereria bancrofti. Life cycle.

of microfilariae in the peripheral blood, e.g. day-biting mosquitoes transmit *Loa loa* and night-biting mosquitoes transmit *W. bancrofti*. When not present in the peripheral blood, microfilarae accumulate in the lungs, specifically in the pulmonary circulation at the junction between arterioles and capillaries. The process of larval nematode accumulation at this location appears to be related to the difference in oxygen tension (DpO_2) between pulmonary venous and arterial blood, e.g. in *W. bancrofti*, a DpO_2 of 55 mm Hg is sufficient to induce accumulation of microfilariae in the lungs, and a drop of DpO_2 to 44 mm Hg will initiate migration of the nematodes to the peripheral blood.

Artificial elevation of the pulmonary DpO_2, by taking exercise or breathing oxygen, will also bring about pulmonary accumulation of microfilariae. It is supposed, but yet to be substantiated, that the sense organs of the microfilariae are chemoreceptors, capable of detecting changes in pO_2 in the host blood. It is not known whether microfilariae possess an endogenous rhythm of their own, for, in the above example, they are clearly responding to an endogenous rhythm of host origin. A similar circadian migration has been described for some species of trypanosomes inhabiting amphibians. One species, *Trypanosoma rotatorium* appears in the peripheral blood of *Rana clamitans* by day and accumulates in the renal blood vessels by night—this trypanosome is transmitted by leeches that feed by day. A second species of trypanosome, parasitizing the same host, shows the reverse pattern of migration and is thought to be transmitted by night-biting insects.

It is not clear whether these migrations are due to an endogenous rhythm of the parasites or of the host, but they are presumably related to the light-dark cycle. There is not evidence for this type of migration in the mammalian trypanosomes, but some malarial parasites appear to be distributed differentially within the host blood system according to a 24 hour cycle. Annual periodicity occurs in many parasites and there is some evidence that this may be related to transmission. It might will be regards as wasteful for a parasite to produce and release infective forms it the potential host animals are not available for infection.

Annual cycles, related to transmission, are evident in some malarial parasite's filarial worms, such as *Dirofilaria immitis* in dogs, *Onchocerca gutterosa* and *O. cervicalis* in cattle and horses, respectively, and in other nematodes, such as *Haemonchus contortus* in sheep. It is probable that circannual rhythms are, in fact, common in many parasites, particularly those inhabiting temperate regions of the world, where climate is seasonally demarcated and host occurrence varies considerably with season.

8

INTERACTION IN ANIMALS

All organism have evolved defence mechanisms for protection against invasion and establishment by micro-organisms and by metazoan parasites. Although ideally we should include the viruses and bacteria in our current discussion, space precludes such a full consideration. Before discussing immunity to parasites, it will be necessary to outline, in the briefest manner, the components of the defence systems—primarily of mammals. The discipline of immunology has developed enormously over the last few decades, and it is often difficult to keep abreast of new developments. Although our knowledge of immunity to parasites is perhaps less complete than that in the areas of tumour and micro-organism immunology, a vast new literature is accumulating on the highly varied immune defence mechanisms adopted by mammals against parasitic infection. Nevertheless, many current textbooks of immunology pay little or no attention to parasite immunology, which perhaps reflects more the difficulties embodies in the subject, rather than the paucity of information.

The immunology of parasitic infections is both an important academic subject, as well as a fundamental applied discipline. It is to the field of immunological achievement that many people from the Third World will be looking for relief from the major parasitic diseases that reduce the quality or length of their lives.

GENERAL PRINCIPLES OF CELLULAR AND IMMUNOLOGICAL DEFENCE SYSTEMS

The essence of the immunological defence system of vertebrates is the recognition of self the ability to distinguish it from non-self, i.e. the host animal must be able to distinguish between its own macromolecules and those of an invading organism. On occasions this system of recognition can fail, giving rise to the so called auto-immune diseases, in which the host's defences act disastrously against its own tissue. Nevertheless, in the formal healthy animal, the ability to recognize non-self material and the development of a specific memory system associated with the recognition is the embodiment of the immune system. Defence mechanisms of vertebrates against invading organisms are divisible into either specific or non-specific responses.

Specific responses are associated with an immunological memory component and are normally directed against secondary and subsequent infections of an organism, whose primary infections initiates the response. Therefore, once a host has been infected with a particular parasite it retains an immunological memory of that infection and responds to eliminate or control any later infections by the parasite. Non-specific responses are more generalized in their nature, and are directed against any invading organism, with no involvement of a memory element. In some cases, specific and non-specific responses cooperative to form a combined defence against invasion.

Non-specific responses tend to have a cellular basis and involve the action of phagocytic cells (macrophages and polymorphonuclear leucocytes), which form the reticuloendothelial system and are present in the circulating blood and in the tissues of the host. Their function is to engulf invading organisms and digest them lysosomally. Engulfment and digestion are sequential processes: first, the invading organism becomes adhered to the surface of the phagoctyic cell; secondly, it is ingested by the formation of a vacuole, becoming trapped within a phagosome; thirdly, a laysosome fuses with the phagosome to form a phagolysosome; and finally, the engulfed organism is destroyed by an array of lyric factors, including enzymes.

Phagocytosis is a common mode of defence against invading micro-organisms, and it can be enhanced by the action of antibodies known specifically as *opsonins*. The complex of serum factors, known as *component*, may also aid the process of phagocytosis by attracting phagocytic cells to the environs of an invading parasite. Other non-specific responses to parasites may take the form of inflammatory reactions and capsule formation.

Inflammation is typified by increased capillary blood supply to a site of invasion, resulting in the accumulation of polymorphs around the invading parasite. Some polymorphs can transform to fibroblasts, which contribute to encapsulation and ultimately, calcification of the parasite; digenean metacercariae and pseudophyllidean plerocercoids are often encapsulated in this way in their intermediate, fish hosts.

Specific immune responses involve the formation of antibodies by the host as a result of stimulation by the presence of foreign material, termed antigens. Antigenic substances are usually polypeptides, proteins or polysaccharides, with a molecular weight in excess of 5,000 Daltons (lipids are only antigenic when combined with other substances, called haptens). Antigens derive from the surface coat of a parasite, i.e. surface antigens, or from secreted or metabolic byproducts, i.e. soluble or metabolic antigens. Substances are only antigenic if they exhibit a degree of foreigness to the host, which then responds by synthesizing antibodies, specific to each antigen. Antibodies are soluble serum proteins that combine, specifically with antigens. They belong to the gamma-globulin fraction of serum proteins, and are referred to as *immunoglobulins* (Ig). They are separable into five distinct classes, depending upon their structure, i.e. IgA, IgD, IgE and IgM.

The molecular structure of the immunoglobulins consists of two heavy peptide chains, linked by disulphide bridges, and two light chains, that are also cross-linked. Papain treatment cleaves the immunoglobulin molecule into two fractions, the F*ab* (antigen binding fragment) and the F*c* (crystallisable fragment) components. Antigen binding takes place at the F*ab* end of the immunoglobulin molecule, and thus, antibody function is retained after papain cleavage of the intact molecule. Immune defence effected by antibodies is divisible into two distinct processes, known as *cell-mediated immunity* (delayed hypersensitivity) and *humoral immunity*. Both types of immunity arise from stem cells produced in bone marrow.

In cell-mediated immunity, the stem cells are processed by the thymus gland to form a population of lymphocytes (T-lymphocytes) which become sensitized upon contact with antigen, in the paracortical region of lymph nodes. These T-lymphocytes are long-lived and carry firmly bound antibody on their surface membranes. In humoral immunity, the bone marrow stem cells are processed in lymphoid tissue (Bursa Fabricius in birds and Peyers Patches in mammals) to give rise to a population of B-lymphocytes, which are short-lived, after contact with antigen. These lymphocytes differentiate into plasma cells which synthesize and secrete antibody into the surrounding blood and tissues. Cellular co-operation may enhance an immune response, and such co-operation can involve contact between macrophages and lymphocytes or between B-cells and T-cells (helper T-cells).

Cell mediated immunity is involved primarily in allergic responses and can take some time to develop (24 hours or longer), whereas humoral immunity is developed much more rapidly. Antibodies react with antigen and effect lysis, agglutination, precipitation or enhanced phagocytosis of the cells producing the antigens. These reasons can form the basis of *in vitro* tests used to detect the presence of specific antibodies to parasite, and they serve to determine whether an individual animal has been exposed to a particular parasite (immuno-or sero-diagnosis). Immunological defence is best developed and most often studied in mammals, and rather less is known for lower vertebrates and invertebrates. In non-mammalian vertebrates, humoral and cell-mediated immunity can be identified, but invertebrates appear to lack antibodies, in the mammalian sense, and rely, to a great extent on, phagocytosis as a defence against invasion. ***Passive immunity*** is passed from a mother to her offspring and reflects in innate resistance to a particular infection. Passive immunity can also be transmitted to the laboratory by the donation of serum or immunologically competent cells to recipient, naive animals. ***Acquired immunity*** depends upon the experience of a primary infection to initiate the specific immunological processes that render the animal insusceptible, or less susceptible, to a second or subsequent infection.

IMMUNITY TO PROTOZOA

Blood-dwelling Protozoa

In this section we shall examine briefly the immunity to malaria, babesias and trypanosomes, which together constitute the major blood-dwelling parasitic protozoans of vertebrates.

Malaria

The general picture of immunity to malarial parasites is remarkably complex due to a range of factors that include strain variation, antigenic variation within strains, and a life-cycle that involves intraerythrocytic stages and exoerythrocytic stages, both in the blood and in liver tissue. Acquired immunity to *Plasmodium* is variable, e.g. premunition is typical of avian malaria, whereas sterile immunity occurs in some rodent malarias, such as *Plasmodium berghei* and *P. vinckei*. In the former case, parasites persist in spite of the development of immunity, while in the latter case immunity is complete and no parasite survives. Premunition is thought to typify human malarial infections, in which relapses an recrudescences are not uncommon.

Protective immunity to malaria is recognized as a collaborative mechanism involving both cell-mediated reactions (T-lymphocytes) and humoral antibodies (B-lymphocytes). Immunoglobulin production increases considerably during the course of a malarial infection, but much of the antibody produced is not protective and does not serve to eliminate the parasite. Recovery from malaria requires protective antibody to the manufactured at a rate sufficient to control the multiplying parasite; drug induced immunosuppression in experimental malarias, can often result in a fatal infection developing from what would otherwise have been a non-lethal, self-limiting condition. There is considerable variation in the levels of the various immunoglobin classes, e.g. IgM levels rise slowly and are maintained for much longer periods.

The roles of IgA, IgD and IgE are not clear in protection against malaria, and this function appears primarily with IgG. This is supported by the observation that, during pregnancy, IgG levels characteristically decrease, and there is often an enhancement of malaria in pregnancy. Cell-mediated reactions, involving lymphokine production, the potentiation of antibody production from B-lymphocytes by helper T-cells and cytotoxicity, may all be involved in the development of immunity to malarias, but current knowledge remains incomplete. Surgical removal of the thymus (thymectomy) in neonatal rats and hamsters renders them more susceptible to *P. berghei* infections than intact controls.

Depression of T-cell function by the administration of anti-T-cell serum, can also exacerbate malaria in experimental animals. There is a number of diagnostic tools, based on immunoglobulin production in response to malarial infection, which serve to identify individuals infected with the parasite, where microscopic examination might prove negative. Such immunodiagnoses include the indirect florescent antibody test, the indirect hemagglutination test and various gel precipitation tests. The application of these methods to human populations can provide considerable and invaluable information on the epidemiology of malaria in endemic regions of the world. Serological monitoring is also proving to an important tool in the prevention of transfusion-induced malaria, which is an increasing problem related to the degree and ease of world travel that typifies the late twentieth century.

Malarial parasites can avoid the immune responses of their hosts to form residual populations, either in red blood cells or within liver tissue, and these residual populations of parasites are able to initiate recurrent relapses or recrudescences. The latter may result either from infected red cells that are situated at deep vascular sites, the plasmodia re-emerging as the effective immunity weakens, or from either the continued existence of merozoites in the circulating blood or sporozoites in the liver.

Antigenic variation, whereby a sub-population of parasites of different antigenic structure from the original invasion, is developed, may also be important in this context. This new antigenic population can now withstand the effects of the antibodies directed against the antigenic variations represents an important problem, hindering the development of a vaccine. The pathology of mammalian malaria is related, in part, to the immunological responses of the host to the parasite—hence the term immunopathology. Red cell destruction (haemolysis) and enlargement of the spleen (splenomegaly) are typical of advanced malaria. Haemolysis may be brought about by the phagócytic activities of spleen macrophages, and both infected and uninfected red cells are destroyed.

Opsonising antibodies enhance this pathological process. Haemolysis in Blackwater fever (*P. falciparum* infections associated with quinine therapy) occurs in the vascular system and is not related to the splenic mechanism. Malarial kidney disease results from antigen-antibody complexes binding to the glomerular basement membrane, producing yet another aspect of malarial immunopathology. An interesting feature of malarial parasites lies in their role as immunosuppressive agents for infections by other parasites. Laboratory animals infected with various species of *Plasmodium*, show a reduced immune response to non-living antigen, such as tetanus toxoid or sheep red blood cells, and in living antigens in the form of viruses, bacteria and several protozoans. It is not known whether this immunosuppressive effects is functionally related to cell-mediated or humoral immunity.

There is god evidence for suppression of antibody production by B-lymphocytes, but the general mechanisms by which immunosuppression to unrelated antigens is effected remain to the elucidated. Toxin production in malaria has been a much debated issue for many years. The key question of whether the parasite releases a toxic substance, which itself is pathogenic, has yet to be answered. There is some evidence that a transferable serum component will induce haemolysis in recipient animals, and that serum elements will induce haemolysis in recipient animals and that serum elements are capable of inhibiting mitochondrial respiration and oxidative phosphorylation in liver cells *in vitro*. Antitoxic immunity, however, has no effect on the levels of parasites in the blood, but may serve to alleviate aspects of the immunopathogenesis.

Babesia

Members of the genus *Babesia* are intraerythrocytic parasites of vertebrates, occurring most commonly in mammals. Several species are of veterinary importance as they cause a variety of tick-borne fevers of domestic animals, e.g. *B. bovis*, red-water fever; *B. equi*, biliary fever. Immunity to many species of *Babesia* is of the premune type, existing only while the host remains infected by the parasite. For some species, however, complete sterile immunity is developed, e.g. *B. divergens*, *B. bigemina* and *B. rhodaini*. Immunity may be highly specific for strains of a single species, such that cattle, immune strain from one locality, can be susceptible to a different strain of the some species from other localities. In other, cases cross-immunity, between different species, is found e.g. *B. rhodaini* and *B. microti* in rodent. The extent of antigenic variation in babesias is not known.

Trypanosomes

Extensive investigations have been made into immunity to both rodent and human trypanosomes. One of the best known model systems is *Trypanosoma lewisi*, a specific rat parasite, which provides a classical demonstration of acquired immunity. Additional interest in *T. lewisi* is generated by the occurrence of an apparently unique antibody, termed ablastin, whose functional significance has proved to be controversial. Typically, the course of a *T. lewisi* infection in a laboratory rat is characterized by an initial phase of rapid multiplication of the parasite, giving a rising parasitaemia. Parasite division is inhibited five days post-infection and a plateau in the number of trypanosomes occurs, but this is of short duration. Subsequently, a primary crisis takes place on or about the tenth day of the infection, during which most of the trypanosomes are killed.

The few parasites that remain survive until approximately the thirty-fifth day, when a second crisis completely eliminates the infection. One explanation of these events involves the production of three distinct antibodies one that inhibits reproduction of the parasite (ablastin) and two lethal (trypanocidal) antibodies, which contribute to the two crises. The name ablastin was coined by Taliaferro, more than fifty years ago, for the antibody that inhibited the reproduction of *T. lewisi* but lacked trypanocidal activity. Since the original experiment which demonstrated ablastic immunity were completed an interesting and controversial literature has arisen. Ablastin is found only in infection of two species of trypanosomes. *T. lewisi* and *T. musculi*, although ablastic-like responses are characteristic of many stercorarian trypanosome infections.

Despite intensive investigation, ablastin has not been unequivocally identified as an immunoglobulin, but it does possess physico chemical properties typical of IgG. However, ablastin is not adsorbed from infected rat serum by trypanosomes *in vitro*, a feature which throws some doubt on its role as an antibody. Furthermore, the trypanosome antigen (ablastinogen) that induces ablastin production, has yet to be identified, though there is some evidence that this antigen is released into the blood of the host during the course of a normal infection. Pregnant female mice are unable to limit infections of *T. musculi*, suggesting that ablastin belongs to the IgG class of immunoglobulins, which are known to decrease during pregnancy. Some authors, however, regard the apparent uniqueness of ablastin as sufficient reason, it itself, for discounting the presence of an antibody that is not lethal but inhibits only parasite reproduction. One alternative explanation that has been advanced, invokes the occurrence of two parasite antigens, one juvenile and the other adult, that stimulate the host to manufacture trypanocidal antibodies.

Antibody against juvenile antigen eliminates juvenile trypanosomes by the tenth day post-infection, while adult parasites survive for longer periods, being—according to this hypothesis—less antigenic. Additionally, it has been proposed that ablastin may act as a metamorphosis-conversion-factor in the development of trypanosomes during an infection, enhancing the change from juvenile to adult form. The first crisis in an infection would then be attributable to the activities of a trypanocidal, anti-juvenile antibody, which may even be ablastin. These views contrast with the classical view of immunity to *T. lewisi*.

Despite considerable speculation, the nature of a ablastin and ablastic immunity remains unresolved. Its apparent uniqueness as an antibody argues for generating research interest rather than for a denial of its unusual role. Studies on immunity to the human trypanosomes, causing sleeping sickness, e.g. *Trypanosoma gambiense* and *T. rhodesiense*, are complicate by the appearance of *antigenic variation*, which typifies these haemoflagellates. Each species has a highly complicated antigenic structure, which gives rise to both strain and variant populations. These render attempts to immunize humans or domestic cattle against trypanosome infections almost futile. The earliest observations on antigenic, variations were made at the beginning of this century, when it was noted that *T. equinum*, isolated from the blood of a monkey, was unaffected by serum antibodies from the same animal *in vitro*, and it was concluded that the trypanosomes had undergone antigenic change during the course of the infection. Similar observations to this have been made on many subsequent occasions.

Antigenic variation can give rise to a *cyclic parasitaemia*, in which the numbers of parasites increase until a crisis occurs which eliminates the majority of individuals, except for a small number of antigenically variant forms which survive and divide to produce a second, rising parasitaemia, followed, in time, by a second crisis. This pattern of increasing and decreasing parasitaemia may continue until the host dies. Experimental observations show the frequency with which new antigenic variants can arise, e.g. *T. congolense* produces a new variant every ninth day in sheep, and *T. vivax* develops new variants every seven days in sheep and calves. The maximum number of different variants arising from a single strain is 23, but this is thought unlikely to represent the upper limit. Only rarely will the same variant appear twice during a single infection.

There is also evidence that antigenic variation is an organised process, with a particular strain of trypanosomes producing the same variants, often in a similar sequence, regardless of host species. Two major subclasses of antigens have been identified from trypanosomes. The first are the bound antigens which remain stable during a relapsing infection. These antigens include enzymes and nucleoproteins and are only weakly antigenic. The second subclass contains the surface-coat antigens, which are unstable entities. The surface-coat antigens are the glycoprotein exoantigens that are present in the serum of infected animals and are responsible for inducing the production of protective antibodies. Antigenic variation in trypanosomes may be regarded as an adaptive mechanism to overcome the immunological defence of the host. Its occurrence makes it most unlikely that successful immunisation will be accomplished in the forseeable future.

Tissue-dwelling Protozoa

Among the major tissue-dwelling protozoans of importance to man from a medical or veterinary point of view are the leishmanias, the coccidians and the amoebae.

Leishmania

Various species of the genus *Leishmania* cause by unpleasant and often lethal diseases of man in many regions of the world. These include oriental sore (*L. tropica*), kala, azar (*L. donovani*) and New World cutaneous leishmaniasis (*L. mexicana* and *L. braziliensis*). The parasite is transmitted to man via the bite of infected phlebotomine flies (sand flies) and, after entry, the amistigote form of the parasite invades and inhabits the cells (such as monocytes and other phagocytic cells) of the reticuloendothelial system. Immunity to *L. tropica* develops after an initial infection, i.e. acquired immunity. The parasite invades the skin macrophages and a superficial sore develops. This sore will heal spontaneously, after which immunity to reinfection is complete. Should the sore be removed prematurely by surgery, however, the individual remains susceptible to reinfection. Circulating antibodies cannot be demonstrated in *L. tropica* infection, whereas humoral components are readily observed in *L. donovani* infections. In general, it is considered that the human immunological response to leishmaniasis is primarily via cell-mediated immunity, with circulating antibodies playing only a minor protective role. A population of specifically sensitized, thymus-dependent lymphocytes develops on contact with parasite antigens, and these cells destroy the infected macrophages.

The actual mechanisms by which the parasite is eliminated is not fully understood, but it may occur by a number of processes, including phagocytosis by activated macrophages, cytoxic effects of lymphokines secreted by sensitized lymphocytes at the infected macrophage surface, or, perhaps by lytic antibody activity. Some degree of cross-immunity occurs in leishmaniasis, e.g. *L. tropica* protects an individual against infection by *L. mexicana* and *L. braziliensis*, but not against *L. donovani*.

Coccidia

Most domestic animals appear to develop immunity to reinfection by species of the genus *Eimeria*, which are obligatory, intracellular, but parasites. Sear of infected animals effectively prevents infection by the invasive stages of *E. tenella*, *E. bovis* and *E. meleagrimitis*, and the presence of circulating antibodies has been demonstrated by a variety of techniques. Rather surprisingly, passive transfer of immunity, by the injection of immune serum into uninfected hosts, has not proved to be particularly successful, and is only effective if the challenge infection is administered shortly after the inoculation of immune serum from the donor animal.

Immunity to *Eimeria* is highly species-specific, so that there is no cross-immunity between the different species. The precise mechanisms of acquired immunity to *Eimeria*, either in chickens or laboratory rodents, is not clear. Humoral antibodies appear to be implicated in the development of protective immunity, while there is some evidence that cellular elements, polymorphonuclear cells, lymphocytes and macrophages, are also involved. Protective antinodies are produced against another important intracellular gut parasite, *Toxoplasma gondii*, but it is thought that the cystic stage of the parasite is resistant to these antibodies.

Amoebae

Entamoeba histolytica is normally a non-pathogenic inhabitant of the human, intestine and, under such conditions, is apparently non-immunogenic. However, the amoebae can occasionally penetrate the gut epithelium and spread throughout the body tissues. When this occurs antibodies are synthesized by a host, and serological tests can be used to identify their presence and thus serve to diagnose amoebiasis. At least two antibody systems have been identified, but it is not certain by which mechanism protective immunity is developed.

Immunity to Helminths

Metazoan parasite differ fundamentally from protozoans in that the great majority do not multiply within the final host, unlike the Protozoa. Thus immunity to helminths will be directed primarily against the individual parasite, rather than its reproductive ability (Cf. *Trypanosoma lewisi*). There is increasing interest in the immunology of mammalian helminth infections, particularly with regard to those species that effect man's health of his economy. While protozoan infections are frequently by an increase in the serum levels of IgG and IgM, helminth infections typically induce the synthesis of reaginic antibody, IgE, which is characteristically abundant in many helminthiases.

Digenea

Information on the immunology of digenean infections depends largely on studies of only two groups, the schistosomes and the liver flukes, e.g. *Fascoila hepatica*. There is some doubt as to the occurrence of acquired immunity to *Fasciola* in cattle and sheep. Cell-mediated reactions may play an important role in the development of protective immunity in domestic animals. In the laboratory, experimental mice treated with peritoneal exudate cells from infected donors support fewer worms than control animals. T-cell responses may also be responsible for effecting repair of the tissue damage wrought by the parasite, since rabbits given antilymphocyte serum, to suppress T-cell responses, did of liver disease due to the parasite. Transfer of lymph node and spleen cells from infected donor rats to naive recipient rats offers a significant degree of protection against a primary challenge infection, but this is not repeatable using sheep. The immunology of schistosome infections, particularly *S. mansoni*, has been examined in considerable detail using model systems in rats, mice and rhesus monkeys.

Experimental studies in man are, naturally, not acceptable, so that evidence for acquired immunity in human infections derives mainly from epidemiological sources. In general, the young are more susceptible to infection than adult humans, and a premune condition develops, in which a primary infection renders the host immune to subsequent reinfection. In schistosomiasis, this condition is termed *concomitant immunity*, since the primary infection is not killed by the immune responses. This situation has led to the suggestion that it may be possible to immunize against infection using attenuated (non-developing) cercariae. There are, however, problems associated with this approach, so the search continues for a dead vaccine that will have fewer attendant risks and be more easily administered. In experimental infections, acquired immunity of the concomitant type has been clearly demonstrated.

Despite the development of a lethal immune response, the primary infection continues to thrive and produce eggs for considerable length of time. In rhesus monkeys, immunity to a challenge infection can be developed within sixteen weeks of administration of a primary burden of 100 cercariae per animal. Mice take much longer to develop resistance to a challenge infection. A *self-cure* type responses has been recorded in heavily infected baboons, in which the primary infection is completely eliminated; this does not accompany light infection doses. The process of concomitant immunity has puzzled researchers for more than a decade, and several explanations for the observed effects have been put forward.

It is generally accepted that concomitant immunity arises through the sharing of common antigens by host and parasite, but it is not clear whether the parasite synthesizes host-like antigens or merely binds antigenic material of host origin to its surface-coat. Support for the latter hypothesis comes from ingenious experiments in which schistosomes are grown to maturity in mice and then transferred to the hepatic portal system of rhesus monkeys that have been immunized against mouse red blood cells. Worms transferred in this way are killed within 24 hours, whereas mouse worms will live in non-immunized monkeys. Similar results have been obtained using worms grown, *in vitro*, in human serum and then transferred to monkeys immunized against human erythrocytes. These results support the notion that schistosomes bind host antigens to their surfaces, and by doing so become essentially disguised, rendering them indistinguishable from the host as far as the immune recognition system is concerned. Challenge infections are thought not to be able to bind host antigen at the rate sufficient to elude the antibodies produced against the primary infection.

Lethal antibodies have been demonstrated in immune monkey serum, killing schistosomula in culture within three to four days. Several antibodies have been isolated from schistosome-infected human and animal serum (many of which are used as sero-diagnostic tools), but it seems unlikely that any are directly involved in acquired immunity. Thus, while schistosomiasis is characterized by an increase in the serum levels of IgG, much of this is non-specific. The general inability to demonstrate the passive transfer of resistance to infection with immune serum has led some to believe that cell-mediated reactions are responsible for acquired immunity in schistosomiasis.

Recent experiments on passive transfer have proved more successful when rats and mice are used, but not so with monkeys. Additionally the occurrence of a lethal serum component, capable of killing schistosomula *in vitro*, strongly suggests that humoral antibodies are, indeed, required for acquired immunity. It may well be that both B-cells and T-cells, working in a cooperative manner, will prove to be the elements involved, but further evidence is required to establish this hypothesis. There is good evidence that cell-mediated reactions are concerned with the immunopathology of schistosomiasis. The disease, as seen in the model system of *S. mansoni* in the laboratory mouse, is characterized by liver and spleen enlargement (hepatomegaly and splenomegaly), by portal hypertension (increased blood pressure in the hepatic portal vein) and by oesophageal varices. Egg production by the worms is a fundamental feature in disease production.

Typical hepatosplenic schistosomiasis does not develop in either unisexual infections of mice or in normal, bisexual infections of mice given drugs to inhibit the production of eggs by the parasites. The formation of granulomata (nodules of inflamed tissue) around eggs has been studied, experimentally, by examining pulmonary inflammation around schistosome eggs that are administered to mice via the tail vein and which become lodged in the lungs. These studies implicate cell-mediated reactions in ganuloma formation. Immunosuppressant drugs will inhibit the production of granulomata around eggs, but there is no evidence to suggest that humoral antibodies are involved in the inflammatory responses to the eggs of either *S. mansoni* or *S. haematobium*. Partially successful attempts have been made to isolate the egg antigens responsible for initiating granuloma formation, which is thought to be due to the secretion of soluble egg antigens through pores in the egg shell. It is probable that soluble egg antigen stimulates the production and secretion of lymphokine by T-lymphocytes, and this attracts mononuclear cells to the egg. Soluble egg antigen includes the enzymes, secreted by the eggs to aid their passage through the host tissues, on their way to the external environment. In contrast with *S. mansoni* and *S. haematobium*, granuloma formation in *S. japonicum* is mediated by antigen-antibody reactions.

Cestoda

Tapeworms live as adults in the lumen of vertebrate intestine and as larvae in the tissues of either invertebrates or vertebrates. The immunological status of lumen-dwelling cestodes tends to be somewhat different from that of the tissue stages, since the latter are in much more intimate with the host's defence mechanisms than the former. Acquired immunity to tissue-dwelling larval cestodes operates either at the transaction phase, during which the oncosphere hatches in the intestine of the host and invades the mucosa, or during the actual tissue phase itself.

Liberated oncospheres are susceptible to immunological attack in the host gut, as shown by studies on *Taenia pisiformis*, *Hymenolepis nana* and *Echinococcus granulosus*. In immune hosts, the oncospheres of these worms either fail to hatch, or hatch but fail to penetrate the gut wall. Immune serum is capable of damaging the oncospheres of *T. pisiformis*, *in vitro*. Rats harbouring the cysts of *T. taeniaeformis* in their livers are resistant to challenge infections for several months, and a similar pattern of immunity occurs in other larval taeniids, e.g. *T. hydatigena*, *T. ovis*, *T. saginata* and *T. pisiformis*. Passive transfer of immunity is accomplished by injecting whole immune serum or lymph node cells from immune hosts. Thymectomy reduces protective immunity of mice to *H. nana*, a tapeworm whose cysticercoids can develop in the intestinal mucosa of the definitive host.

Protection against reinfection is restored by implantation of thymus tissue. In *H. nana* infections, passive protection is also associated with serum antibodies, particularly IgG. Studies on the hydatid stage of *Echinococcus granulosus* suggest that a complete protective immunity does not develop, since challenge infections are numerically less successful than the primary, immunizing infection, but they still survive. Although contact with host tissues tends to be less intimate in many of the lumen-dwelling adult cestodes, they are, nevertheless immunogenic. Current interest has focussed on *Hymenolepis diminuta*, *H. microstoma*, *Raillietina cesticillus* and *Echinococcus granulosus*. *Hymenolepis diminuta* develops very differently when grown in laboratory rats and mice. In the mouse, this parasite grows normally for ten days, then growth ceases and destrobilation occurs. Transfer of destrobilated worms to rates restores normal growth.

A challenge infection given to immune mice (infected and cleared of worms by anthelminthic treatment) dies more rapidly than a primary infection. Additional evidence for the immunological basis of these events derives from the use of immunosuppressants, e.g. cortisone, methotrexate and antilymphocyte serum prevent both the rejection response and destrobilation. The rat, on the other hand, is the natural host for *H. diminuta* and will support low density infections, i.e. containing between one and ten worms, indefinitely. Destrobilation and worm loss is a feature of higher density infections, but it is not clear whether this is a function of the immune response of the host or if it is due to the physiological alterations associated with crowding of worms in the intestine. Recent evidence supports the former proposition; challenge infections of 50 or 100 worms survive less well than primary infections, an effect alleviated by cortisone treatment. Antibodies of various classes are bound to the tapeworm surface in the rat gut and it may be that this binding interferes with the absorption of essential nutrients.

Hymenolepis microstomoa, which normally develops in the bile duct or mice, fails to grow or survive in the rat, but it has not been established that this is due to the immune response of the host. In the mouse, *H. microstoma* infections are accompanied by increased levels of circulating antibodies (IgA, IgE and IgG), but these do not appear to play a protective role. In general, the early developmental stages of cestodes are more immunogenic than mature worms, so that the intermediate vertebrate host develops a strong protective immunity. Acquired immunity to lumen-dwelling adult cestodes may play a controlling rather than an eliminating role. Larval cestodes that develop in invertebrate animals are exposed to the haemocytic defences, which are either phagocytic or encapsulating processes. There is some evidence that tapeworm larvae, living in the haemocoel of insects, can either evade this defence or disguise themselves so that the host haemocytes fail to identify them as foreign, e.g., *H. diminuta* in four bettles.

Nematoda

Interest in the immunity of mammals to nematode parasites is well established, and has focussed particularly on the pathogenic species of economic or medical importance to man, or on convenient laboratory systems. Nematodes differ from other parasites in the presentation to the host of a cuticular surface, which is only weakly immunogenic.

Furthermore, it is unlikely that antibodies or cellular defences will be able to exert any lethal effect by attaching to the nematode cuticle, and only the anal, oral and glandular openings of the parasite will be vulnerable. Most nematode infections, and indeed helminth infections in general, stimulate the host to produce reaginic antibodies (IgE), but these antibodies do not appear to be exclusively responsible for limiting an infection. The parasite antigens that stimulate IgE production are termed *allergens*, and these have been identified in some nematodes.

There is a relationship between IgE production, eosinophilia and immuno-pathology in many nematode infections, and this may be mediated via anaphylaxis, involving the synthesis and secretion of histamine from mast cells at the site of invasion or habitation of the parasite. The phenomenon called self-cure is probably related to the release of histamine in the gut of hosts infected with lumen-dwelling nematodes, and this causes leakage of plasma proteins into the gut, bringing about expulsion of the worms. Self-cure occurs in *Haemonchus contortus* infections of lambs and *Nippostrongylus brasiliensis* infections of rats, and may prove to be a common expulsive mechanism for gut-dwelling nematodes.

Haemonchus contortus is an economically important parasite, causing pronounced anemia and often killing young lambs. Self-cure of haemonchosis was first described almost 50 years ago. It is characterized by a sudden decrease in the numbers of parasite eggs present in the faeces of lambs between 12 and 13 weeks post-infection. This immunity is precipitated by the ingestion of additional larval nematodes from the pasture and occurs in lambs over four months old. Vaccination against *Haemonchus* is partially effective using irradiated third-stage larvae, that can migrate but do not develop to maturity. Variation in response by different breeds of sheep diminishes the usefulness of this approach to artificial protection.

It has been proposed that the antigens responsible for stimulating self-cure arise from the exsheathing fluid of the third moult. Self-cure is typical of *Nippostrongylus brasiliensis* infections in the laboratory rat. In heavily, infected rats, parasite egg production rises between six and ten days post-infection and then decreases rapidly, accompanied by the expulsion of the majority of the worms. The rats are not highly resistant to a challenge infection. By contrast, worms given to rats as repeated low doses, called *trickle infections*, fail to stimulate any protective immunity and these worms will survive and produce eggs for extended periods, being described as "adapted" and capable of surviving transfer to an immune host.

Self-cure in *Nippostrongylus* was originally thought to be due to two distinct events; antibodies inhibiting worm acetylcholinesterase activity and immunopathological effects upon the rat gut epithelium rendering it an unsuitable environment for the worms. The latter is related to IgE production, stimulating histamine release from mast cells. Other immunoglobulins, including IgA and IgG, are thought to play some role in self-cure. It is apparent, on examining the immunological response of the rat to heavy infections of *Nippostrongylus*, that several aspects of the immune system must operate coincidentally to effect worm expulsion. Both antibodies are sensitized lymphocytes are required for the rejection process, whereas IgE, mast cells and eosinophils are not regarded as non-essential components. Protective antibodies belong to the IgG class of immunoglobulins but not to IgA. The former have been shown to interfere with worm metabolism and cause damage to the worm's alimentary tract.

Trichinella spiralis, causing trichinosis in man, is a good model parasite for the study of acquired immunity using mice, rats and guinea pigs as laboratory hosts. Trichinosis in man is not accompanied by a noticeable increase is serum antibody levels, but repeated infections of experimental animals results in a greatly increased immunoglobulin (precipitin) titre. In rabbits, IgM levels rise early in an infection, followed later by elevated levels of IgG. IgA remains consistently high throughout an infection, and forms a localized mucosal response to the parasite.

Trichinosis is characterized by the production of homocytotropic antibodies, which stimulate the release of pharmacologically active substances from the cells to which they bind. This type of immediate hypersensitivity can be detected by active and passive anaphylaxis. In the mouse, homocytotropic antibodies are of two types; one is a reaginic antibody (IgE) and other is IgG_1. Inflammatory, cell-mediated immunity occurs in experimental trichinosis and it is passively transferable with the peritoneal exudate cells of infected mice. Acquired immunity to *Trichinella* is developed following a primary infection, so that the larvae of the challenge infection rapidly become eliminated in the immune host gut due to local inflammatory responses.

Anti-inflammatory agents, such as cortisone, markedly depress the elimination of a challenge infection, as does whole-body-irradiation of hosts. Both anti-lymphocyte and anti-thymocyte sera reduce the resistance of immune mice to a challenge infection, implicating T-cells in acquired immunity. Although the exact nature of the mechanisms of acquired resistance in laboratory animals to *Trichinella* remains to be elucidated, current opinion is that the parasite is expelled from the gut of the host by specific inflammatory mechanisms. It is possible that non-specific processes that alter the environment of the gut, rendering it unsuitable for the worms, are also involved. In trichinosis, protective, immunity is directed essentially against invading larvae rather than established worms.

Ascaris suum in pigs and laboratory animals serves as a convenient model for human ascariasis (*A. lumbricoides*), the immunology of which is poorly understood. Immunity to *A. suum* possibly involves both humoral and cell-mediated elements, since passive transfer of immunity can sometimes be accomplished with anti-serum, but cannot be effected with lymphoid cells. Again, protective immunity acts against invading larvae, though self-cure has been recorded for adult ascarids in pigs. During an infection, IgM levels are the first to increase, followed by IgE, IgG and IgA. It is not known to which class of immunoglobulins the protective antibodies belong.

In human infections with *Ascaris*, by contrast, the immunoglobulin titre does not always increase. Experimental ascariasis is typified by both inflammatory and cell-mediated responses directed, particularly, against larval worms in the tissues of the host. Granulomata are formed around parasite eggs, probably as a result of cell-mediated reactions. The third-stage larvae are the most active immunogenically and some characterization of parasite antigens has been accomplished by gel precipitation techniques. Rather little is known about immunity to other nematodes that parasitize humans.

Ancylostoma caninum, a dog hookworm, has been studied experimentally and is similar to *Nippostrongylus brasiliensis* in the rat, in the immune responses it invokes. Immunity to the human hookworms has not been examined in any detail. The immunology of human and experimental filarial infections has received some attention using *Litomosoides carinii* in cotton rats, *Dirofilaria immitis* in dogs are *Brugia pahangi* in cats as model systems. Infection in these systems produces increased IgE and IgG levels in the respective hosts.

There is a significant degree of common antigenicity between filarial worms of different genera and other helminths. Many nematodes like some protozoans and platyhelminths, are capable of living in a host which has responded to the parasite by humoral and cellular reactions. The avoidance of the immune responses is a characteristic feature of many parasitic infections, but little is known about the mechanisms involved. Additionally, there are unexpected interactions between nematodes and micro-organisms at the immunological level, e.g. mice infected with *Plasmodium berghei*, *Trypanosoma brucei*, on various babesias, lose their protective immune responses to *Trichuris muris* and *Trichinella spiralis*. No explanation of these events is, at present, forthcoming.

Immunization Against Parasitic Diseases

The development of vaccines to combat any of the major parasitic diseases of man and domestic animals is of fundamental importance to the medical well-being and economy in much of the world, particularly the tropical and subtropical regions. Unfortunately, vaccine production has not yet met with any real success since so many parasites are capable of evading the immune responses of the host (antigenic variation in malaria and trypanosomiasis, and concomitant immunity in schistosomiasis are example of the problem). Nevertheless the search continues and it is a widely held view that a complete knowledge of the immune responses to each parasite is a prerequisite for the development of artificial immunization.

As we have seen already in this chapter our depth of knowledge is far from complete for the immunology of any parasitic disease, but despite this, there have been encouraging results with immunization against some parasites. A variety of different approaches has been adopted to produce immune hosts by artificial treatment and these include the use of attenuated parasites, homogenized parasite tissue, dead parasites, soluble parasite antigens, heterologous protection, controlled live infections and non-specific measures. Perhaps the most successful immunization yet achieved has come through the use of attenuated larvae to protect cattle against the nematode lungworm, *Dictyocaulus vivaparus*, and there is now a commercially available vaccine of irradiated larvae which confers a long-lasting immunity. Similar success has been achieve with *D. filaria* in sheep and *Ancylostoma caninum* in dogs. The use of larvae attenuated by irradiation has been found to offer protection in experimental infections of a number of nematode parasites, including *Brugia malayi*, *Dirofilaria immitis*, *Capillaria obsignata*, *Haemonchus contortus*, *Litomosides carinii* and *Trichostrongylus colubriformis*, and in some schistosome species (*S. mattheei*). An effective live vaccine has been developed against *Babesia argentina* in Australia.

The parasite is obtained from splenectomized calves and therefore has a lowered virulence and will not infect the tick vector. The use of killed parasites has not found much success and it has proved more efficacious to use soluble parasite antigens to induce immunity. Highly immunogenic substances have been isolated from a number of parasites, e.g. *Ascaris suum*, *Trichinella spiralis*, *Trichostrongylus colubriformis* and larvae of *Taenia saginata* and *Taenia ovis*. One of the more promising recent development in the quest for suitable anti-parasite vaccines derives from tumour immunology. Following treatment with BCG (*Bacillus Calmette Guérin*) vaccine, non-specific immunity is developed for a wide range of parasites, e.g. experimental success has been achieved with rodent babesias, malarias, *Trypanosoma*, *cruzi*, *Leishamania tropica* and *L. donovani*, *Echinococcus granulous* and *E. multilocularis*. It is remains to be seen whether the potential of non-specific immunization with BCG vaccine and other such substances can be realized as a practical approach in the field.

Pathogenesis of Parasitic Infections

Immunopathology

It is well recognized that the pathology due to infection with many species of parasite is related to the immunological of the host to the parasite. Immunopathology is classified into four subclasses, that together constitute the allergic mechanisms often responsible for the production of disease symptoms.

Type I mechanisms

This class of allergic reactions includes immunopathology due to immediate hypersensitivity, in which parasite antigens react with host tissues that have been passively sensitized by antibodies, thus stimulating the release of pharmacologically active substances, which can, among other effects, increase blood supply and stimulate smooth muscle contraction in the host. The general term *anaphylaxis* is applied to the effects of these substances. Antibodies capable of sensitizing tissues for anaphylaxis are called *homocytotropic* antibodies and belong to the IgE class; IgG can also be involved. Included among the pharmacologically active substances that are released are histamine, serotonin, "slow reacting substances of anaphylaxis" and peptides called kinins (e.g. bradykinin). Immediate hypersensitivity reactions have been identified in a number of parasitic infections including *Leishmania brazililiensis*, *Trichomonas foetus*, and *Trypanosoma cruzi*, as well as in many helminth infections, the latter involving both short-lived (IgG) and long-lived (IgE) homocytotropic antibodies.

The prevalence of immediate hypersensitivity reactions in helminth infections provides a useful immuno-diagnostic tool for the identification of infected individuals, e.g. skin reactions, of the wheal and flare type, are often regarded as diagnostic for *Ascaris*, *Trichinella*, *Echinococcus*, *Schistosoma*, hookworms, filarial worms and many others.

In ascariasis, pathological conditions, including bronchitis, bronchial asthma and pulmonary eosinophilia, are associated with type I reactions. In trichinosis, the reactions include fever, skin, rash, oedema and eosinophilia. Penetration of schistosome cercariae into sensitized mammalian skin is accompanied by urticaria ("nettle rash"), subcutaneous oedema, leucocytosis, eosinophilia and, occasionally bronchial asthma. Infection with filarial worms is characterized by eosinophilia while local swellings arise a result of worm migration. Type I reactions, occurring as they do in individuals with a previous experience of an infection (necessary to cause sensitization), are probably not involved directly in protective immunity. They may, however, affect the growth and development of parasites and the ability of the parasite to migrate within the host.

Type II mechanisms

This class is of reactions involves the, usually, lytic effects of antibodies against antigens that are fixed to cell or basement membranes. The antigen may be an integral part of the membrane or it may simply be adsorbed to it. Complement and lymphocytes are both involved in type II reactions. It is thought that haemolytic anemia in malaria results from this class of allergic reactions, as indeed may the anaemias associated with schistosomiasis, trypanosomiasis, leishmaniasis and babesiasis.

Type III mechanisms

Immunopathology associated with this class of reactions is either inflammatory (arthus) or systemic (serum sickness), both of which result from the formation of antigen-antibody complexes. Complement is also involved in type III reactions. Arthus lesions are characterized by oedema, infiltration of neutrophils, necrosis of arteriole and venule walls, and thrombin containing platelets and leucocytes. Serum sickness, due, for example, to the treatment of humans with animal serum, is rare nowadays. Pathological processes associated with animal serum, is rare nowadays. Pathological processes associated with this class of events include glomerulonephritis in malaria and schistsomiasis, and the nephrotic lesions in some malarias.

Type IV mechanisms

These reactions include cell-mediated responses or delayed hypersensitivity. Delayed skin reactions have been demonstrated in human trypanosomiasis, leishmaniasis, toxoplasmosis, trichomoniasis, ascariasis, toxocariasis and schistosomiasis mansoni. In laboratory animals, many parasites induce delayed hypersensitivity reactions. The sores developed in cutaneous leishmaniasis (*Leishmania donovani*) and hepatosplenic disease (due to the formation of granulomata around tissue-bound eggs) in schistosomiasis are classical examples of type IV reactions.

General Pathology

Apart from the immunopathology that we have discussed briefly above, many parasites cause damage to their hosts, resulting in the production of disease symptoms. Protozoan parasites multiply within the final host, whereas helminths do not. Therefore, any density-dependent effects of the former in the process of pathogenesis will depend upon the rare of division or reproduction and in the latter on the rate of invasion, longevity and size of the parasites. These factors are of particular importance in pathogenesis. It must be made clear, however, that a great many parasites do not, under natural conditions, appear to cause their hosts any ill effects, i.e. an undetermined number of wild animals may be parasitized without ever showing outward or overt physiological signs of infection.

It is rare, in the author's experience, to find unparasitized fishes, amphibians, birds or small mammals, yet these animals are living, until capture, apparently in harmony with their parasite fauna and it seems probable that it is only under certain circumstances that disease conditions arise from an infection. Such conditions would include the intensive monoculture of fishes, domestic fowl or cattle. Hence the view that a parasite is, by definition, a disease-causing organism is patently not true, and we must exercise care when defining the pathogenic nature of parasites, since only a relatively small number of species are disease-causing in natural environments. It has been an understandable tendency among parasitologists to focus attention on the pathogens and relegate the innocuous species to a lower order of significance. Nevertheless man, domestic and wild animals can be seriously affected by their parasite fauna. The ways in which pathology can arise are varied and are briefly outlined below.

Mechanical injury

There are a number of different ways in which parasites inflict mechanical injury to their hosts, e.g. during penetration and migration through the tissues, during feeding, through the action of the organs of attachment and cellular injury wrought by the presence of intracellular parasites within host cells. Many of the skin-invading digeneans produce localized damage at the site of entry, along with allergic reactions. Mechanical damage due to invasion and subsequent migration is often related to the number of invading larvae. The feeding habits of helminths, particularly nematodes, are responsible for considerable local damage to host epithelia, e.g. hookworms bite into the intestinal mucosa to obtain a blood and tissue meal and in doing so may leave open wounds with resultant bleeding and loss of tissue fluids. Mechanical damage attributable to parasite attachment organs is usually associated with the digeneans and cestodes.

The unarmed tapeworm scolex, lacking hooks (e.g. Pseudophyllidea) may or may not inflict damage to the host mucosal tissue, whereas the presence of hooks on the scolex tends to be more injurious. In general, attachment of tapeworms to the intestinal mucosa is associated with inflammatory lesions and characteristic fibrosis. Digeneans, such as the liver flukes, produce pathological conditions by the mechanical disruption of host tissue, attributable to the body spines, attachment organs and the physical presence of the parasite in the narrow ducts of the liver. Anemia, resulting from the ingestion of host red blood cells (e.g. *Schistosoma mansoni*, *Fasciola hepatica* and *Haemoncus contortus*) or from haemolysis (malaria) is not a common form of mechanical injury. Heavy infections of duct-dwelling parasites may cause occlusion of the ducts occupied, with its associated pathology, e.g. *Fasciola* in the bile duct, malaria in hepatic venules, and cestodes in the intestine. The severity of mechanical injury due to parasitic infection will usually be directly related to the number of parasites present, the size and behaviour of the parasite, or its multiplication rate if it is a protozoan.

Toxic effects

The most potent toxins known to man are produced by bacteria and include tetanus, diphtheria and botulism. By contrast, the majority of parasites do not release toxic substances into the blood and tissues of their hosts. Although toxins are thought to the produced by *Ascaris*, some digeneans (*Schistosoma*), some cestodes (*Moniezia*) and several protozoans (*Plasmodium* and *Sarcocystis*), there is no unequivocal evidence to support this notion. In malaria, for example, a toxin is postulated to inhibit mitochondrial respiration and oxidative phosphorylation in host liver cells, but this substance has not been isolated or identified.

Effects on host cell growth

Several parasites are responsible for inducing altercations in host cell and tissue growth patterns. Such effects are most commonly hyperplastic, i.e. cell growth without cell proliferation, or neoplastic, i.e. new cell growth, as in tumour production. Hyperplasia accompanies a number of infections, including *Eimeria*, *Schistosoma haematobium*,

Paragonimus westermani, *Fasciola hepatica* and many adult tapeworm infections. Neoplasia, or tumour development, has been described for infections with *Taenia taeniaformis* in rodents, and *Schistosoma haematobium* in man.

Effects on the reproductive system of the host

There are a small number of well documented cases in which the host's reproductive system is altered, or its tissue destroyed, by the presence of a parasitic infection. The general term for this phenomenon is *parasitic castration*. The best known examples are *Sacculina parasitica* (a cirripede crustacean) that causes sex reversal in male crabs and sterility in females, and *Ligula intestinalis* (a pseudophyllidean cestode plerocercoid) which causes gonad regression and sterility in cyprinid fishes. In neither of these examples is the mechanism of castration understood. It seems probable that the parasites release compounds that mimic the activity of the host sex hormones, but extensive investigation has proved negative to date. The larvae of some digeneans are also known to castrate their molluscan intermediate hosts, e.g. *Himasthla secunda*, *Cercaria emaculens*, and *Cercaria lophocerca* in *Littorina littorea*; *Cercaria milfordensis* in *Mytilus edulis*; and *Leucochloridium paradoxum* in *Succinea putris*. The mechanisms involved here are also known.

Metabolic effects

Many parasites may bring about altercations in the various metabolic pathways of their hosts, but few cases have been documented. Among the best known are the pernicious anemia and vitamin B_{12} deficiency characteristic of some human infections with the cestode *Diphyllobothrium latum*, and the growth-stimulating effects of larval *Spirometra mansonoides* (Pseudophyllidea) on laboratory rats and mice. In the latter case, the growth effects is due to the parasite releasing a substance (known as sparganum growth factor) that mimics the activities of mammalian growth hormone.

This has questionable significance, however, since the larval parasite, the sparganum or plerocercoid, normally infects water snakes and is grown in rodents as a laboratory model only. Similar growth-stimulating effects are found in natural infections of molluscs with larval digeneans, and the phenomenon of molluscan parasitic gigantism is well-known. Molluscs parasitized by some larval digeneans may show the reverse effect, i.e. stunting. Explanations for these interactions are not yet forthcoming.

9

INTERACTION IN PLANTS

Plant parasitism by nematodes is confined to three orders: Dorylaimida (class Adenophorea), Tylenchida and Aphelenchida (class Secernentea). There is vast difference in the nature of parasitism by genera and species of these orders. Parasitism is established when the nematode pierces the cell wall and starts ingesting cell contents. The injection of saliva during feeding may only help in ingestion of food or it may additionally induce changes in the host. The differences between nematode groups in their parasitism are seen during this process.

ORIGIN AND EVOLUTION OF PLANT PARASITISM

Plant parasitism in nematodes is multiple in origin and may have evolved from free-living, stylet-bearing soil nematodes. There is no evidence to suggest any major role of adaptation in the evolution of parasitism. Although local adaptation may possibly be responsible for some differences as, for example, the local races of a nematode breaking resistance of a resistant host variety, there is little possibility of a major pest evolving through adaptation. This can be attributed to a low number of generations per year through which the more specialized parasite passes. Two main trends in the development of plant parasitism among nematodes are considered.

First, there is a trend from ectoparasitism to migratory endoparasitism to sedentary endoparasitism. With the development of sedentary endoparasitism, which is restricted to the females, there is also morphological adaptation leading to a high rate or reproduction. Second, there is a trend from indiscriminate feeding on the epidermis and cortex with accompanying cell necrosis and tissue damage to descrete feeding sites with specific plant responses and a suppression of necrosis. Exceptions are genera like *Ditylenchus* and *Aphelenchoides* which are both ecto- and endoparasites.

Some ectoparasites become semiendoparasites but the host response remains primitive, that is, necrosis occurs. It appears that the plant parasites belonging to the order Dorylaimida, especially the Trichodoridae (*Trichodorus*, *Paratrichodorus*), are most primitive forms followed by other Dorylaimids (*Longidorus*, *Xiphinema*). The Tylenchids are more highly evolved and in this group the sedentary parasites of roots and the cereal seed gall nematode, *Anguina*, can be considered as most highly evolved parasites. They establish good synchronization between their life cycle and that of the host and often have an elaborate system of deriving nutrients from the host.

Proof of Pathogenicity

To establish that a particular nematode causes a particular disease, the Koch's postulates must be proved. The four postulates for proof of pathogenicity are:

1. Reproduction of the disease by inoculation with a pure culture of the organism.
2. Reisolation of the organism from the inoculated diseased host, and identification of it with the original inoculant.
3. Constant association of the suspected organism with this disease.
4. Isolation association of the suspected organism with the disease.

The Koch's postulates were originally developed for diseases caused by bacteria which were easy to isolate and grow on synthetic media. Literal application of these postulates to nematode disease is difficult for two main reasons. One reason is that plant parasitic nematodes generally cannot be grown on artificial media. The second reason is that in many cases the nematode forms a complex with other plant pathogens (fungi, bacteria and viruses). Since the second postulate cannot be proved the third naturally becomes inoperative. However, since the second law simply stresses the necessity of obtaining aseptic monospecific inoculum, methods have been developed to separate the nematodes from infested plants, free them from contaminants, grow a host in aseptic conditions and inoculate it with the nematode. The inoculation should result in the specific disease symptoms unless association of other organism is essential for the symptoms.

Mechanisms Involved in Parasitism

The phytonematodes use their stylet to puncture the epidermis and the cell walls and for displacing the cells while moving intercellulary.

During feeding and causing mechanical damage with their stylet, they secrete hydrolytic enzymes from their esophageal glands which dissolve cell walls or act as digestive enzymes. Most common enzymes in nematodes are cellulase, protease, and amylase. Enzymes detected in nematode secretions and homogenates and listed below:

Pectinmethylesterase:	*Ditylenchus triformis, D. myceliophagus, D. dipsaci* (alfalfa race)
Non-specific esterases:	*Meloidogyne hapla, M. javanica*:
Alkaline phosphatase:	*Meloidogyne* species.
Acid phosphatase:	*D. triformis, Meloidogyne*, spp., *Tylenchulus* semipenetrans.
Amylase:	*Aphelenchoides sacchari, T. semipenetrans, D. dipsaci, D. destructor, D. triformis, Heterodera schachtii, Globodera rostochiensis, Meloidogyne, javanica.*
Cellulase	*Aphelenchoides avenae, A. sacchari, D. dipsaci, D. myceliophagus, D. triformis, Radopholus similis, Helicotylenchus nannus, Pratylenchus penetrans, P. zeae, Heterodera trifolii, H. schachtii, M. arenaria, M incognita, T. semipenetrans.*
Polygalacturonase	*A. avenae, P. zeae, D. dipsaci, D. myceliophagus, D. destructor.*
Invertase	*A. sacchari, M. arenaria, R. similis, P. penetrans, P. redivivus.*
B-glucosidase:	*Aphelenchoides fragariae, P. penetrans, Globodera rostochiensis.*
B-galactosidase:	*Globodera rostochiensis*
Pectinase:	*D. dipsaci, D. destructor, R. similis, P. penetrans, M. arenaria, M. hapla.*
Proteinase:	*G. rostochiensis, H. schachtii, D. dipsaci, D. destructor, D. allii, D. triformis, A. ritzemabosi, Pratylenchus redivivus.*
Chitinase:	*D. dipsaci, D. myceliophagus, D. destructor.*

The enzymes not only dissolve cell walls and help in ingestion and digestion of food, they include metabolic changes in the host also. Enzyme activities in nematodes are influenced by parasitic habits. In migratory endoparasite *Pratylenchus penetrans* the cellulase activity is seven times more than in the sedentary endoparasite *Heterodera trifolii*.

The root parasitic, nematodes, whether primitive or highly evolved in the mode of parasitism, establish a modified feeding site by disrupting or stimulating normal root cell function to satisfy their nutritional demands for growth and reproduction. Two main strategies to meet these demands are the creation of large volumes of cytoplasm near the feeding site and the modification of this cytoplasm for the efficient withdrawal of nutrient. The establishment of specific feeding sites is closely linked with feeding strategies.

Parasitism of the Dorylaimida

The phytonematodes of this group are migratory root ectoparasites which either browse along the root surface (*Trichodorus*, *Paratrichodorus*), or feed for longer periods at specific sites from deeper root tissues (*Longidorus*, *Xiphinema*). Apart from causing direct damage to roots, these nematodes are of great economic importance as vectors of several plant viruses. Species of *Trichodorus* and *Paratrichodorus* transmit Tobraviruses or NETU (nematode transmitted tubular viruses) whereas *Longidorus* and *Xiphinema* transmit the Nepoviruses (nematode transmitted polyhedral viruses).

Trichodorid nematodes

These are also known as stubby root nematodes. The members of Trichodoridae usually aggregate near the tips of rapidly growing roots where they often feed gregariously on epidermal cells and root hair, occasionally on subepidermal cells. When root growth is retarded due to massive destruction of the epidermal layer, the parasitism is transferred to the apical region of the root. Roots that stop growing lose their attraction. The nematode then moves away in search of new feeding sites, usually emerging in lateral rots which when heavily attacked give rise to the characteristic stubby root symptoms. In all Trichodorid nematodes the stylet is a ventrally curved mural tooth with a solid tip.

Obviously, this is only to ensure piercing of cell wall, not for food ingestion. The nutrients are removed from the cell through a self-made feeding tube composed of hardened saliva. This feeding tube is left in the host cell after each feeding cycle which takes only a few minutes. After achieving perforation of cell wall with rapid stylet thrusts, the nematode first inserts only the tip of the stylet repeatedly through the hole during salivation. The saliva hardens to form the feeding the tube along the stylet. Saliva is injected into the cytoplasm also and a massive accumulation of cytoplasm occurs at the perforation site. The host nucleus is drawn to this aggregation. It swells and soon becomes optically empty. When sufficient cytoplasm has accumulated at the perforation site, the nematode starts making deeper thrusts of stylet (with feeding tube) for ingestion of the accumulated cytoplasm. Probably the tonoplast is also ruptured and cytoplasm and cell sap mix together. This liquefaction of cell contents facilities ingestion through the feeding tube. Cell death soon becomes evident by the coagulation of the remaining cytoplasm by action of enzymes.

Longidorid nematodes

These nematodes are relatively of large size and possess a hollow axial stylet, about 100 to 300 microns long. Thus, they can feed deep within the plant root. The two parasitic members, *Xiphinema* (dagger nematodes) and *Longidorus*, differ in their parasitic specificity. Some species of *Xiphinema* may feed for many hours or even days from cells with the differentiated vascular cylinder while others like *X. index* and *X. diversicaudatum* prefer root tips as their feeding site. Terminal galls on root tips are seen in the attack of *X. index*. These galls appear to be attractive for the nematode and provide nutrients for egg production. The galls contain much enlarge multinucleate cells which arise by synchronous mitosis without cytokinesis. These cells are metabolically highly active and seem to be indispensable for reproduction of the nematode.

The non-host plants react by necrosis of the cells. The modification of cells or the necrotic reaction are triggered by enzymes in the saliva of the nematode. Species of *Longidorus* attack only root tips which show terminal galls as in the attack of *Xiphinema*. In contrast to the latter, which in most cases, feed for several minutes at a time from several layers within the root tip galls. *Longidorus* species insert their stylet nearly fully protected into the root tips before they start salivation and feeding at the same spot for several hours. Hypertrophy of root tip cells in the initial stage of gall formation is followed by hyperplasia and then by secondary hypertrophy at the actual feeding site. However, the root tip modification induced by these species is not uniform for the whole genus. Some species induce formation of multinucleate cells in root tips similar to those in the attack of *Xiphinema*.

Parasitism of the Tylenchida

The Tylenchids are more advanced than Dorylaimids. They show great diversity in parasitic behaviour and adaptation which culminate in sedentary endoparasitism among the root parasites (*Meloidogyne*, *Globodera*, *Heterodera*, *Tylenchulus*) as well as in some parasites of aerial parts (*Anguina*).

Root parasites

The types of parasitism and related modifications amongst the root parasitic Tylenchida are given below:

Migratory ectoparasitism

The genera *Tylenchus*, *Tylenchorhyncus*, *Paratylenchus*, *Hemicycliophora* and *Belonolaimus* are important members of this group. These nematodes remain vermiform throughout their life. *Tylenchus* species and some species of *Tylenchorhynchus*, with their rather short stylet feed from surface cells for a few minutes or longer. *Pratylenchus* species feed for many hours or even days at the same spot on epidermal cells or root hair without inflicting any apparent damage. Some are equipped with long stylet for food withdrawal from deeper tissues along the root (*Belonolaimus*) or from root tips (*Hemicycliophora*). The root tips are transformed into terminal galls.

Sedentary ectoparasitism

Some genera of the family Paratylenchidae (family Criconematidae, subfamily Paratylenchinae) such as Cacopaurus and *Glacilacus* show sedentary ectoparasitism. These nematodes possess long stylets and feed from deeper tissues of the root. The adult females become slightly swollen and remain attached (sessile) to the feeding site.

Migratory ect-endoparasitism

The nematodes remain vermiform. Not only the stylet but the entire head region of the body is inserted into the cortical cells from which nutrients are removed (species of *Macroposthonia*, *Rotylenchus*, *Seutellonema*).

Migratory endoparasitism

This type of parasitism is common in members of the family Pratylenchidae (or subfamily Pratylenchinae of the family Hoplolaimidae) such as in *Pratylenchus* (lesion nematode), *Radopholus* (burrowing nematode), and *Hirschmanniella* (rice nematode). The nematodes remain vermiform and may invade or leave the roots at any stage of their development. They usually feed cortical cells which are then destroyed by intracellular migration of the nematodes. In the attack of the root lesion nematode, *Pratylenchus*, species, the consistent host response is necrosis which is both due to mechanical damage to the cells as well chemical damage during feeding. In peach roots attacked by *P. penetrans* it has been shown that the enzyme β-glucosidase secreted by the nematode in its saliva hydrolyses amygdalin in the root cells and releases phytotoxic compounds benzaldehyde and hydrogen cyanide. These compounds cause necrosis of tissues. The burrowing nematode, *Radopholus*, penetrates the epidermis and moves through the rot cortex by apparent dissolution of the cells which are ingested by the nematode. The nematode, thus, forms tunnels and cavities in the cortex and stele. A large number of them can be seen accumulated in the phloem-cambium ring which is often destroyed by necrosis. The females lay eggs in the cavities.

Sedentary endoparasitism

The nematodes showing this type of parasitism are considered highly evolved root parasites. They invade the root, partially or completely, at a definite developmental stage, for example, as second stage juveniles in *Meloidogyne*, *Heterodera* and *Globodera* or as young females as in *Rotylenchulus reniformis*. The invading nematodes migrate to specific sites where they induce the formation of highly specialized nurse cell systems. These cells provide a continuous source of nutrients until the parasite dies. Female nematodes become saccate, remain completely inside the host tissue or their posterior portion may partly protrude on the outside, and usually produce several hundred eggs. Males may never feed as in *Tylenchulus semipenetrans* and *Rotylenchulus reniformis* or feed at an early juvenile stage as in root knot and cyst nematodes. In *Meloidogyne*, the second stage juveniles enter roots and move quickly through the region or cell elongation and establish themselves to a permanent feeding site usually with the head adjacent to or protruding into the vascular cylinder. The body of the nematode lies entirely within the cortex in various positions but usually parallel to the root axis and close to the stele with the anterior end pointed away from the root tip. After sometime they become sedentary and feeding is limited to a few cells that constitute the nurse cell system.

A similar pattern is seen in *Heterodera* and *Globodera* but while the root knot nematodes induce gall formation the cyst nematodes do not produce galls although they also feed from a system of nurse cells. In *Tylenchulus semipenetrans*, the second stage juveniles penetrate epidermal and hypodermal cells and feed on the hypodermis and other layers of cortical parenchyma. They also establish a nurse cell system in the cortex and the developing young females remain attached to this site.

In *Rotylenchulus reniformis*, young females penetrate the roots some distance away from the root tip. They move inter and intracellularly to the feeding site which may be cortex as in cowpea, phloem as in cotton and castor roots, or the pericycle as in some dicotyledonous crops. Whatever is the feeding site the nematodes induce a nurse cell system for their nutrition. In the above five kinds of root parasitism specialized nurse cell systems are found only in the sedentary endoparasites. In this group also the different genera vary in their histopathological effects. These are described later. Parasitism by some other phytonematodes not included in the above five types in given below:

Leaf, Bulb, and Stem parasites

Parasitism by anguina

The genus includes species that are fully mobile ectoparasites for a part of their life and then become endoparasites and lose mobility. They induce formation of organization galls and exhibit good integration of their life cycle with that of their host plants, thus showing characters of most highly evolved nematode parasites of aerial parts of plants. The galls produced in leaves, steams and inflorescences are highly organized structures geared primarily to provide resistance to desiccation. There is no evidence to suggest that the gall tissue provide nutrients to the nematode. The galls are small, pronounced, localized swellings arising from the leaves, on the stem, or in the inflorescence. Internally they contain several layers of cells.

Parasitism by ditylenchus

The stem nematodes inhibit the intercellular spaces of parenchymatous and cortical tissues of leaves and stem where the enzymes injected by them cause hypertrophy and hyperplasia of cells resulting in swellings. These swellings are spongy and soft. The middle lamella is dissolved (digested) by pectic enzymes of the nematode and cells become spherical. Large air spaces are seen in the swellings. The swellings and giant cells induced by *D. dipsaci* are similar to galls formed by the root knot nematode. Dissolution of middlc lamella is associated with reproduction of the nematode. In resistant varieties necrosis develops around the invading nematode and it dies or fails to mature. There is no cell separation and hypertrophy.

Parasitism by aphelenchoides

The leaf and bud nematodes, *Aphelenchoides* spp., feed either internally or externally upon the leaves of their hosts. All stages of their development are seen in leaf tissues. They are fully mobile in their adult stage.

In the infection of chrysanthemums by *A. ritzemabosi* the first reaction of the leaf is the production of patches of an opaque brown substance within the cells near the nematodes. The browning then becomes more extensive and the number of chloroplasts in the cells is reduced. Eventually this browning extends to the epidermis; few chloroplasts remain and cells in the mesophyll are ruptured producing large cavities. During necrosis the mesophyll and palisade tissues are equally affected but only in the final stage of decay do the vascular bundles and collenchyma break down. In mature leaves there is no hypertrophy or hyperplasia but in young leaves such reactions probably occur. The browning is attributed to chlorogenic and isochlorogenic acid.

Development of Nurse Cell Systems of Sedentary Root Endoparasites

This aspect of pathogenesis has been reviewed by many workers. The nurse cell systems of the sedentary endoparasites have, as a common feature, a high degree of metabolic activity. There is pronounced increase in cytoplasmic density in ribosomes, polyribosomes and cell organelles. Nuclei are modified to increase the capacity for nuclear-cytoplasmic exchanges. Once introduced the nurse cell systems continue to grow and are probably maintained by the nutrient sink situation imposed by the continuous nutrient demand of the developing and reproducing parasite.

Metabolic activity is most pronounced when the females start laying eggs. At this stage the nurse cells are packed with cytoplasm. When egg laying is complete, there are signs of disintegration of the system. Walls of the nurse cells, during active metabolic condition, especially those adjacent to the vascular elements (xylem) exhibit remarkable ingrowth leading to amplification of plasmalemma. These are "transfer cells" meant for short distance transport of solutes to the area of nutrien sink. When ingrowths are not present, solutes move through plasmodesmata, especially in the pit fields. The nurse cell systems are nematode specific. These have been classified into four categories:

(a) Groups of uninucleate, not enlarged, descrete parenchyma cells located in the cortex or the vascular cylinder near the mouth of the parasite. These are seen in the infection of *Tylenchulus semipenetrans*.

(b) The nurse cell is a single, uninucleate giant cell. It is usually initiated in the pericycle and expands as it matures into the vascular cylinder which may become greatly distorted. Transport of solutes may occur through cell wall ingrowth but more commonly through plasmodesmata in the numerous pit fields.

(c) Probably the most common type of nurse cell system develops by the expansion of several cells at the feeding site. Wall expansion is accompanied by partial wall dissolution, mainly at pit fields where the wall is thinnest. This causes mixing of contents of several cells and a multinucleate syncytium is established. Mitosis is not stimulated in the syncytia. The nematode feeds from cell which draws nutrients from neighbouring cells. The syncytia induced *Rotylenchulus reniformis* are confined to the pericycle. Beyond the syncytial cells proximity to the nematode, other pericycle cells are metabolically stimulated but they remain discrete and uninucleate. Such syncytial nurse cell system is also seen in *Naccobus*, *Globodera*, *Heterodera* and *Atalodera* but the method of solute transfer differs. Extensive transfer cells with wall ingrowths around the xylem elements are common in *Globodera* and *Heterodera*. In others solute transfer occurs through plasmodesmata in pit fields adjacent to xylem vessels.

(d) The fourth category comprises nurse cell systems that consist of a variable number of discrete multinucleate giant cells which develop within the differentiating vascular cylinder by the expansion of individual cells accompanied by synchronous mitosis without cytokinesis. Wall ingrowths develop adjacent to vascular elements for nutrient transfer. Such nurse cells are known to be induced and maintained by only *Meloidogyne* species which also induce copious gall formation around the infection site hence called root knot.

The nurse cell system not only profoundly affect morphology and physiology of the root, they also increase the susceptibility of the plant to other soil-borne fungal and bacterial pathogens.

Some Biochemical Aspects of Parasitism by Sedentary Root Endoparasites

Numerous reports have been published on biological changes associated with infection by sedentary root endoparasites. Growth regulators both in the affected tissue and in the parasite, amino acids, proteins, nucleic acids, and phenolic compounds have been studied. As has been stated earlier the nurse cells are metabolically very high active. In syncytia there is intense RNA and DNA synthesis. Walls of the giant cells contain cellulose and pectin but no lignin, suberin or ninhydrin positive substances. The protoplasm contains carbohydrates, fat, RNA and large amounts of proteins. The gall tissues contain increased amounts of cellulose, sugars, keto, acids, free amino acids, nucleic acids, phosphorus, and nitrogen as compared to healthy tissue but starch is absent. The mechanism of gall formation in *Meloidogyne* infections has not been thoroughly investigated. Probably growth regulators play an important role. Changes caused by some nematode species in cells of a susceptible host are similar to those caused by exogenous indole acetic acid, that is, hypertrophy, hyperplasia, adventitious roots, nuclear division without cell division, and the break down of the cell wall.

Probably, the mechanism of auxin action depends upon the induction of RNA, synthesis. Indole compounds detected in galls caused by *M. javanica*, *M. hapla* and *M. incognita* are: indole-3-acetic acid (IAA), indole-3-acetic acid ethylester (IAE), indole-3-acetonitrile (IAN) and indole-3-butryic acid (IBA). Two possible source of these compounds in the roots galls have been suggested. First, nematodes inject through their saliva the enzymes glycosidase and protease into the host cells and release free auxins from the complexes in the host. The proteases breaking down proteins may also release tryptophan, the IAA precursor, and also amino acids such as phenylalanine, alanine, histidine, and serine, which promote auxin synthesis. The second possibility is that the nematode itself releases auxins during feeding. Perhaps, indole compounds are formed in nematodes as end products of metabolism and are excreted by endoparasites into the plant tissue or by ectoparasites into the root region. IAA, IAN, IAE, and IBA have been detected in larvae and egg masses of *Meloidogyne* species. IAA in larvae of *Heterodera schachtii* and indole-3 acetic acid methylester (IAM) in larvae of *D. dipsaci* have also been reported.

Host Specificity

Phytonematodes are always specialized parasites, that is, they can reproduce only when they feed on certain kinds of plants. The range of such plants may be very large for such nematodes as *D. dipsaci* and *Meloidogyne incognita* or very restricted as in *G. rostochiensis* and *H. avenae*. While *H. schachtii* has several hundred known hosts, *H. glycines*, a closely related species on soybean, has less than 20 hosts. The hosts of nematode species are often botanically related but many nematodes such as *Meloidogyne* species have hosts in several hundred plant species belonging to widely separated genera and families. The factors which determine whether a plant will be a suitable host for a nematode or not are largely unknown.

Generally, the structural make up of the plant does not affect nematodes. There is evidence that some plants are not hosts or are even immune to attack of nematodes because they contain or produce toxins that inhibit nematode development. Others possibly lack some chemical essential for the nutrition of the parasite. The fact that nematodes can penetrate even non-hosts or resistant hosts but fail to develop and reproduce indicates that host specificity is determined to some factor(s) within the plant. However, in potato cyst nematode it appears that specificity is influenced by host chemicals external to the plant since in *Globodera rostochiensis* only the hatching factor from the host roots can stimulate larval emergence from cysts. On the other hand, root diffusates from even non-host wild beet plants stimulate larval emergence in *H. schachtii*. Root exudates or emanations that often attract nematodes to roots are non-specific.

Resistance of Susceptibility

In the successful host-parasite relationship enters the host, feeds, metabolizes, grows and produces. These steps are accompanied by alterations in host morphology and physiology that manifest in disease symptoms, damage or injury, and loss. When the host does not restrict these processes and the parasite is free to carry on its life activities in or on the host, the latter is susceptible. If the host puts restrictions to any of these activities is shows signs of resistance. When vital activities of the parasite such as growth and reproduction are checked the host is considered resistant, and if all steps mentioned above are prevented the plant is immune. Thus, from susceptibility to immunity there can be many grades of resistance. Resistance of plants to nematodes has been variously interpreted as "lack of ability or failure of juveniles to enter the host", or as "any characteristic of plant or any interaction between host and parasite which retards or prevents the occurrence of parasitic relationship between the nematode and its host.

Veech defines resistance as the absence or inhibition of disease development upon challenge by a pathogenic agent. In other words, anything that prevents, retards, or restricts disease development contributes to host resistance. Feeding and the resultant growth and reproduction of the nematode parasite are the main criteria to decide resistance and susceptibility. Based on the degree of plant resistance one can group the plants into different categories according to ability of the nematode parasite to reproduce on them. Taylor had given the following convenient and practical classification:

1. Moderately resistant: plants in which reproduction is 10 to 25 per cent of that in susceptible plants.
2. Very resistant: reproduction is one to 10 per cent of the susceptible plants.
3. Highly resistant: plants in which reproduction is less than one per cent of that in the susceptible plants.
4. Immune: plants in which no reproduction of the nematode occurs.
5. Susceptible: plants in which nematode reproduction is normal.

6. Slightly resistant: plants in which reproduction is about 25 to 50 per cent of that in susceptible plants.

In slightly or moderately resistant plants the life cycle of the nematode may be longer than in susceptible plants; while the resistant or highly resistant plants only a small proportion of the nematodes may complete a life cycle or produce eggs. Nematodes may enter an immune plant and pass through several stages of development but they do not produce fertile eggs. Crop plants may also be *tolerant* to nematode invasion. Tolerance is not related to resistance in genetic terms. The plants are able to make satisfactory growth and produce a fair yield even though they come under susceptible category. In the root-knot disease such plants may form only small galls or the roots system is so copious that it compensates for the root damage caused by the nematode.

Mechanism of resistance

Mechanisms of resistance to nematodes are complex and involve both the host and the nematode parasite. The mechanisms may operate from outside the host, on surface of the host, or within the host. There is no role of physical and chemical barriers in the host in resistance. The plants may contain or produce toxins that kill the nematodes. These have been sometimes called the pre-existing-nematode compounds which provide resistance outside or within the host. *Asparagus officinalis* var. Mary Washington is resistant to the ectoparasite *Trichodorus christiei*.

On young plants the nematodes are able to feed but populations decline at the same rate as in follows soil. In the presence of older plants in which fleshy storage roots have formed, the population decline is more rapid than in fallow soil. Obviously, the nematotoxic principle develops in the fleshy roots and is released in the soil. Other host plant roots in the vicinity of asparagus roots are also protected. The roots contain a water soluble glycoside with a low molecular weight aglycone which has been found to kill nematodes. In addition to *T. christiei*, the compound is toxic to many other nematodes such as *Pratylenchus penetrans*, *Tylenchorhynchus claytoni*, *Helicotylenchus nannus*, *Paratylencus projectus*, *Belonolaimus gracilis* and some species of *Meloidogyne*.

The effect of the toxic compound is related to the function of acetylcholinesterase in the nematode nervous system which is neutralized. A similar role of French and African marigold (*Tagetes patula* and *T. erecta*) has been proved against several nematode species.

Root knot nematodes fail to induce giant cell formation in marigold roots. When grown between tomato and potato, marigold has been shown in protect these crops against root knot. Used as a cover crop, marigold is reported to give a high degree of control of *Pratylenchus penetrans* and *Tylenchorhynchus dubius* but other nematodes such as *Paratylenchus* and *Criconemoides* are not effected. Two nematicidal compounds have been isolated from marigold roots. These are alphaterthienyl and a derivative of bithienyl.

The resistance and nematicidal effect of *Eragrostis curvula* against root knot nematodes is due to the presence of high level of pyrocatechol in roots. There are many graminaceous plants including millets that disinfest the soil. However, the concentration of the toxic compounds from their roots at which they can produce the nematicidal effect can be reached only after a prolonged stand of the plant (three to four months). It was thought that since many nematodes are attracted to host roots along a gradient of root exudates or emanations, resistance may be linked to this phenomenon of attraction. However, it is now conclusively proved that in resistant as well as in susceptible plans attraction of an penetration by juveniles occurs almost equally well. It is after penetration that a resistant plant challenges the parasite.

Similarly, there are views of some workers emphasizing that a resistant plant may lack a substance required by the nematode for its development of the substance may be short supply. There is no evidence to prove that some components of the nematode diet are lacking in resistant plants. However, indirectly the resistance is believed to the linked to feeding of the nematode. To grow and reproduce, the nematode must have access to uninterrupted supply of food. Like in most of the advanced parasites the nematodes induce changes in the host plant that enable the nematode to ensure uninterrupted supply of host cell cytoplasm. In resistant plants this part of host reaction is absent or suppressed resulting in interruption of food supply to the nematode which fails to develop and mature. The changes in the host induced by the nematodes are nurse cell systems (sycytium, giant cells) and galls.

Resistance is provided to the plant through failure of these systems. There has been extensive work in recent years to find out the mechanisms that suppress or inactivate the nurse cells and galls. Hypersensitive host reaction has been considered since long as possible mechanism of resistance within the host. The importance of this reaction in resistance has been emphasized in studies of root knot and cyst nematodes. The hypersensitive reaction (necroses) prevents development of syncytium or giant cells or renders them non-functional, thereby interrupting food supply to the nematode. Resistance to root infection by root knot nematode usually manifests itself at an earlier stage than in cyst nematodes.

The development of *Meloidogyne* species is often blocked by a rapid hypersensitive reaction which may be associated with the accumulation of nematostatic phytoalexins at the right time and place. Phenolics are also known to stimulate indoleacetic acid-oxidase activity thereby inducing necrotic reaction. The hypersensitive reaction resulting in formation of a protective necrosis instead of the normal giant cells at the feeding site was first studied by Paulson and Webster. In addition, the reaction may be linked to cyanide-insensitive respiration. In resistant tomato infested with *Meloidogyne incognita* the cyanide sensitive and cyanide insensitive respiration both increase but in susceptible plants there is no alteration in respiration.

The necrosis in cells around the feeding site prevents the nematode to develop into adult females. The juveniles die at an early stage of development. The necrotic response of the plant is genetically controlled. In roots resistant to cyst nematode, the syncytium is usually induced but disintegrates a few days later so that males with limited food requirement develop. Although hypersensitive type of reaction occurs around the body of the juveniles it does not appear to inhibit initiation of feeding sites by the nematode and may not be part of resistance mechanism at this stage. Changes occurring after the initiation of syncytia can be directly implicated in resistance mechanisms.

There are two types of responses after syncytium initiation. The major gene mediated response results in isolation of the developing syncytium by necrosis of the surrounding cells and rapid degeneration of the syncytium while in polygene mediated response the development of transfer cells (wall ingrowths) in inhibited and there is no necrosis. In this case also the syncytium degenerates because in both types of responses flow of nutrients to the syncytium is stopped. The developing nematodes are deprived of food supply resulting in either death or preponderance of males.

Roles of phenolics in resistance

Phenolic compounds are the best known factors involved in susceptible-resistant responses of plants. There is a definite correlation between the degree of plant resistance and the phenolics present in the plant tissue. Most phenols occur in plants in the bound form as glycoside (polyphenols) of low physiological and chemical activity.

Activation requires their decomposition to free phenols. Nematodes are able to do this by secreting β-glycosidase into the host cells. Four kinds of host responses invasion in which phenolics are known to play the main role were suggested by Giebel:

1. IAA-oxidase stimulation favouring auxin decomposition and formation of necroses in plant resistant to sedentary endoparasites.
2. Browning and slow formation of wide necroses in plants susceptible to migratory ectoparasites.
3. IAA-oxidase inhibition thus favouring auxin accumulation and giant cell and gall formation in plants susceptible to sedentary endoparasites.
4. Quick browning and formation of non-expandable necroses in plants resistant to migratory nematodes.

The presence of chlorogenic acid (a phenolic) is thought to the cause of browning and resistant reaction of chrysanthemums to *A. ritzemabosi* and of root knot resistant tobacco and tomato cultivars to *M. incognita*. Artificial introduction of β-glycosidase causes necrosis in potato resistant to *Globodera restochiensis* and formation of syncytium in susceptible potato.

Biological Races in Nematodes

A biological race can be defined as a segment of nematode species differing from the rest of the species in some physiological characteristic such as pathogenicity. In practice, biological race or biotype is usually distinguished by its ability to attack a certain plant usually considered highly resistant or immune to the same nematode species. Many biological races of nematodes have been reported. Some of the most important are those found in root knot nematodes, *Meloidogyne* species, bulb and stem nematode, *Ditylenchus dipsaci*, and the burrowing nematode, *Radopholus similis*. It is however, believed that many races occur in other nematode species such as *H. avenae*, *H. schachtii*, *Ditylenchus radicicola*, etc.

Two biotypes were identified in the golden nematode of potato (*Globodera rostochiensis*). The original race was called biotype. A and the resistance breaking race was called biotype B. The latter is now described as a separate species, *G. pallida*. In *Melaidagyne*, four races of *M. incognita*, two of *M. arenaria*, and one each of *M. javanica* and *M. hapla* have been identified Seinhorst distinguished 11 biologic races in *D. dipsaci* on the basis of their pathogenicity on nine plant species. In *R. similis* two races are known. One parasitises banana but not citrus (the banana race) and the other parasitises banana as well as citrus (the citrus race). The biologic races are important in breeding for resistance to nematodes and in deciding proper crop sequence in rotations for nematode control. They are also useful to nematologists who rely to some extent on the pathological responses of the plant to identify the nematodes. Race identification is based on pathogenicity of populations on different hosts.

Symptoms of Nematode Infection

A successful host-parasite relationship results in appearance of symptoms on the host. Symptoms of nematode infestation are mostly non-specific. Poor growth of the plant stunting, patchiness of the crop stand, and discoloured foliage are such abnormalities that can be attributed to many other causes such as attack of fungi, bacteria, and viruses as well as soil factors including nutritional disorders and unsuitable soil physical conditions. While some nematodes are primary pathogens, others, by feeding on the host plant, may allow other pathogens or secondary invaders which cause the damage and visible symptoms. In many diseases, there is clear association between nematodes and other pathogens.

Above ground symptoms

1. *Discolouration of foliage:* Leaf chlorosis often accompanies stunted growth of crops and die-back in fruit trees. Yellowing in indicative of nutritional deficiency caused by root destruction and immobility of nutrients from root to shoot. The discolouration may range from light yellow to deep red, purple, or even black. Sometimes foliar symptoms are very specific. *Aphelenchoides besseyi* causes pale yellow or white tips or white tips or rice leaves.
2. *Reduced growth:* Stunting and slow growth are common symptoms of nematode attack. Such plants appear in patches due to non-random distribution of nematodes in the field or introduction of nematode infested seed in nematode-free soil. Stunted growth is common in the attack of *Pratylenchus penetrans* and *P. pratensis* in different crops, *Globodera rostochiensis* in potato and *H. glycines* in soybean. In fruit trees stunting is accompanied by decline and die-back such as in the infestation of citrus by *R. similis* and *T. semipenetrans*. Generally, stunted growth occurs when nematode populations are high. Sometimes, low populations enhance plant growth.
3. *Distorted and abnormal growth:* Distortion of aerial parts is caused by the stem and leaf feeding nematodes such as *Ditylenchus*. *Aphelenchoides*, and Anguina species. These distortions include

swellings on leaves and stems, twisting of leaves and replacement of grain with cockles.

Below ground symptoms

1. *Root galls and cysts*: Galling of roots is the most characteristic symptom of attack of root knot nematodes. However, many others nematodes such as *Xiphinema* and *G. rostochiensis* also cause swellings on underground parts of their hosts. In the attack of cyst nematodes, the presence of white to brown cysts projecting on the root surface is a characteristic symptom.
2. *Reduced root system*: Depletion of the root system is the most common below ground symptom of nematode attack. *Trichodorus* spp. and *Tylenchorhynchus claytoni* inhibit root growth without producing other recognizable symptoms. Short stubby and malformed roots are symptoms of attack of *Trichodorus*, *Belonolaimus gracilis*, *Xiphinema diversicaudatum*, etc. The roots may show lesions as in the attack of *Pratylenchus* and *Radopholus*.
3. *Root proliferation*: Some species of nematodes do not cause general decay of roots. The injury induces the plant to grow more roots, in clusters, especially behind the damaged portion. *H. avenae*, *G. rostochiensis*, *H. glycines* and *H. schachtii* on cereals, potato, soybean, and sugar-beet, respectively, as well as *Trichodorus christiei*, *Belonolaimus gracilis* and *Meloidogyne* spp. produce such symptoms.

Relationship of Nematodes with Other Plant Pathogens

In the preceding pages it was mentioned that nematodes are often involved in disease complexes in which nematode is the primary pathogen and some fungus or bacterium is the secondary pathogen. Both can cause injury and disease independently but when in association there is synergistic effect and damage is much more along with alteration in symptoms and host-parasite relationship. Apart from synergistic effect, nematodes generally provide avenues for entry to those pathogens which normally cannot penetrate the intact host. In some cases there is obligatory relationship also. This aspect of host-parasite relationship has been reviewed by many nematologists.

Nematode-Fungus Associations

The possibility of involvement of nematodes in aggravation of a fungal disease was hinted as early as in 1892 when it was found that cotton wilt (*Fusarium oxysporum* f.sp *vasinfectum*) was more severe in the presence of the root knot nematodes. Since then considerable attention has been paid to disease complexes involving phytonematodes and fungal parasites of plants.

Association with root pathogens

Sedentary entoparasites

Root knot nematodes are the most extensively studied nematode parasites that influence fungal root diseases. They are known to predispose the host plants to attack of *Fusarium*, *Pythium*, *Phytophthora*, *Rhizoctonia*, *Sclerotium* and *Verticillium*. These fungi are independent root pathogens and can infect the host without the aid of nematodes but when the nematodes are present the effects are aggravated. Of the numerous references describing effects of root knot nematodes on soil-borne fungal parasites of roots, the common in the reference to vascular wilts caused by *Fusarium* spp., especially on varieties resistant to the fungus. The incidence of wilt of cotton (*F. oxysporum* f.sp. *pisi* race 1), sweet potato (*F. oxysporum* f.sp *batatas*) and many others have been found to increase when *Meloidogyne* spp., especially *M. incognita* have invaded the roots.

The increase in wilt incidence is related to the number of nematodes infesting the plant before invasion of the fungus, race of the nematode, and resistance of the host to the fungus, the effect being more pronounced in resistant varieties. It would appear that wounds caused by the nematode are directly responsible for increase in wilt incidence but the interactions are much more complicated and less understood. Propagule counts of *Fusarium* in the rhizosphere of tomato infested by root knot nematode increase and actinomycete counts decrease suggesting changes in root exudates. Root galls of tobacco are much more rapidly attacked by *Fusarium* than healthy roots. Breakdown of resistance or inconsistency in the pattern of inheritance of resistance to wilt are most important effects of root knot nematode invasion. Changes in host physiology due to activities of the nematode are given as a reason for resistance breakdown. The fact that an association with nematodes for sometime is essential for an increased effect of the fungus suggests a change in host physiology.

Resistance to vascular wilt is often related to formation of gel and tyloses (occlusion bodies) in the xylem with which the sedentary endoparasitic nematodes are also closely associated. It is possible that activities of the nematodes upset the mechanism by which the host forms these bodies. *Verticillium albo-artum* and *V. dahliae* also cause vascular wilt but in their cause root knot nematode is not as important as other nematode species. Increased susceptibility of cotton to this fungus in the presence of root knot nematode has, however, been reported.

Black shank, a destructive stem and root disease of tobacco caused by the fungus *Phytophthora nicotianae* var. *nicotianae* is greatly enhanced by earlier invasion of *M. incognita*. The fungus needs no assistance to infect susceptible tobacco cultivars. However, where the interaction between the fungus and the nematode occurs resistant varieties suffer heavy losses. There is extensive colonization of the hypertrophied and hyperplastic regions of the root galls and the fungus even colonizes the anterior and posterior regions of the female nematode body as well as the egg masses. Interaction between *Meloidogyne* and *Rhizoctonia solani* has been reported on cotton, soybean, peas, tobacco, tomato, and okra.

The root knot nematode predisposes the plants to severe damage from certain fungi normally harmless to them. *R. solani* and *Pythium ultimum* are only minor pathogens of tobacco, especially after the plants have grown for some time and have attained tissue maturity. However, if plants, even those approaching maturity, have been previously invaded (about four weeks earlier) by *M. incognita* the two fungi are capable of invading roots and causing rapid necroses. No damage or decay occurs if roots are inoculated with either of the two fungi in the absence of the nematode. Little damage occurs if nematodes and fungi are inoculated simultaneously. Tobacco varieties resistant to root knot do not show this response to the fungi. Tomato plants also respond similarly to such nematode-fungus interactions. In tomato, *R. solani* appears to be more aggressive than *P. ultimum*. The host tissues (galled as well as non-galled) in nematode infested plants become good substrate for colonization by the above two fungi. *Curvularia trifolii*, *Botrytis cinerea*, *Aspergillus ochraceus*, *Penicillium martensii* and *Trichoderma harzianum* are common soil inhabitants not parasitic on tobacco roots.

When any one of these fungi is added to roots that have been previously exposed for three to four weeks to *M. incognita*, invasion is so severe that root destruction may be more than that caused by *R. solani* and *Pythium* under similar conditions. Although *Meloidogyne* and *Heterodera* (*Globodera*) arc similar in their host-parasite relationship except for the nature of nurse cell system, reports of *Heterodera* being involved in root disease complexes with fungi are not many. Some associations between *Heterodera* and species of *Fusarium*, *Thielaviopsis*, *Aphanomyces*, and *Colletotrichum* have been reported without convincing proof of interactions. Miller has reported interaction of *H. tabacum* and Verticillium and Fusarium wilts of tomato. On the other hand, in sugar-beet, when *H. schachtii* and *F. oxysporum* both were present there was antagonistic relationship. The damage was less than what only nematode was present. The complex of *H. schachtii* and *R. solani* (root rot) is sugar-beet is convincingly proved. Giant cells and adjacent areas provide a highly suitable substrate for extensive growth of the fungus.

Migratory endoparasites

The role of nematode enzymes in fungal pathogenesis through change in host physiology was further demonstrated in the *Ditylenchus dipsacibotrytis allii* complex of onion collar rot. When water containing enzymes excreted by the nematode was injected into the plant in the presence of the fungus 100 per cent plants showed collar rot against 50 per cent when only the fungus was present or when only plain water was injected. There is evidence that root decay phase of spreading decline disease of citrus (*Radopholus similis*) is accelerated when fungal pathogens such as *F. solani*, *R. solani*, *S. rolfsii* and *Thielaviopsis basicola* are present. In banana wilt (*F. oxysporum* f.sp. *cubense*) disease expression is aggravated and incubation period shortened when the plants are invaded earlier by *Radopholus similis*. *Pratylenchus penetrans*, the root lesion nematode, aggravates eggplant wilt (*V. dahliae*), which through a vascular wilt fungus, is not affected by the presence of root knot nematode. In this case the interaction appears be the dependent on root lesson and necrosis rather than the highly specialized host response in the case of root knot. Similar *Pratylenchus Verticillum* associations are reported in tomato, peppermint and *Impatiens balsamina*. Charcoal rot of sorghum (*Macrophomina phaseolina*), *Cylindrocarpon radicicola* on potato, carrot, and red clover, and black shank of tobacco are reported in increase when *Pratylenchus* is present. *Trichoderma* spp. are not important plant pathogens but in the presence of *P. penetrans* they can cause significant damage to alfalfa and celery.

Sedentary semiendoparasites

In some nematode fungus interactions invasion by nematode selectively encourages certain fungal pathogens but not others. The citrus root nematode, *Tylenchulus*, *semipenetrans* aggravates the root decaying action of *F. solani* but does not encourage *F. oxysporum*. Increased severity of cotton wilt (*F. oxysporum* f.sp. *vasinfectum*) and okra root rot (*R. solani*) in the presence of *Rotylenchulus reniformis* has been reported. These studies have also shown that rhizosphere microflora influence the nematode-fungus interaction. The damage caused by *R. reniformis* and *V. dahliae* is less in unsterilized than in sterilized soil.

Ectoparasitic nematodes

Ectoparasitic nematodes feed on cells of the root epidermis or in the outer cortex, or sometimes deeper. *Hoploalimus uniformis* is associated with *F. oxysporum* f.sp. *pisi* in early yellowing of peas. Singly the two pathogens cause only slight discolouration, lesion formation or cracks in the cortex. Jointly, they cause extensive dark brown discolouration along the roots and complete decay and the prior of inoculations. Rice plants infected with *H. indicus* are predisposed to infection of *Sclerotium*. The stunt nematode, *Tylenchorhynchus claytonii* causes only slight increase in black shank disease of tobacco cultivars resistant to the fungus but significantly increases Fusarium wilt in susceptible varieties. The severity of root rot of peas (*Aohanomyces euteiches*) is increased by *T. martini*. The increase is directly related to nematode population.

Saprobiotic nematodes

These nematodes are omnivorous, not feeding on living plants, and are rarely considered accomplices in plant disease complexes. However, some reports point to their indirect role in bacterial and fungal diseases. As their diet these nematodes ingest bacterial cells and fungal spores which remain viable and are protected from deleterious effects of fungicides and nematicides. These nematodes usually seek wounds, lesions or necrotic tissues as sites for feeding or microbes. In this process they deposit the living fungal spores and bacterial cells for infection of plants.

Suppression of nematodes by secondary pathogens

In some interactions, the secondary invaders suppress the primary invaders. It depends on how early the secondary invader gains entry into the host. A gray sterile fungus is reported to reduce penetration of roots by *G. restochiensis*. In such combinations probably the destruction of the feeding sites by the fungus is the cause of suppression of the nematode. Some reports suggests that mycorrhizal fungi hinder penetration of roots by nematode. The bacterium *Psudomonas solanacearum* so heavily colonizes the giant cells induced by *M. incognita* that these feeding sites quickly degenerate and the nematode is starved.

Nematode-Bacterium Associations

Investigations on nematode-bacterium interactions in plant diseases are comparatively fewer then the nematode-fungus interactions. Among root diseases major studies have been carried out with *Pseudomonas solanacearum* which causes bacterial wilt in potato, tomato, egg-plant, banana etc. The root knot nematodes, predispose potato, tomato, and egg-plant in infection by this bacterium. Similar increase in wilt is obtained if roots are cut and exposed to a thick suspension of bacterial cells. This suggests that the interaction is mainly the effect of wounding permitting easier entry of bacteria. In resistant varieties the only path of entry of bacteria is through injuries on roots. However, some reports suggests a role of structural and physiological changes also in these interactions.

Sitaramaiah and Sinha have demonstrated that the egg-plant wilt caused by race 3 of this bacterium is more severe when *M. javanica* is present than when it is absent. The mass severe wilt membrane was observed when the bactenum was introduced two to three weeks after nematode inoculation. The bacterium showed a preference for the giant cells which degenerated following bacterial invasion. *M. hapla* is also reported to increase the incidence of crown gall caused by *Agrobacterium tumefaciens* on raspberry and wilt of alfalfa caused by *Corynebacterium michiganense* subsp. *insidiosum*. Similar interactions between *D. dipsaci* and bacterial wilt of cotton and between *Helicotylenchus nannus* and *Pseudomonas solancearum* is tomato have been period.

Cauliflower disease or fasciation disease of strawberry is a good example of how the nematode-bacterium interaction is essential for full expression of symptoms. In this disease the interactions involves the foliar nematode *Aphelenchoides ritzemabosi* and the bacterium *Rodococcus* (*Corynebacterium*) *fascians*. The nematode alone produces localized feeding sites on the crown buds. When the bacterium is also present, proliferation of axillary buds on affected crown occurs and leaf galls are formed.

It is believed that the nematode acts as a vector of the bacterium to carry it to the crown buds which it could not enter by itself. One of the best studied interactions is the '*tundu*' or yellow slime or eat rot disease of wheat in which the nematode *Anguina tritici* and the bacterium *Calvibacter* (*Corynebacterium*) *tritici* are involved. The nematode causes ear cockle or seen gall disease. The bacterial yellow ear rot cannot develop unless the nematode is also present. Survival of the bacterium is carrying in soil, on surface or inside the galls. The nematode acts as a vector carrying the bacterium as the surface contaminant from the galls. There is involvement of an attractive forces of considerable magnitude between the cuticle surface of the nematode and the bacterial cell well.

Mechanisms Involved in Interactions of Nematodes with Fungal and Bacterial Pathogens

A simple mechanism does not supply to all cases of interpathogen interactions. Each interactions has a mechanism of its own and distinct from another involving a different pathogen or a different host. Broadly, the mechanisms could fall under the following categories:

Wounding by the nematode (primary pathogen)

The mechanism is simple. Where the secondary pathogen (bacterium or weak fungi) is not strong enough to break down mechanical barriers of the host, nematode punctures of the host epidermis not only provide avenues of entry, they also expose a greater root surface area for infection and colonization by the secondary pathogen. Indirectly, the root wounds increase root exudation which may alter the rhizosphere effect and the secondary pathogen may be stimulated or inhibit. Root wounding involves for more tissues than the epidermis alone. When cortical cells are damaged by the nematode, the dead cells or necroses may provide a good substrate for the secondary pathogens. The nematodes may externally carry the propagules of the secondary pathogen to these deeper tissues.

Physiological changes

During pathogenesis the nematode injects into the host cell several enzymes including proteases and β-glycosidases. Breakdown of proteins releases free amino acids the concentration of which increases in the host cell and also in the root exudates. The amino-acid tryptophan is metabolized to the auxin IAA and the resulting hormonal imbalance is not only confined to the feeding sites but also affects farther tissues. The physiological changes may affect pathogenesis by the secondary pathogen in one or more of the following ways:

1. The toxin-oxidizing or inhibiting mechanisms in the plant may be destroyed thus enabling the wilt fungus to break the host resistance.
2. Characteristics of the rhizosphere microflora may change and the microflora antagonistic to the secondary pathogen, such as actinomycetes, may be suppressed. *Streptomyces* are effective antogonists of *Fusarium* species.
3. Tissues affected by nematodes or the nurse cell system and galls formed by nematodes are metabolically highly active containing denser cytoplasm. These tissues provide good substrate for nutrition of the secondary pathogens.
4. Resistance to vascular wilts often lies in the endodermis which is normally avoided by fungi. The endodermis contains large quantities

of antifugal substances like phenols, naphthols, and anthrols which, in later stages, give rise to benzoquinones, naphthoquinones and anthroquinones, compounds highly toxic to fungi and bacteria. The giant cell system of the root knot nematodes originating in the pericycle does not allow proper differentiation of the endodermis. The absence of the endodermis removes the barrier against invasion of xylem by the vascular wilt pathogens.

Interactions between Nematodes and Plant Viruses

Although nematode-virus interactions are better known through the role of nematodes as vectors of soil-borne plant virus, there are reports of associations in which the viruses increase or suppresses nematode populations or results in enhanced damage. The exact mechanism are not known.

Stimulation or suppression of nematodes and enhanced damage

Multiplication of *A. ritzemabosi* and *D. dipsaci* is increased in tobacco infected with arabis mosaic. Tomato mosaic virus increase multiplication of *M. javanica*. The tobacco mosaic virus suppresses population of *A. ritzemabosi* and *D. dipsaci* in tobacco. Increased damage due to *M. incognita* in the presence of leaf curl virus in tomato, maize mosaic virus in maize, potato virus Y in tobacco, and cowpea mosaic virus in cowpea has also been reported.

Nematodes as vectors of plant viruses

Many virus diseases of plants were found to arise even when the plants were raised from clean propagation material and were protected from the visit of insect vectors. It was suspect that such viruses were soil-borne and were transmitted to plants from the soil. Cadman divided such viruses into two groups: those losing infectivity when the soil was allowed to dry at 20°C for a week or more, and those which remained infective on drying to soil for several weeks or years. These differences were found to be related to the transmitting agent, and not to properties of the viruses.

The first group, in which the soil becomes non-infective on drying, contains the viruses transmitted by nematodes. Drying destroys the nematodes and hence no transmission of the virus. The second group contains the viruses transmitted by soil fungi. Drying cold not destroy resting spores of the fungi. Hewitt and associates were the first to positively demonstrate that a nematode could transmit a plant virus when they found that the fanleaf virus of grapevines was transmitted by the nematode *Xiphinema index*. Now, four genera of nematodes, all from the same group, Dorylaimida, in the order Adenophorea, are

Table 9.1. Nematode species transmitting plant viruses

Nematode species	*Plant viruses transmitted*
1	2
Xiphinema americanum	Tobacco ringspot, Tomato ringspot.
X, coxi	Tobacco ringspot, Arabis mosaic (Type strain), Cherry leaf roll.
X diversicaudatum	Arabic mosaic (Type strain and Hop strain), Cherry leaf roll, Strawberry latent ringspot, Brome mosaic and Carnation mosaic.
X index	Grapevine fanleaf.
X italiae	Grapevine fanleaf.
X vuiltenezi	Cherry leaf roll.
Longidorus attenuatus	Tomato black ring (English and German strains).
L. elongatus	Tomato black ring (Scottish strain), Raspberry ringspot (English and Scottish strains), Carnation mosaic.
L macrosoma	Raspberry ringspot (English strain), Brome mosaic, Prunus necrotic ringspot.
Trichodorus cylindricus	Tobacco rattle (European strain).
T. primitivus	Pea early browning (English isolate). Tobacco rattle (European isolate).
T. minor	Tobacco rattle (European isolate).
T. similis	Tobacco rattle (European isolate).
T. viriliferus	Pea early browning (European isolate). Tobacco rattle (European isolate).
Paratrichodorus anemone	Pea early browning (English isolate). Tobacco rattle (European isolate).
P. allius	Tobacco rattle (American isolate).
P. christiei	Tobacco rattle (American isolate).
P. nanus	Tobacco rattle (European isolate).
P. pachydermus	Pea early browning (Dutch isolate), Tobacco rattle (European isolate).
P. porosus	Tobacco rattle (American isolate).
P. Teres	Pea early browning (Dutch isolate). Tobacco rattle (European isolate).

known to be vectors of many plant viruses. These genera are *Xiphinema* and *Longidorus* of family Longidoridae and *Trichodorus* and *Paratrichodorus* of family Trichodoridae. The Longidoridae and *Trichodorus* and *Paratrichodorus* of family Trichodoridae.

The Longidorids are relatively large nematodes and transmit the isometric (polyhedral) NEPO viruses (nematode transmitted polyhedral viruses). The Trichodorids transmit the tubular viruses of the Tobraviruses group, also known as NETU viruses (nematode transmitted tubular viruses). Transmission of viruses by nematodes resembles the transmission by insect vectors. Viruses are picked up from the infected host and injected in the healthy host by the stylet during feeding process. There is ac sition feed time and inoculation feed time and the viruses are retained in the body of the nematode for different durations. Arabis mosaic virus in *X. diversicaudatum* and tomato ringspot virus in *S. americanum* can be acquired after one hour of feeding and transmitted in a similar period of feeding. *S. diversicaudatum* retains arabis mosaic virus for at least 31 days when kept in fallow soil and for at least eight months when kept on virus-immune cultivars of raspberry. The site of virus retention varies in different nematodes.

In *Longidorus* the virus particles are found in the cuticular lining of the guide sheath of odontostylet. In *Xiphinema* the particles accumulate in lumen of odontosty and in the esophagus while Trichodorus has the virus particle in the entire oesophagus. Specificity of the transmission also occurs. Different strains of the same virus are often transmitted by different species of the same nematode genus. For example, *Paratrichodorus christiei* transmits. American isolate of Tobacco rattle virus while the European isolate is transmitted by *P. anemone*. Site of retention or accumulation of the virus in the vector is responsible for some degree of specificity but more important is specific absorption of virus particles by linings of the nematode body.

10

PARASITIC IMMUNITY

When a parasite or other 'foreign' body (e.g. a splinter) invades a host, the latter is able, in some way not understood, to recognize that 'non-self' material is present, and respond accordingly. How this recognition is carried out is not understood, but presumably recognition must occur at the molecular level on or near the surface of susceptible cells and requires contact between the invading material and the recognizing cell. In some cases, possibly associated with a phenomenon termed *molecular mimicry* the host is unable to 'recognize' the parasite as 'non-self' and consequently does not respond to its presences; this phenomenon is discussed later. Vertebrate hosts respond to the presence of 'non-self' material by bringing into action mechanisms which can be broadly grouped into two kinds; these compliment each other to some degree:

(*b*) *specific responses*: these depend on the prior exposure to 'non-self' material and the subsequent recognition and reaction to this material specifically, during later invasions;

(*a*) *non-specific responses*: these are selective to the extent that they can differentiate 'non-self' from 'self' but they are not dependent on *specific* recognition (i.e. of some very specialized type of 'foreign' molecule).

IMMUNITY (RESISTANCE) AND INSUSCEPTIBILITY

It is usual to speak of specific responses as *immune* or *allergic* responses and the field of study incorporating these phenomena as *immunology*. In many situations, these responses do not actually confer immunity on the host, and it has been recommended that responses or reactions should not be designated 'immune' until shown to be concerned with immunity the 'uncommitted word "allergie" is much better and presupposes nothing as to their role or significance'. Ideally, it would be desirable to clarify usage of the terminology in this way, but the terms 'allergic' and 'immune' have been widely used as being synonymous.

The term immune is generally used throughout this text, since the word allergic is derived from the noun *allergy*, often considered to be synonymous to hypersensitivity. In a similar manner, the terms 'immunity' and 'resistance' have been widely used as synonymous, although the former term is now, perhaps, more frequently used. Theoretically, the word 'resistance' implies that the host is not completely successful in repelling invading organisms, whereas 'immunity' suggests complete rejection of such organisms.

Insusceptibility

Although the words 'resistance' and 'immunity' are widely used, in many instances an alternative term, *insusceptibility* may be more appropriate. Insusceptibility implies that certain features related to the morphology, behaviour, ecology, physiology or biochemistry of the host are unsuitable for the establishment of a particular species. For example a parasite with enzymes adapted to a cold-blooded host is unlikely to establish itself in a warm-blooded host. Again, the physiological conditions in a gut—such as pH, PCO_2, PO_2 and Eh—may be such that it provides an unfavourable site in relation to the metabolism of particular species of parasite. In the gut too, the size and shape of the villi and/or the composition of bile may be important. For example, it has been shown that the tegument of the cestode *Echinococcus granulosus*, a parasite of carnivores is rapidly lyzed by deoxycholic acid but not by cholic acid. Since the former occurs in high concentration in herbivore bile and the latter in carnivores, this would be one factor in restricting the host specificity of this parasite.

Williams (1960) has shown that the structure of the gut in fish is an important factor in the host specificity of Tetraphyllidea. Numerous other examples involving behavioural or ecological insusceptibility could be quoted. Advances in our understanding of the immune process evolve with bewildering rapidity and it is not possible here to give more than the most generalized account of the current position. Useful general texts on immunology are those of Bellanti (1971), Gold & Peacock (1970), Humphrey & White (1970), Roitt (1971), Ward (1970); a valuable glossary of immunological terms (and their origins) is also available (Halliday, 1971). Immunity to animal parasites in particular has also been the subject of a major (two volumes) review by Jackson, Herman & Singer (1969—1970) and several symposia.

Natural and Acquired Immunity

If a parasite makes contact with a potentially susceptible host but fails to become established in it, the host is said to possess *natural* or *innate* immunity or resistance to that parasite. If an initial infection induces a degree of resistance to subsequent infections of a parasite, the phenomenon is termed *acquired* immunity or resistance. Acquired immunity may develop in two ways:

(*a*) *artificially*, as a result of the deliberate inoculation (= *vaccination*) of the live or dead organism or some material associated with it (i.e. an extract, homogenate, secretion, excretionor metabolic product). Passive immunity may be induced artificially by injection of antiserum i.e. serum containing antibody.

(*b*) *naturally*, as the result of a natural infection with a parasite of the same (i.e. homologous) species or a different (i.e. heterologous) species. Some degree of immunity in newly born mammals can occur due to transfer of antibodies from the mother to the offspring either via placental passage or in milk.

These various types of immunity are discussed further below.

Cell Types in the Immune System

In vertebrates, the effector system concerned with immune mechanisms is known as the *lymphoid-macrophage* or *lymphoreticular* system. The cellular elements are found throughout the body but especially in the thymus, lymph nodes, spleen, bone marrow and in the respiratory, gastrointestinal and urinogenital systems. The cells chiefly involved are the *macrophages*, *granulocytes*, *plasma cells* and *lymphocytes*; they all share the property of being activated by contact with non-self material. Lymophoctes, in particular, play a central role in the specific immune response and an understanding of their origin and properties is central to an understanding of immunology. In the non-specific response, on the other hand, the phagocyte is the major cell involved.

Non-specific Responses

Although the reactions below are considered under the heading of 'non-specific' responses, they are all capable, under certain conditions of being influenced by, or interacting with, certain components involved in the specific immune response, such as antibody or complement.

Phagocytosis

The first reaction of the host to the invasion of small particles of foreign material, such as bacteria, is to attempt to engulf them by

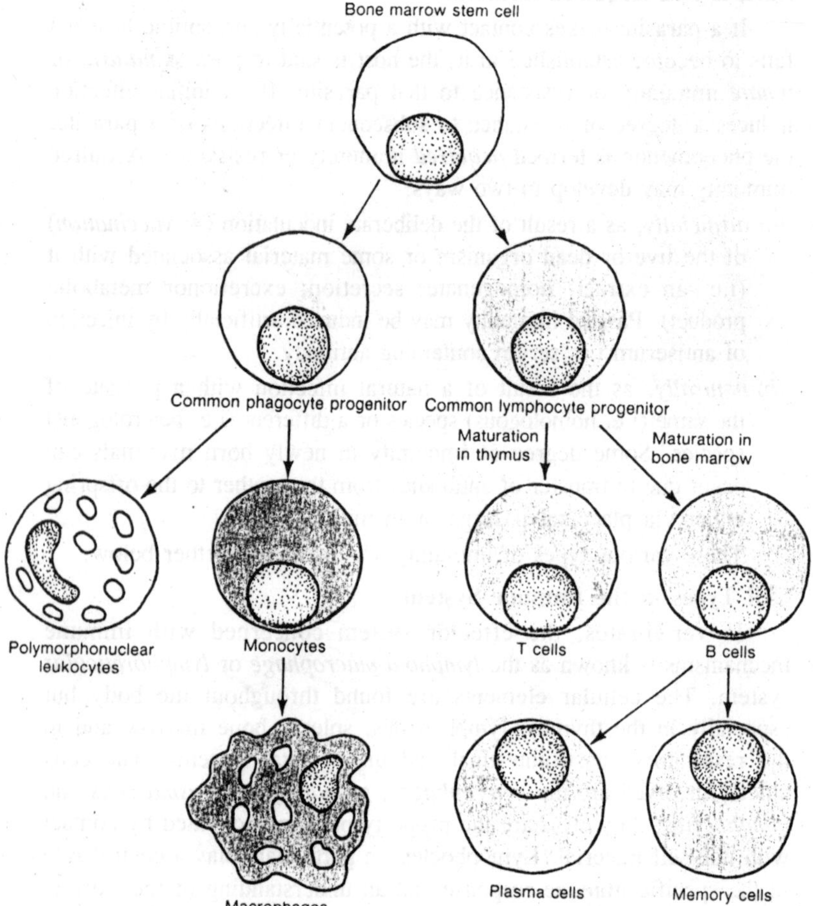

Fig. 10.1. Origin of major cells involved in the immune response. There are two major lines, one generating phagocytic cells, the other generating lymphocytic cells.

phagocytosis. Two types of phagocytic cells are involved in this process, *macrophages* and *polymorphonuclear leucocytes* sometimes referred to as *microphages*. Macrophages in the circulating blood are known as *monocytes*, those in the connective tissues as *histiocytes*, and those in the spleen, thymus and lymph nodes as reticulum cells. *Sinuslining* macrophages occur in liver sinusoids, capillaries etc.

Phagocytes belong to the *reticulo-endothelial system*; the latter is now often called the *lymphoid-macrophage system*. The role of phagocytes is to engulf particulate matter and digest it, where possible, by lysosome action, or if undigestible (e. g. carbon particles) to store it so that it no longer acts as an irritant. Although phagocytoiss can occur in the absence of antibody, the process can be greatly facilitated by the action of *opsonizing antibodies* or *opsonins* (especially, IgG and IgM antibodies) which, in some way not understood, make the antigen more readily ingested.

Phagocytosis is also enhanced by a serum component known as *complement* in the presence of antibody. Macrophages can also pass on antigens or their products to lymphoid cells and thus can act as an important link between the non-specific and specific immune responses.

Inflammatory Response

If the amount of foreign material invading the host is small it may be surrounded by phagocytic cells and gradually immobilized by the deposition of collagenous tissue around it. If the amount is large, the reaction is more severe and *inflammation* occurs. This shows itself as a swelling (*oedema*) of the tissue brought about as a result of a localized dilation of the capillaries (*vasodilation*) which results in an increase in the blood supply to the affected area. This is accompanied by the passage of protein from the blood into the tissue fluids together with polymorphonuclear leucocytes which collect (by chemotaxis), in large numbers, at the invasion site. The onset of vasodilation is brought about by a number of factors which include (*a*) a local nervous reaction and (*b*) the release of an active substance termed *histamine* from mast cells. In addition to histamine, other pharmacologically active substances, 5-hydroxytryptamine (=serotonin in USA), bradykinin and SRS-A (slow reacting substance of anaphylaxis) are released as a result of antigen-

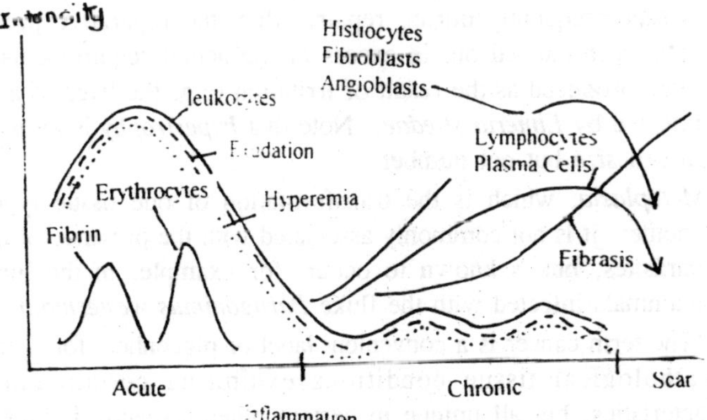

Fig. 10.2. Schematic representation of the tissue reactions to inflammatory stimuli.

antibody reactions. The role of these substances is not fully understood (see *hypersensitivity*). The passage of leucocytes is followed by a migration of lymphocytes, some of which may transform into mononuclear cells or *fibroblasts*. The latter also take part in the formation of the fibrous *capsules* which surround many issue parasites — especially larval helminths (e.g. *Cryptocotyle lingua*, *Trichinella spiralis*). If capsules persist for long periods, calcification may occur.

Abnormal Growth Responses to Parasites

One of the most interesting features of tissue reactions, often associated with the presence of parasites, is a change in the growth pattern of parasitized tissue. This change of pattern may be reflected in a number of ways as follows:

(*a*) *Neoplasia*, which is the growth of new kind of tissue from existing cells, and such growths are commonly referred to as *tumours*. Any definition of a tumour can only be regarded as temporary and provisional. 'A neoplasm or tumour may be defined as a new growth arising from pre-existing tissue, independent of the needs of the organism, and subserving no useful purpose, but on the contrary, often acting deleteriously. It is autonomous, i.e. it follows its own line of growth, and is not regulated by those governing the tissues of the body in which it is growing. . . the life history. . . of a neoplasm appears to have no typical termination, the tumour cells seeming to be endowed with the power of continuous, aimless and unlimited proliferation.' A malignant tumour is popularly referred to as a 'cancer.'

(*b*) *Hyperplasia*, that is an increase in the rate of cell division. In this condition, the cells increase in number but not in size. This is seen frequently during repair, when the reparative processes have been carried out in excess of the actual requirements. It is often produced as the result of irritation (e.g. the liver of a rabbit infected by *Eimeria stiedae*;. Note that *hypertrophy* is an increase in cell *size* but not number.

(*c*) *Metaplasia*, which is the transformation of one tissue type into another. It is not commonly associated with the presence of animal parasites, but is known to occur, for example, in the lungs of mammals infected with the fluke *Paragonimus westermani*.

'The term cancer is a convenient label or pigeonhole for a number of pathological tissue conditions exhibiting certain common characteristics, but all unique in certain other respects. Indeed, it is not a disease at all, but an assemblage of many different diseases which have certain symptoms in common. A century ago, medical men spoke of "fever" as a disease. Now they no longer do so. Something of this sort is happening with cancer during the present century.'

In medical terminology, however, the term 'cancer' is used synonymously for a *carcinoma*, a malignant tumour of epithelial tissue. The term *sarcoma* refers to a malignant tumour of connective tissue. It is significant that helminth parasites are amongst the very diverse agents associated with the production of neoplasia; the literature has been reviewed by Schwabe (1955). A number of species, has been incriminated. Of these, only the larvae of the cestode *Hydatigera taeniaeformis* (*Cysticercus fasciolaris*) in the liver of rodents, and the adults of the nematode *Spirocerca lupi* in the oesophagus of dogs have been unequivocally shown to be associated with sarcoma formation in these sites. The other cases listed are doubtful ones.

The presence of *Schistosoma haematobium* has long been associated with cancer of the bladder in man, and this question has received much attention. The evidence has been reviewed by Jordan & Webbe (1969) who conclude that *S. haematobium* is apparently associated with bladder cancer in some areas of Africa but not in others. It has been suggested that this variation in effect in different areas may be nutritional in origin or possibly related to the absorption of a carcinogenic substance from the different cooking pots used in these areas!

Table 10.1. Parasites believed to be associated with tumour formation (neoplasia) in mammals. Only in the case of those marked has the relationship been unequivocally established.

Group	*Parasite*	*Host*	*Site of tumour*
Protozoa	*Eimeria stiedae*	rabbit	liver
Trematoda	*Schistosoma mansoni*	man	intestine
	Schistosoma haematobium	man	bladder
	Schistosoma japonicum	man	intestine
	Paragonimus westermani	tiger	lung
	Clonorchis sinensis	man	lung
Cestoda	*Hydatigera taeniaeformis*	rats	liver
	Echinococcus granulosus	man	lung
Nematoda	*Gonylonema neoplasticum*	rat	tongue
	Spirocerca lupi	dog	oesophagus

Cysticercus fasciolaris

Hepatic sarcoma in the liver of the rat induced by the presence of the larva of *Hydatigera taeniaeformis* (*Cysticercus fasciolaris*) has received much attention since its initial discovery by Borrel in 1906. To date, more than 7500 cysticercus tumours have been developed experimentally. The majority of these have been sarcomas, but a few have been of other histological types. In general, malignant tumours arise only following an infection of from 250 to 675 days' duration. The tumours appear to arise not from the liver tissue itself, but from the fibrous capsule formed by the host around the larva. Attempts to isolate a specific active agent from *Cysticercus fasciolaris* have been unsuccessful; there is some evidence that it may be associated with the calcareous corpuscles normally present in the body fluid of cysticerci. The extraordinary difficulties inherent in attempts to obtain unequivocal results in experimental work on tumour formation are reflected in the case of *Gogylonema neoplasticum*. This nematode occurs in the tongue, oesphagus and forestomach of rodents. In these sites, it was thought to induce malignant tumours and an intensive research programme was carried out on this organism, especially by Fibiger and his co-workers. Critical experiments have shown that, although this nematode can act as chronic irritant, it does not induce malignancy. It was shown that the so-called maligancy described by early workers was related to the unintentional use, in the original experiments, of Vitamin A deficient diets for the experimental rats used.

The parasite was first encountered in rats in a sugar refinery where the abundance of sugary food available resulted in the natural diet being a vitamin A deficient one, although this was unrecognized at the time. When the infections were established and maintained in laboratory rats, it happened that a vitamin A deficient diet consisting of white bread and water (not milk) was again used. Examination of the records of the millers suggests that the bread used in Norway at that time was deficient in vitamin A. Thus, by a series of coincidences, the abnormal diet conditions occurring in the sugar refinery were essentially reproduced.

Spirocerca lupi

The association of this nematode and the occurrence of sarcomas in dogs has been much studied and presents an interesting problem in host-parasite relationships. The adult worm lives in a nodular mass in the wall of the oesophagus of dogs and other carnivores; there is a small passage from the lumen of the nodule through which eggs pass.

Numerous species of dung beetles act as intermediate hosts and animals such as birds, hedgehogs, lizards, mice and rabbits serve as transport hosts. Infective larvae ingested by a dog are freed in the stomach, penetrate the mucosa and eventually, after a migration, end up in the oesphageal wall. A reactive granuloma develops around the worms and, under conditions not understood, becomes transformed into a sarcoma; the genetic disposition of a dog to the neoplasm is probably an important factor. Research on this intriguing problem has been reviewed by Bailey (1963, 1972).

SPECIFIC IMMUNE RESPONSES: BASIC CONCEPTS OF IMMUNOLOGY

Antigens and Antibodies

In addition to the reaction of phagocytic cells described earlier, the invasion of a vertebrate host by a parasite or other 'foreign' (i.e. non-self) material is associated with a number of serological (or humoral) reactions, so-called because they occur primarily (but not always) in the serum of the infected host. A substance which invokes these reactions is said to be *antigenic* and the molecules it releases or exposes to the host are known as *antigens*. The introduction of antigen provokes the appearance, in the blood and tissue fluids, of a soluble protein entity known as *antibody* which has the property of combining with that specific antigen. The nature of antigens and antibodies is considered further below.

Antigens

Antigens are macromoleules of protein or complex polysaccharides. In order to act as an antigen, a substance must (*a*) have a minimal molecular weight (rarely less than 5000), and (*b*) exhibit 'foreignness', i.e. must contain chemical grouping or configurations which are forcign to the animal. Lipids are not directly antigenic but may act as haptens, i.e. 'partial antigens,' which contain at least one determinant group of an antigen. When coupled to proteins or complex polysaccharides they can dictate specificity of the coupled 'antigen.' The term *allergen* has recently been introduced for antigens which provoke *allergic* reactions (see hypersensitivity).

Antibodies

As mentioned above, the introduction of an antigen results in the appearance in the serum of a soluble protein entity termed *antibody*. Antibody has the property of combining specifically with the antigen which provoked its production, but with that antigen only, so that the molecular structure of antibody is in some way related to that of

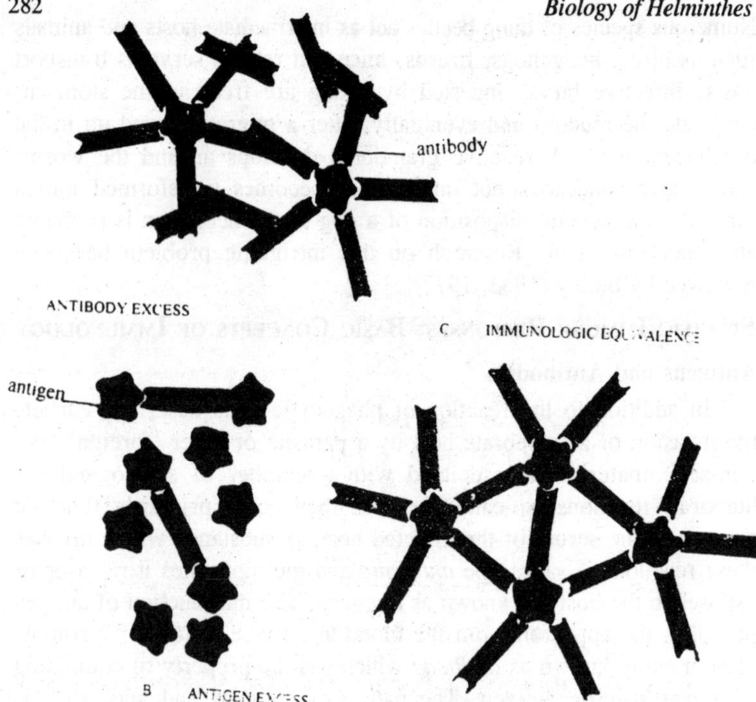

Fig. 10.3. Combination of antigen and antibody in different proportions.

antigen. *In vitro*, the presence of antibodies in serum is revealed by the ability to react with their corresponding antigens. A number of complex antigen-antibody reactions are known and these are discussed later.

Theories of Antibody Synthesis

Although antibody appears to originate in cells of the lymphoid series, or their derivatives (e.g. plasma cells, the method whereby antibodies are actually synthesized is one of the most controversial problems in immunology. Detailed consideration of this problem is beyond the scope of this book; the position may briefly be summarized as follows. There are two main theories: (*a*) the instructive theory, and (*b*) the selective theory; several modifications of these are known.

The instructive theory postulates that antigen enters the site of serum protein synthesis and acts instructively as a template around which an immunoglobulin chain could be moulded. This then takes on a configuration which is complementary to one of the determinant groups of the antigen. On separation from the template, the antibody molecule would thus have a specific antigen-combining site.

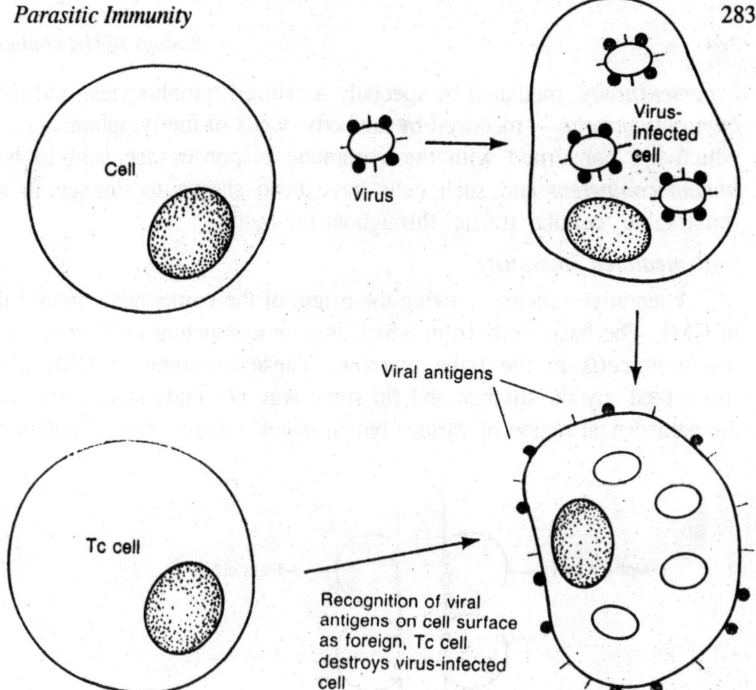

Fig. 10.4. Manner in which a T cell recognize a foreign antigen.

The selective theory was first put forward by Jerne (1955) and later modified by Burnet (1959) who offered a 'clonal selection theory.' This postulastes that before lymphoid cells reach maturity large numbers of variants arise each capable of producing one specific type of antibody. When an antigen enters the body it makes contact with the cell (s) which have the corresponding reactive sites on their surfaces and combines with them. This stimulates the cell to multiply by repeated cell division, producing a clone of similar cells all producing the same antibody. The validity of this theory in the light of modern work is discussed by Gold & Peacock (1970). According to either of these views, for an antigen to stimulate the formation of a reacting antibody, its specific molecular pattern must not be destroyed or altered to any significant extent. This probably explains why some antigens used as vaccines (e.g. bacteria) are only active in a living state, while others may remain active after having been killed by formalin or other treatment.

Cellular Processes in Antibody Formation

Specific immune responses are brought about by two types of effector mechanisms: (*a*) *cell-mediated immunity* (CMI) or *delayed hypersensitivity*, mediated by specially sensitized lymphocytes, and (*b*) *humoral immunity*— mediated by antibody. Cells of the lymphoid series which are concerned with these immune responses are said to be *immunocompetent* and such cells have been shown to engage in a remarkable 'cellular traffic' throughout the body.

Cell mediated immunity

A tentative scheme showing the origin of the components involved in CMI. The basic cells from which immunocompetent cells arise are the *stem cells* in the bone marrow. These involved in CMI are 'processed' by the thymus and (in some way not understood) pass to the paracortical region of various lymph nodes, where, after stimulation

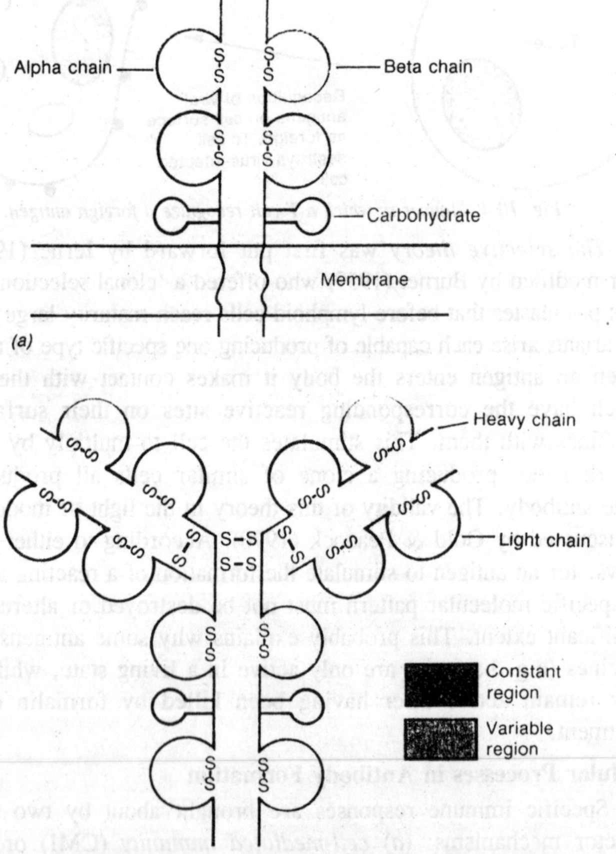

Fig. 10.5. Structural comparison of the T-cell receptor with an immunoglobulin.

by antigen, they proliferate to produce a population of 'sensitized' lymphocytes (thymic dependent cells or *T-cells* or *T-lymphocytes*); these carry surface-bound antibodies directed against the antigen which stimulated their proliferation.

T-lymphocytes are especially active against 'fixed' antigens or substances (such as drugs) which persist and which are themselves not immunogenic but become so through combination with tissue protein. They are thus largely concerned with so-called *allergic* reactions. The cellular events involved in CMI are slow to develop, usually requiring at least 24 hours—hence the expression 'delayed hypersensitivity.' In contrast, humoral reactions (see below) can occur within minutes (*anaphylaxis*) or hours. T-lymphocytes are long lived.

Humoral immunity

The production and secretion of serum antibody is dependent on a population of B-lymphocytes which, like T-lymphocytes, develop from stem cells in bone marrow. These B-lymphocytes develop into cells known as *plasma* cells which in birds are 'processed' by the Bursa of Fabricius, a lymphoid organ connected to the cloaca. Plasma cells are characterized by possessing a well developed rough endoplasmic reticulum and they are actively engaged in the synthesis of protein which is secreted into the bloodstream. The equivalent of the avian Bursa of Fabricius has not been identified in man although it is thought possible that gut-associated lymphoid tissue, such as Peyer's patches, the tonsils and the appendix may act as the processing sites.

Structure and Function of Antibodies

The proteins in plasma may be separated by electrophoresis into five groups—albumin, and four globulins a_1, a_2, b and *y*. Antibodies were formally thought to be associated with the *y* globulin fraction, but this is now recognized to be an oversimplification, for some antibody has been detected in other components also. The more general term *immunoglobulins* is now used to describe the closely related proteins involved in antibody activity. In each vertebrate species, immunoglobulin molecules can be subdivided on the basis of the structure of their 'backbone'. In man, for example, five classes of immunoglobulins have been recognized. These are referred to by the letters G, A, M, D and E after the symbol *y* (indicating their electrophoretic mobility as *y* globulins), e.g. *y*G *or* after the abbreviation Ig (indicating their immunoglobulin function), e.g. IgG; the latter terminology is used here. IgG, IgM and IgA are the major immunoglobulins and may be detected by immunoelectrophoresis.

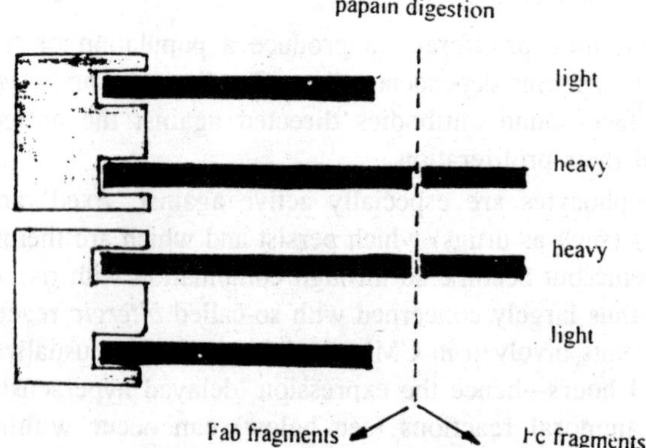

Fig. 10.6. Structure of an immunoglobulin molecule.

Table 10.2. Some properties of human immunoglobulins.

	IgG	*IgA*	*IgM*	*IgD*	*IgE*
Mean serum concentration (mg/100 ml)	1240	280	120	3	0.03
Sedimentation constant ($S_{20,}$ W) (Svedberg units)	6.5-7	7-15	18-19	6.2	8
Molecular weight	150 000	170 000	890 000	150 000	1960 000
Electrophoretic behaviour	γ_2 (slow)	γ_1 (fast)	γ_1 (fast)	γ_1 (fast)	γ_1 (fast)
Heavy chain designation	γ	α	μ	δ	ε
Light chain designation	$K_1\lambda$	$K_1\lambda$	$K_1\lambda$	$K_1\lambda$	$K_1\lambda$
Antibody activity	Crosses placenta	Secretary antibod	Agglutinins	None known	Reaginic antibody
	Complement fixation	Antitoxins	Lysins		
	Anaphylaxis	Some	Opsonins		
	Antitoxins	reagins	Bactericidins		
	'Late' antibody' valence = 2		'Early' antibody; valence = 5		

IgG

This is the most abundant of the immunoglobulins and is believed to provide the bulk of immunity in the tissues to invading organisms including bacteria, viruses and fungi, as well as parasites. It accounts for more than 85 per cent of the total. It has a molecular weight of

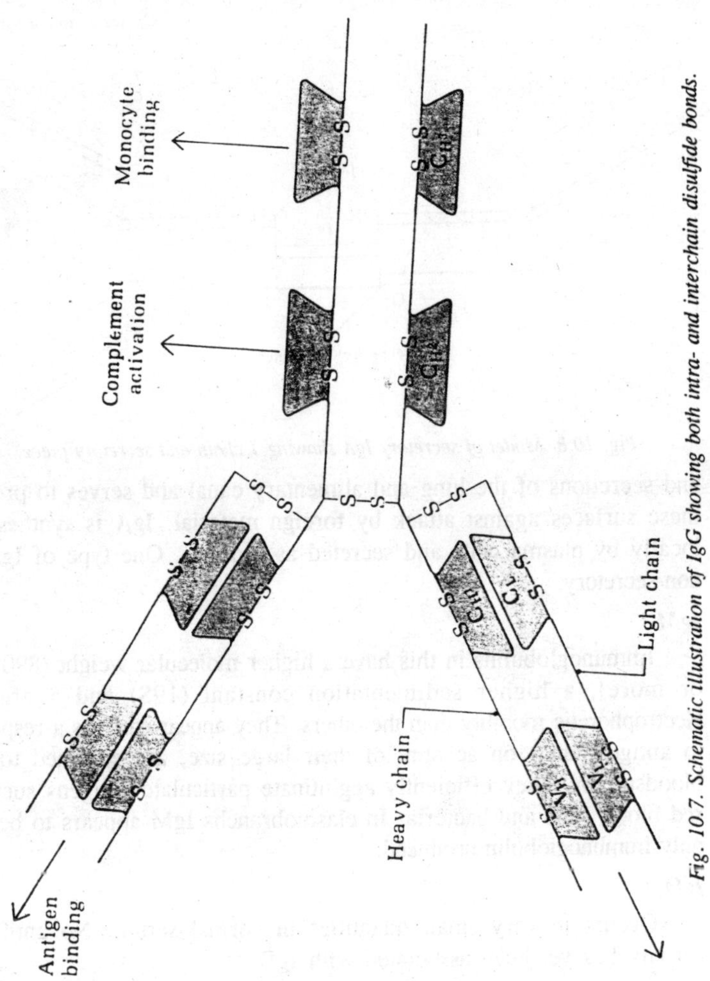

Fig. 10.7. Schematic illustration of IgG showing both intra- and interchain disulfide bonds.

about 150 000 and a sedimentation constant (in Svedberg units) of 7S. It can cross the placenta and can fix antigen. It achieves significant concentration in the internal body fluids, particularly in extravascular sites. Most antibodies concerned in serological tests for helminths occur in this class. The complexes of bacteria with IgG antibody are able to attach to the phagocytic cells because some of these cells have specialized receptors for sites on the heavy chain of the Fc portion of IgG.

IgA

The second most abundant immunoglobulin. It appears especially in the external sero-mucous secretions such as tears, nasal fluids, saliva

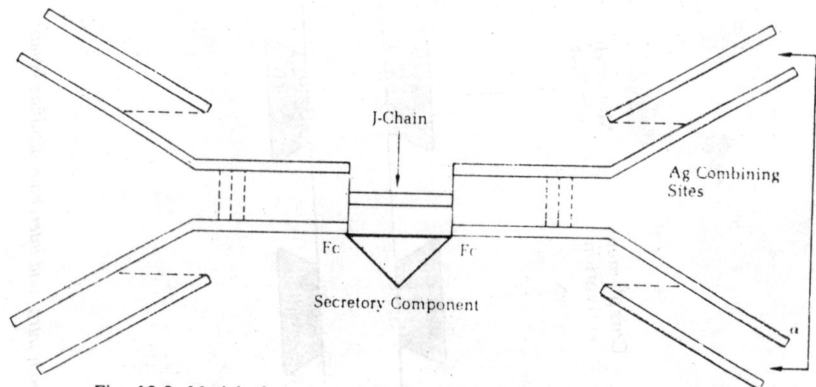

Fig. 10.8. Model of secretory IgA showing J chain and secretory piece.

and secretions of the lung and alimentary canal and serves to protect these surfaces against attack by foreign material. IgA is synthesized locally by plasma cells and secreted as a dimer. One type of IgA is non-secretory.

IgM

Immunoglobulins in this have a higher molecular weight (890 000 or more), a higher sedimentation constant (19S) and a greater electrophoretic mobility than the others. They appear early as a response to antigen, and, on account of their large size, are confined to the bloodstream. They efficiently agglutinate particulate antigens such as red blood cells and bacteria. In elasmobranchs IgM appears to be the only immunoglobulin produced.

IgD

Occurs in very small quantities in normal serum. No antibody activity has yet been associated with IgD.

IgE

Present only in trace amounts in serum; known as *reagin*, *reaginic* or *cytotropic antibody*. It has the ability to remain firmly fixed when injected into human skin, probably being bound to mast cells. Contact with antigen leads to release of mast cell granules (degranulation with release of vasoactive amines, e.g. histamine hypersensitivity reaction, Type III. It has been suggested that the release of histamine initiated by parasites coming in contact with mast-cell bound IgE antibody facilitates the rejection of parasites (e.g. *Nippostrongylus brasiliensis*, by the gut. Reagins probably play an important role in the resistance to several helminths.

Subclasses

Immunochemical analysis of the above main classes of immunoglobulins has revealed that they can be further divided into subclasses such as IgG_1, IgG2 etc. Serum of an immunized animal (containing antibodies) is termed *antiserum* or less aptly *immune* serum and if some of this serum is injected into a nonimmunized animal, this second animal also will usually exhibit immunity. Immunity produced by injection of antiserum is termed *passive immunity* since the recipient animal has not taken part in the development of the immune state. Duration of passive immunity is short, perhaps a few weeks, and depends partly on the amount of antibody injected. In contrast, the kind of immunity developed after the injection of antigen is termed *active immunity*, since the infected animal body is itself actively responsible for its production. The degree and duration of active immunity depends on a number of factors; in some it may last for years; in others it may wane sharply after the antigen has disappeared.

ANTIBODY-ANTIGEN REACTIONS

General Principles

In vivo. At the cellular level, combination of antibody and antigen may cause (*a*) agglutination (clumping), or lysis of cells containing antigen or (*b*) more ready ingestion of such cells or antigens by phagocytes or (*c*) damage by the antigenatibody complexes forming precipitates.

In vitro. Many of the above reactions can better be studied *in vitro*. For laboratory methods and detailed descriptions of these reactions see Campbell *et al.* (1963), Crowle (1961), WHO/UNESCO (1964), Ackroyd (1964), Weir (1973) and Kwapinski (1972). The interactions between antibodies and antigens are. Antigens (multivalent) combine with the antibodies (generally bivalent) to form three-dimensional lattices which aggregate and precipitate. However, (at least *in vitro*) the proportions in which antibodies and antigens are mixed have a marked effect on the composition of the resulting precipitates:

(*a*) *Immunological equivalence.* The combining sites are used to the full and virtually all the antigen and antibody precipitate together and neither can be detected in the supernatant.

(*b*) *Excess antibody.* The resultant precipitate contains more antibody than antigen; the complexes formed are insoluble.

(*c*) *Excess antigen.* Resultant precipitate contains more antigen than antibody and the precipitate tends to dissolve due to formation of solube complexes.

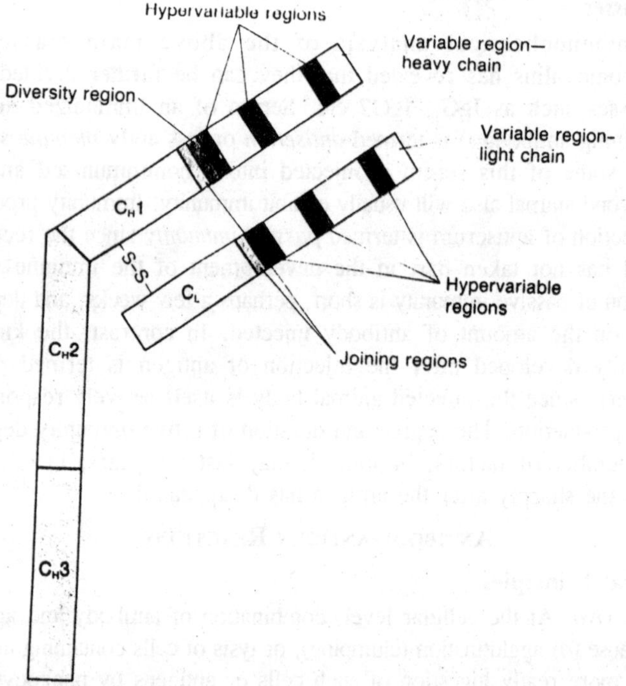

Fig. 10.9. Detailed structure of variable regions of immunoglobulin light and heavy chains.

Tests for antigen and antibody

(a) *Precipitation.* In the simplest test, antigen is carefully overlayed on the antiserum in a tube. A white line of precipitate forms at the interface if an antibody-antigen reaction has taken place.

(b) *Agglutination.* The basic mechanism of agglutination is essentially the same as that of precipitation except that the antigens are larger and either particulate or cellular (e.g. bacteria, yeasts, erythrocytes). Agglutination of erythrocytes is known as *haemagglutination*. The use of the haemagglutination reaction has been extended to produce a delicate test for antibody in serum. Erythrocytes are treated with tannic acid after which antigens can be absorbed on their surface and used to test for the presence of antibody. Other particles—such as polystyrene, latex, sephadex or bentonite can similarly be used as 'carriers' and coated with antigen. Particles or cells treated in this way are exceedingly sensitive and can detect as little as 0.003μg of antibody (measured as nitrogen).

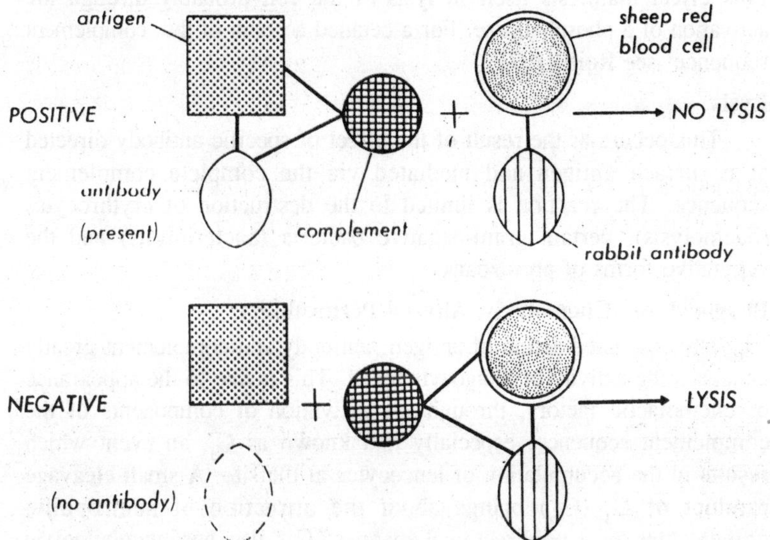

Fig. 10.10. The complement-fixation reaction.

(c) *Titre*. Antibody concentration is expressed as the *titre* of a serum sample. Titre is the reciprocal of the highest dilution of serum which gives a detectable antibody-antigen reaction and is expressed as the number of (arbitrary) antibody units per unit volume of undiluted serum.

Complement-dependent Reactions

The term *complement* was originally applied to a component of normal serum which, when acting upon an antibody-coated cell, led to lysis of that cell (cytolysis). This activity is destroyed by heating serum to 56°C for 30 minutes. Guinea-pig serum is particularly rich in complement. Reactions which are largely dependent on complement include cytolysis (cytotoxicity), phagocytosis-promoting activity (opsonization), chemotaxis, and altered permeability. Although 'complement' was originally thought to be one substance, it is now known to consist of at least nine functional entities (or eleven discrete proteins) which act in sequence.

When complement is first 'activated' by an immune complex (e.g. antibody bound to a red cell) it develops the ability to activate the next component, which in turn activates the next and so on. The process involves a series of complex reactions. Hence the 'triggering' of a single C_1 molecule can lead to activation of many thousands of the lat components, the whole process thus producing an *amplification* effect. This effect manifests itself in lysis of the cell probably through the activation of a phospholipase. For a detailed account of the 'complement sequence' see Roitt (1971).

Lysis

This occurs as the result of the effect of specific antibody directed at a surface antigen and mediated via the complete complement sequence. The reaction is limited to the destruction of erythrocytes (haemolysis), certain gram-negative bacteria (bacteriolysis) and the vegetative forms of protozoans.

Phagocytosis, Chemotaxis, Altered Permeability

The combined action of antigen, antibody and complement greatly enhances the activity of phagocytic cells. This is due to the appearance of chemotactic factors, through the activation of components of the complement sequence, especially that known as C_3, an event which results in the accumulation of leucocytes at the site. A small cleavage product of C_3 (C_{3a}) brings about the attraction of neutrophilic granulocytes in a unidirectional manner. C_{3a} also has *anaphylatoxin* activity which results in histamine release from mast cells. As well as causing smooth muscle contraction, histamine release results in increased vascular permeability. The overall effect of the latter, together with neutrophil reaction, is to produce *acute inflammation*. Phagocytosis is also enhanced by the action of *opsonins*— antibodies which adhere to the surface of micro-organisms which are thus said to be 'opsonized.'

Complement Fixation

Antigen-antibody complexes have the ability to 'fix' or bind complement and this may be used as a test for antibodies using a known antigen or vice versa. A haemolytic system must be used to make the effect visual. Thus, serum (inactivated by heating to destroy its own complement) which is suspected of containing a particular antibody is mixed with the corresponding parasite antigen and added to fresh guinea-pig serum (as a complement source). 'Sensitized' sheep red blood cells (i.e. those mixed with (heated) antiserum prepared against them in a rabbit) are added, and if haemolysis occurs, the complement level has remained high and the serum is therefore negative for parasite antibody. If no haemolysis occurs, the complement has been bound by the parasite antigen antibody complex and the serum under test is positive for antibodies to the parasite.

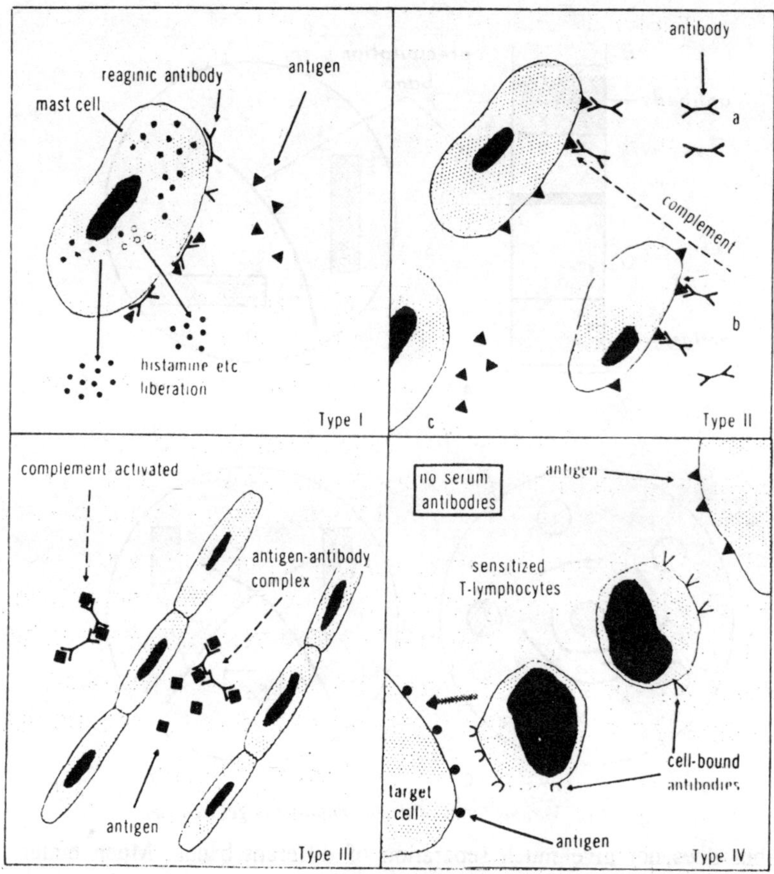

Fig. 10.11. The four chief types of hypersensitivity reactions; Type V omitted. Sites of involvement of complement are indicated by dotted lines.

Immunodiffusion: In vitro Detection of Antibodies and Antigens

Double diffusion. Detection of antibody and antigen has been greatly facilitated by the fact that the precipitation reaction between them can be visulaized in gels, usually agar. In immunodiffusion techniques, antigen and antibody are allowed to diffuse towards each other in agar and – according to the number of types of antibodies and antigens present – one or more precipitation bands are formed.

In the simplest techniques, the antibody and antigen are mixed with agar, placed in tubes separated by a further layer of pure agar, and allowed to diffuse towards each other through the pure agar. Precipitation band(s) form at the interface. This method is sensitive

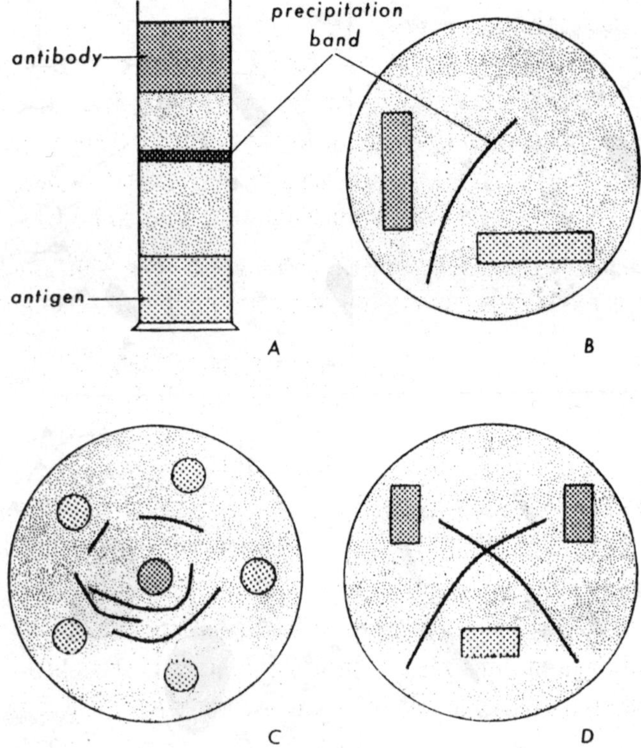

Fig. 10.12. Various forms of double diffusion-in-gel techniques.

but does not give much separation of different bands. Much better 'resolution' is obtained in plates, in which antigen and antibody are allowed to diffuse towards each other from troughs or wells cut in agar. Such plates are known as Ouchterlony plates, after the inventor.

Immunoelectrophoresis

In immunoelectrophoresis, the antigen, placed in a circular hole in agar, is subjected to electrophoresis. The components of the antigen separate at rates depending on the net charge in their molecules. At the alkaline pH's used, the proteins, which carry a negative charge, might be expected to move towards the anode. However, agar itself carries a negative charge and since it is not able to move towards the anode, pressure (endosmosis) on the liquid in the gel makes the protein fraction move in the other direction towards the cathode (endosmosis). Antibody is then placed in the trough and allowed to diffuse into the agar towards the separated antigen components. Where antigen and antibody meet, precipitation arcs develop.

Hypersensitivity

In addition to immunity, an antigen-antibody reaction may give rise to a state of *hypersensitivity*; this term is widely used as being synonymous with *allergy*. Hypersensitivity is essentially a heightened response to an antigen, in an animal which has already been immunologically primed, or *sensitized* to that antigen. Thus, a guinea-pig injected with a small quantity of ovalbumin becomes 'sensitized' to that substance. If a further challenging dose of ovalbumin is given, 2—3 weeks later, the animal suffers a violent general systemic reaction (*anaphylaxis*) which may result in death within a few minutes. This would be an example of *immediate hypersensitivity*. By restricting the antigen to a small area of skin, local hypersensitivity may be produced. Intadermal reactions based on this procedure are of value in the diagnosis of certain diseases of helminth origin.

Table 10.3. Comparison between humoral and cell-mediated immunity.

	Humoral	*Cellular*
Sensitizing material	proteins, polysaccharides, lipids	proteins or protein-hapten
Reaction time'	immediate: minutes to hours	'delayed': 24-48 hours
Initiating event	union of antibody with antigen	Reaction of 'sensitized' lymphocytes with antigen
Transfer	circulating antibody in in serum	cells and cell derived 'transfer' factor
Effector mediators	complement; vasoactive amines (e.g. histamine)	secreted soluble factors (lymphokines)

Other hypersensitivity reactions in man include asthma, hayfever, urticaria, eczema or (more rarely) anaphylaxis. 'Delayed' hypersensitivity (= cell mediated immunity) has already been dealt and its main characteristics are mentioned briefly below. One fundamental difference between immediate and delayed hypersensitivity is that whereas immediate hypersensitivity (a humoral reaction) can be transferred with serum antibodies (which can be detected by the common serological techniques of precipitin tests, complement fixation etc.), delayed hypersensitivity (since it is a cell mediated reaction) is transferable only with *lymphoid cells* or their derivatives. Hypersensitivity is a complex subject, and the principles on which it operates can only be summarized briefly here. For detailed consideration see texts such as Roitt (1971), Bellanti (1971) and Gold & Peacock (1970). The following five types of hypersensitivity have been recognized:

Type I Anaphylactic-type hypersensitivity (anaphylaxis)

Antigen reacts with a special type of antibody known as *reaginic* (*homocytotropic*) antibody bound to mast cells or circulating basophils. This results in degranulation of the mast cells and release of vasoactive amines (e.g. histamine). SRS-A (slow reacting substance of Anaphylaxis) is also released during anaphylaxis, but its origin is unknown. Reaginic antibodies are characterized by their high affinity and permanent attachment to skin. They are best tested for by utilizing their high affinity for cells in *passive cutaneous anaphylaxis* (PCA) with several days' interval between intradermal injection of antibody and intravenous injection of antigen. Serum from an infected animal is injected into the skin of a normal, parasite-free animal of the same species.

Several days later, a mixture of antigen and Evans blue is injected intravenously. Within 20-30 minutes a discrete blue area 10-20 mm in diameter appears at the initial site of injection in the case of a positive serum. In this way, reaginic antibodies are distinguished from other antibodies which do not have an affinity for cells and which diffuse away from the site of injection. Reaginic antibodies appear to be characteristic of certain helminth infections, e.g. *Nippostronylus brasiliensis*.

Type II Cytotoxic-type hypersensitivity

In this type, either phagocytosis or lysis of a cell takes place. This is due to the action of antibodies (IgG or IgM) against antigens on the surface of target cells (e.g. blood cells). Lysis takes place when the operation of the full complement system can occur.

Type III Complex-mediated hypersensitivity

In this type, complexes or microprecipitates formed between antigen and humoral antibodies can lead to (*a*) cell damage, (*b*) activation of the complement system, (*c*) platelet aggregation and (*d*) activation of the Hageman factor. A series of complex reactions follows including vasoactive amine release, polymorph attraction and activation of the *kinin* system. The latter involves the release of small peptides known as kinins with potent pharmacological properties. Kinins have been known to be released in various *Trypanosoma* infections. Examples of Type III hypersensitivity are Arthus' reaction, serum sickness and malarial nephrosis.

Type IV Cell-mediated (delayed type) hypersensitivity

As already discussed in this type, T-lymphocytes (thymus 'processed'), with specific antibody on their surface, on contact with antigen, are stimulated to release factors which induce delayed-type hypersensitivity. A specific inflammatory reaction develops in the area where the antigen is localized. The reaction takes 24-48 hours-hence the term 'delayed.'

Type V Stimulatory hypersensitivity

Antibodies directed against certain cell surface components may stimulate cellular activity.

Hypersensitivity reactions in parasite diagnosis

Intradermal reactions may be useful in the diagnosis of certain diseases of helminth origin. The test is carried out by injecting an extract of parasite or placing some powdered parasite on a small area of scarified skin. A positive result is indicated if a weal develops in the test region within 10—20 minutes.

Immunity in Lower Vertebrates

The immune response in lower vertebrates has been relatively little investigated; Rowlands (1969) has reviewed the literature in this field. In general, antibodies take longer to appear after the antigen invasion, but persist longer, than in mammals. Their development is also very dependent on temperature. Only a very brief account is given here. The phylogeny of immunity has been reviewed by Smith *et al.* (1969).

Amphibia

The immune response and immunoglobulins in amphibia more nearly approach those of mammals than fish. Two immunoglobulins (18S and 6.7S) have been identified.

Reptilia

Humoral antibodies have been little studied 18S, 11S, 7S and 4S components of serum have been identified.

Fish

Lampreys possess immunoglobulins and are capable of antibody response. This antibody activity appears to be confined to immunoglobulins with sedimentation coefficients between 17S and 8S. Higher fish have more complex immune systems and immunoglobulins have been detected in elasmobranchs and teleosts. Antibody has also been detected in fish mucus. In general, however, the immune response is weak and poorly protective. Snieszko (1970) and Hoar & Randall (1970) have reviewed the problems of fish immunology.

Immune Reactions in Invertebrates

Immune reactions have been very little studied in invertebrates, the relevant literature begin reviewed by Tripp (1969) and Huff (1940). Most work in this field has been carried out on arthropods (especially insects). Invertebrate body fluids do not appear to contain substances comparable to antibodies of vertebrates. Nevertheless, substances which agglutinate or haemolyse cells have been reported from the body fluids of some invertebrates. The main reaction of invertebrates to foreign material is to remove it by phagocytosis, although large particles may be surrounded and encapsulated. Little information appears to be available on the factors necessary for phagocytosis in invertebrates and this is a field which would undoubtedly reward further study.

11

ANTIBODY PRODUCTION IN PARASITES

Few fields in parasitology have expanded so rapidly or have aroused such widespread interest as that dealing with the immunology of parasitic organisms. Much of this interest has been concerned with some of the world's worst diseases of parasitic origin, such as malaria, schistosomiasis, or hydatidiasis and has centred on attempts to develop vaccines or to improve diagnostic methods of the causative organisms. At a more basic level, the evolution of antigenic patterns in host and parasite has received some attention. It has been found that parasites possess antigens which may have similar molecular configurations to those of (*a*) closely related species, (*b*) species from other classes of parasites and (*c*) their hosts.

This last phenomenon appears to provide some protection for the parasite since the host is unable to recognize the parasite surface as 'non-self.' This *host unresponsiveness* has aroused great interest among parasitologists. Immunity to parasitic organisms, although showing many features in common with microbiological infections, is, in many cases, made immensely complicated by the occurrence of several hosts in the life cycle. Thus, immunity is rarely a simple reaction to a single stage in the life cycle of a parasite, but frequently one involving responses to stages of the parasite in perhaps several invertebrate and/or vertebrate hosts - stages which may show a continually changing antigenic pattern. Moreover, in some infections - particularly protozoan infections - completely 'sterile' immunity is not achieved. Instead relatively small numbers of parasites persist at a 'latent' or 'subclinical' level - a phenomenon referred to as *premunition*.

PARASITE ANTIGENS AND FUNCTIONAL IMMUNITY

It has been pointed out earlier that invasion of a vertebrate generally results in local tissue reactions and the appearance of antibody in the host serum. It must be emphasized that the appearance of antibody in the serum does not necessarily imply that thereby the resistance of the host to the parasite has been increased. A degree of immunity will only have been developed if the antibody acts in such a way as to inhibit the development of the parasite. This inhibition (or death) of a parasite can result from:

(*a*) stimulation of host tissue reactions (especially cell mediated reactions) resulting in phagocytosis or lysis of the parasite or encapsulation of it;

(*b*) production of precipitates which either interfere with metabolic processes (perhaps by interacting with enzymes), especially those taking place at surfaces, or which block functional openings such as the mouth or excretory pores and prevent their normal operation;

(*c*) reactions interfering with glandular secretions from the parasite.

Antibodies which confer some measure of protection on a host are known as 'protective' antibodies and the antigens which stimulate their production are sometimes referred to as 'functional' or 'essential' antigens. Thus these antigens are essential to the parasite and if the processes in which they are involved are modified in some way by antibody, the parasite cannot grow and develop normally. A complex terminology has grown up around parasite antigens although all authors do not necessarily use these terms in quite the same way. Thus antigens of the tissues are sometimes referred to as *somatic* antigens — also known as *endogenous*, *structural*, *bound* or *internal* antigens.

The terms *surface* or *external* antigens are sometimes used for those presumed to be present on the surface. Antigens released during growth, development or tissue penetration as metabolic products (especially secretory or excretory materials) are referred to as *metabolic* antigens - also known as *exogenous antigens* or *exoantigens*. The distinction between somatic and metabolic antigens, however, may not always be clear.

Cell Mediated Immunity and Reaginic (homocytotropic) Antibodies in Parasitic Infections

Much of the early work on immunity to parasitic organisms was concerned with the effects of humoral antibodies, and cellular phenomena were generally neglected. It has now become increasingly apparent that cell cediated immunity and reaginic (homocytotropic) antibodies may play an important role in the immune processes in parasitic infections. In considering immunity to a limited extent e.g. *Schistosoma mansoni*. Knowledge of these topics, which is expanding rapidly, has been comprehensively reviewed by Soulsby (1972b) and Sadun (1972).

Protozoa

Haemoflagellates

Immunity to trypanosomes in general has been reviewed by Neal, Garnham & Cohen (1969), Gray (1967), Goble (1970), Desowitz (1970), D' Alesandro (1970) and Lumsden (1972a).

Trypanosoma lewisi

This species presents one of the classical cases of acquired immunity and one of the earliest established for protozoa. The life cycle has already been described. Typically, in a new infection, the trypanosome multiplies in the peripheral blood of rats for five or six days, after which time retardation followed by complete cessation of reproduction occurs. The organisms agglutinate in rosettes and a 'crisis' occurs on or about the tenth day when the majority are destroyed. A second 'crisis' occurs about the 35th day characterized by complete disappearance of all organisms from the bloodstream.

It is thought that three antibodies might be involved in the development of immunity (D'Alesandro, 1970); (*a*) a reproduction inhibiting antibody, termed *ablastin*, (*b*) a trypanocidal antibody (IgG) specific for division forms and (*c*) a trypanolytic antibody (IgM) responsible for termination of infection by killing the adult forms. All these effects, however, could be explained also on the assumption that only one antibody, ablastin, is produced. According to this view, reproduction is inhibited by ablastin which inhibits nucleic acid synthesis, and the organisms become sticky and agglutinate.

The agglutinated organisms are filtered out by the reticulo-endothelial system, especially by the liver and spleen, with the result that the majority suddenly disappear. This accounts for the first 'crisis.' As the agglutinated organisms disappear, it becomes increasingly difficult for those remaining non-agglutinated to meet, and the *complete* removal of all trypanosomes is accomplished by normal phagocytic mechanisms. An increase in the number of granules in *T. lewisi* just before the onset of the 'crisis' has been noted but the association between their appearance and the onset of immunity is no understood; these granules appear to consist of a phospholipid. That the '10-day crisis' and the subsequent elimination of parasies is due to an immune reaction has been dramatically demonstrated by splenectomizing rats before infection — an operation which removes the most active site of antibody formation. In these splenectomized rats a very long parasitemia without death developed. Although the peak infection (294 000 per cm^3 $\pm$ 38 700) developed on the tenth day - as in the non-splenectomized controls - by the 25th day there was still a very high parasitemia (125 690 per cm^3 $\pm$ 20 000) and, even after 60 days, 6 per cent of the experimental rats showed some parasitemia (1250 per cm^3).

Human trypanosomiasis

Immunity to *T. rhodesiense* and *T. gambiense* which cause sleeping sickness in man or related forms in cattle such as *T. brucei* has been studied extensively. Such species have an exceedingly complex antigenic structure made up of somatic, surface and metabolic antigens. Moreover, each species contains numerous 'strains' and each strain gives rise to numerous 'variants,' up to 20 being described in one instance. Occurrence of such a high degree of antigenic variation is responsible for the chronic relapsing course of trypanosomiasis and the appearance of distinct variant specific antibodies in the host serum. Moreover, there is a tendency for variants to appear in a characteristic order. The mechanism host. In some species (e.g. *E. tenella*) in chicks, a single infection is sufficient to provide complete protection against reinfection. The mechanism of immunity is not understood. It appears that penetration of cells can occur in some hosts but further development is prevented.

In other host species, penetration may be prevented by the secretion of an antibody against an unknown proteolytic enzyme utilized by the penetrating sporozoite. This view could receive support from the experiments of Leathem & Burns (1971) who found that stages of *E. tenella* recovered from the epithelial cells of immune hosts at 12-24 hours were still infective when transferred to susceptible chicks, whereas those from a 48-hour infection were not. This has been interpreted as showing that within 24-48 hours after invasion, development had been irreversibly inhibited by the immune mechanism. Alternatively, of course, the 48-hour sporozoites may have developed beyond the infective stage. Humoral antibodies have been shown in many host species infected with coccidea, and may, in some cases, be protective. Passive immunity can be induced by the inoculation of serum from immune hosts, but only if the intervals between serum and parasite inoculations are short.

Helminths

Helminths offer a number of advantages over protozoans, bacteria or other micro-organisms for basic investigations in immunology, e.g. (*a*) with a few exceptions (e.g. *T. crassiceps*) they do not multiply in the body (in contrast to bacteria or protozoa) so that precise control of an effective dose is possible, (*b*) they tend to mature in a specific site within the body, or, if they undergo a tissue migration, the route is well defined, (*c*) they are relatively enormous in size, compared with a single bacterium or protozoon, so that individual tissues can be isolated and antigens prepared from them, (*d*) immunological effects, such as the formation of an envelope around a cercaria in immune serum, can sometimes be directly observed and followed under the microscope.

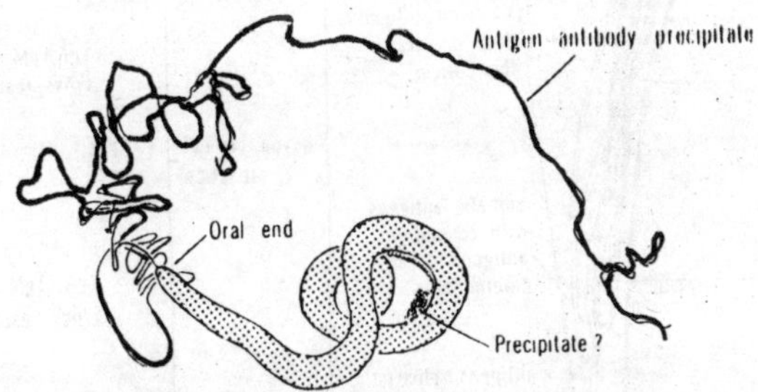

Fig. 11.1. Third stage larva of Ascaris lumbricoides var. Suum removed from the lungs of an experimentally infected mouse and incubated for 25 days in immune guinea-pig serum; a strand of precipitate originates from the oral opening

Although complete sterile immunity against helminths may be developed, an immune response may be evidenced in other ways such as stunting, retardation and inhibition of reproduction. Since many helminths are large, immune phenomena such as precipitation or phagocytosis may have no effect on them. Much, however, depends on the type of worm. Thus, an antibody reacting with the antigenic cuticle of a nematode may virtually be without effect, as this is a non-absorbing surface. On the other hand, the tegument of a cestode is highly absorptive and transport activities could be greatly affected by the presence of antibody precipitated on the surface.

In nematodes, however, precipitations may occur in immune serum at the oral or excretory pores and these may be inhibitory with some protozoa, many antibodies to helminths appear to be directed against metabolic antigens; in the case of *Nippostrongylus brasiliensis* released lipases and esterases may serve as metabolic antigens. The realization that metabolic products of helminths can serve as antigens has given

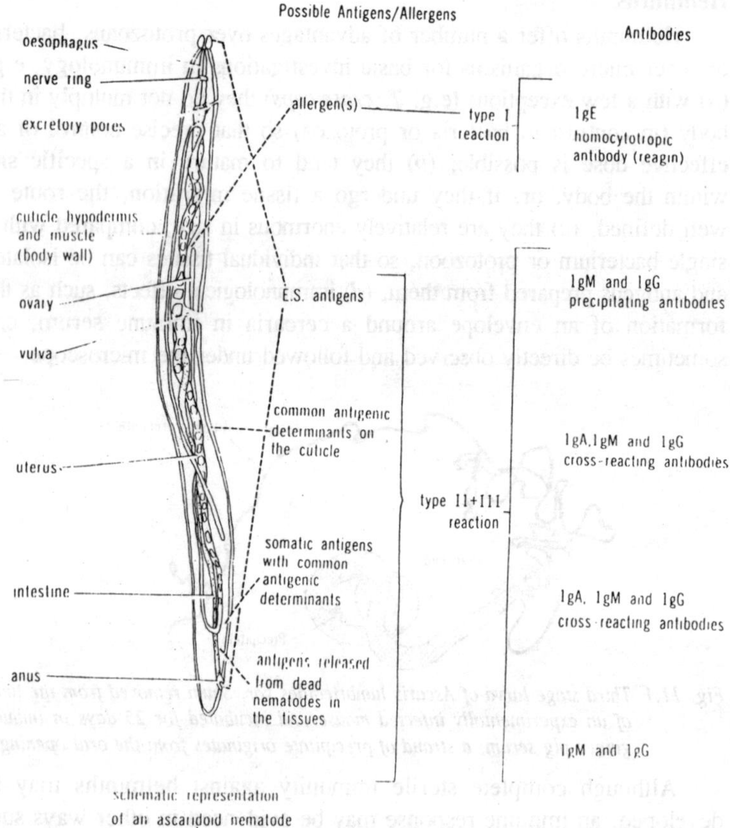

Fig. 11.2. The possible site of some nematode antigens/allergen and their antibody responses.

new importance to the elaboration of techniques for the *in vitro* cultivation of helminths.

Irradiated larvae as antigens

A number of attempts have been made to utilize extracts of worms as antigens, but although these invoke the production of antibodies the degree of immunity developed is generally of a low order. Immunity develops much more strongly in the presence of living whole worms, presumably because of the production of metabolic antigens. Means have, therefore, been sought to treat worms in such a way that they would be attenuated and yet still retain their ability to invoke somatic and metabolic antibodies. It was found that when treated with X-rays or radiation from cobalt 60, some species of larval worms became sterilized and yet retained their ability to induce immunity.

Much of the early work was carried out on larval nematodes as many of these undergo extensive migration in the definitive host, and although sexual maturation was usually prevented by irradiation, the migratory phase (during which the immune reaction is evoked) was comparatively little affected. This has resulted in the development of vaccines which can produce some degree of immunity against several nematodes of economic importance. This approach has now been attempted with a number of other helminths of economic importance, e.g. *Schistosoma mansoni*, generally with only limited success; this work is discussed further below.

Trematodes

Immunity to trematodes in general has been reviewed by Kagan (1966), Platzer (1970) and Smyth (1966).

Fasciola hepatica

There appears to be some conflicting evidence as to whether acquired resistance to this species ever occurs. Thus, based on experiments with normal and X-radiated metacercariae, Boray (1967) concluded that acquired resistance to this species does not develop in sheep or cattle. Nor can passive immunity be demonresistance has been claimed in rabbits and mice although only relatively small numbers of experimental animals were used. On the basis of lymphocyte infiltration into the liver and the histopathology of the latter it was concluded that delayed hypersensitivity may play a role in the degree of immunity developed. This view is supported by the fact that the transfer of lymphoid cells from rats infected with *F. hepatica* has been claimed to confer some immunity on isogenic hosts subjected to primary infections.

Schistosomes

Immunity in schistosomiasis has been extensively investigated, various aspects of the field being reviewed by Brujining (1967), Jordan & Webbe (1969), Smithers & Terry (1969a, b) Smithers (1968, 1972) Kagan (1966) and Lewert (1970).

Antigen-antibody interactions

Antibodies against both adult and larval schistosomes have been detected in serum and a number of diagnostic tests based on the usual antibody manifestations have been developed. These include precipation, flocculation, haemagglutination, complement fixation and skin tests. In addition, three unusual reactions have been developed. These are (*a*) the *Cercarienhullenreaktion* = CHR), of Vogel & Minning (1949), (*b*) the *miracidial immobilization* test and (*c*) the *circumoval precipitin* test (= COP).

(*a*) *The COP test* is based on the fact that when a live schistosome egg is placed in antiserum, a globular precipitate, contiguous with the edge of the egg-shell, appears. The precipitate is best seen after 24 hours in serum at 37°C but can be detected as early as two hours. Host serum becomes positive about the 40th day after infection at the time the eggs are produced. It should be noted, however, that even serum from *unisexual* infections, whether males or females gives a weak COP reaction. The source of the antigen in eggs is not known, but it is probably of metabolic origin, for if eggs are allowed to hatch in a small volume of water, the metabolic products are strongly antigenic and will produce a weal if injected intradermally into an infected host or will cause a precipitate in a tube of antiserum.

(*b*) *The CHR test* is of special biological interest. It is manifested by the appearance of a transparent membrane around the tail of a schistosome cercaria when placed in immune serum. This reaction is destroyed by heating the serum to 56°C for 30 minutes. Electrophoretic studies have revealed that the CHR membrane develops only in the gamma and alpha fraction of the serum. The

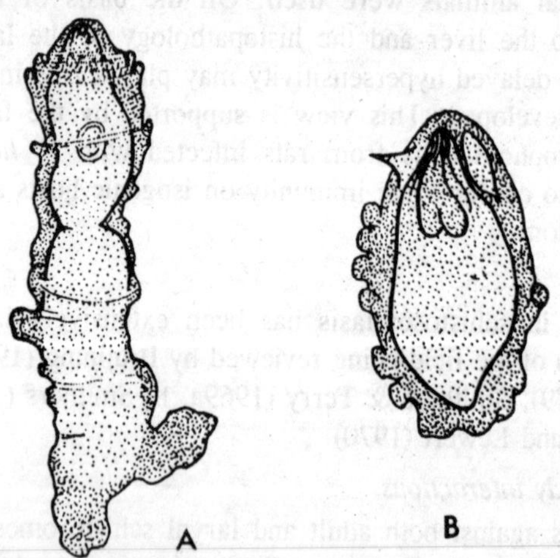

Fig. 11.3. Immunological reactions of Schistosoma mansoni in antiserum. A—Cercarienhullenreaktion (CHR). B—Circumoval precipitin test (COP).

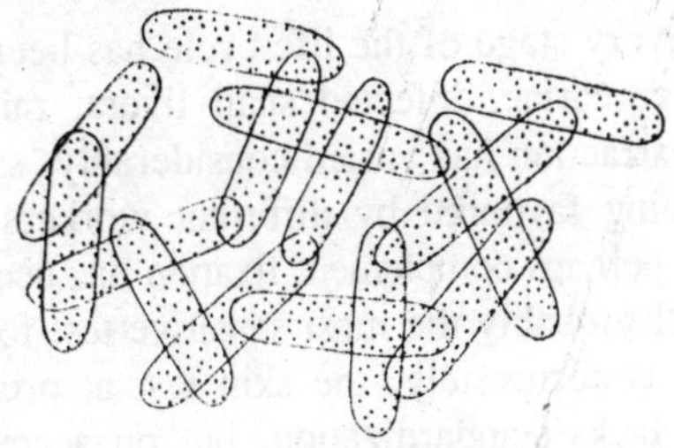

(a) Antiserum against somatic antigen

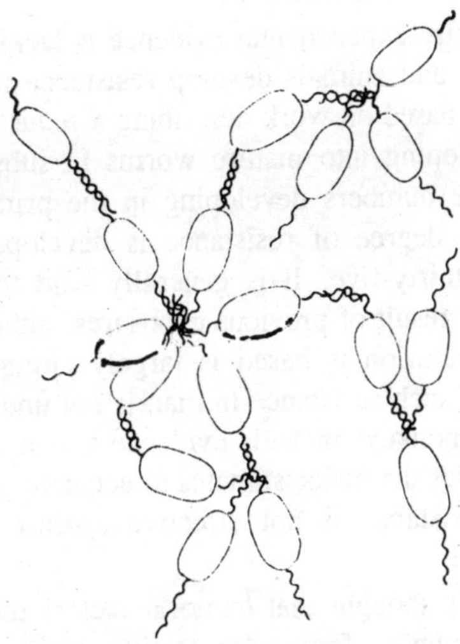

(b) Antiserum against flagellar antigen

Fig. 11.4. The antibody as agglutinin. (a) Appearance of agglutinated bacterial cells when antibody is directed against somatic, or O, antigen. (b) Appearance of agglutinated bacterial cells when antibody is directed against flagellar, or H, antigen.

CHR reaction has been applied to both *S. mansoni* and *S. douthitti* with rather similar results. As a diagnostic tool the method has obvious limitations. It is not as sensitive as some of the other reactions and also necessitates maintaining living cercariae in the laboratory.

(c) *The miracidial immobilization test* is based on the immobilization of miracidia in immune serum, and it has been found to be an exceedingly sensitive one. The method has not yet been widely used. All stages of schistosomes have been found to produce immobilization antibodies.

Practically every stage of the life cycle has been used for antigen - adult worms, cercariae, infected snail livers, miracidia and eggs. The method of extraction has varied considerably, saline, alcoholic or ether extracts being favoured by different workers. Of all the tests mentioned above perhaps complement fixation has been the most widely used and it is still probably the most sensitive test for the diagnosis of early infections. Unfortunately, the skin test at present developed is not specific and lacks standardization, but on account of its relative simplicity would be a diagnostic tool of value. Kagan (1958) gives an excellent review of these tests.

Acquired immunity to schistosomes

Although precise experimental evidence is lacking, it is generally accepted that man and animals develop resistance to schistosomiasis. This conclusion is based on work describing a reduction in the number of cercariae developing into mature worms in subsequent infections compared with the numbers developing in the primary infection. In man, a very high degree of resistance is developed in adults over about the age of thirty-five. It is generally held that this resistance has developed as a result of previous exposures, although the evidence on which this conclusion is based is largely circumstantial in man. The mechanism of such resistance in man is not understood, but from experiments with monkeys there is evidence that it is the living *adult* worm which provides the major stimulus to acquired resistance although apparently this resistance is not effective against the adult worms already established.

It was formerly thought that humoral factors did not play a part in schistosome immunity, for no 'protective' action of the numerous antibodies reported could be identified. *In vitro* experiments have shown that hyperimmune sera from rhesus monkeys (i.e. animals exposed to two to four infections of *S. mansoni* cercariae) contain an antibody (IgG) lethal to schistosomula. Moreover, this antibody is dependent on a labile factor in serum—probably a component of the complement system. It is thus likely that, in hyperimmune monkeys, invading schistosomula would be rapidly killed. If, however, schistosomula are cultured *in vitro* in normal serum or are maintained for the same period in a mouse they are almost completely protected against the action of the lethal antibody *in vitro*. A possible explanation of this is that a worm could 'acquire' a layer of host antigens and hence in its adult condition be 'protected' against host antibody.

Newly penetrated schistosomula would not, of course, have such a protection. These results clearly need confirmation in other host species, and more supporting evidence is needed before we can conclude, unequivocally, that protective antibodies operate in schistosome immunity. It is significant to note, however, that a specific schistosomal antigen has been detected in hamsters infected with *S. mansoni*.

Vaccination against schistosomiasis

Numerous attempts have been made to develop vaccines against schistosomes using either dead worms, extracts from them or irradiated cercariae. Dead materials are useless as antigens but an initial infection with normal cercariae induces immunity in experimental hosts. Results have been very variable, especially if the initial exposure is less than 100 cercariae or of the challenge occurs less than 16 weeks after the initial infection. In some cases, a high degree of resistance has been induced with as few as 25 cercariae. In contrast to nematodes irradiated cercariae even in doses of 25 000 may induce no immunity.

Concomitant immunity to schistosomes

The existence of an infection of an adult worm in a host induces a condition known as *concomitant immunity* in which such a host is immune to a second infection. The term arose first intumour transplantation to describe a condition in which an animal bearing one tumour was sometimes resistant to a second graft. The question must be asked: why it is that the established worms are not destroyed by the immune response? It appears, as suggested above, that, in some way not understood, adult schistosomes accumulate host antigens on their surface which serve to 'protect' them from antibody attacks. This question is discussed further under *molecular mimicry*.

Cell mediated immunity in schistosomiasis

Cell mediated immunity appears to operate in relation to the immunopathological events associated with the presence of eggs in the tissues. Thus, in these tissues, the typical granulomas which form around the eggs appear to be a form of delayed cell mediated hypersensitivity. This reaction has been shown to be not only species-specific but also stage-specific. Thus, there is no cross sensitization between *Ascaris suum* and *S. mansoni* eggs and little between the latter and eggs of *S. haematobium* and *S. japonicum*. Furthermore, sensitization is transferable by cells but not by serum.

Cestodes

Various aspects of immunity to cestodes have been reviewed by Gemmell & Soulsby (1968), Smyth (1969), Weinmann (1966, 1970) and Gemmell & MacNamara (1972). Since the body surface of adult and larval cestodes presents a virtually 'naked' cytoplasmic surface, it may be expected that in host tissue sites, at least, cestodes would be highly immunogenic. In general this appears to be the case and, in mammalian hosts, even a few larvae may induce a high and durable level of resistance. Moreover, acquired immunity may develop very rapidly as early as one to two days in *Hymenolepis nana* or *Hydatigera taeniaeformis* in rodents.

It is important to stress that in cestodes, in particular, the strain of both host and parasite appear to be of special importance in relation to the degree of immunity developed. Thus, cysts of *Echinococcus multilocularis* when injected intraperitoneally into different strains of mice will develop into hydatid cysts in some but not in others. In the best known cestodes of man and other mammals (i.e. *Taenia* spp.) the host/parasite contacts established during the life cycle are (*a*) the oncosphere, during the initial stages of penetration of the intestinal mucosa of the intermediate host, (*b*) the larva in its final tissue site (often the liver) and (*c*) the scolex of an adult worm attached to the intestinal mucosa. Most of the work on cestode immunity has centred on species of economic importance (such as *Taenia* spp.) or those readily maintained in laboratory animals (e.g. *Hymenolepis* spp).

Taenia* (= *Hydatigera*) *taeniaeformis

Much of the early work on cestode immunity centred around this organism and it was shown that even a few cysts in the rat liver were sufficient to prevent the establishment of further infections when challenged with this organism. This protection persists for at least two months, even after encapsulated larvae are removed from the liver. In common with many cyclophyllidean cestodes, oncospheres must be eliminated either at the site of penetration of the intestine or during migration, a process known as *early* or *pre-encystment* immunity or at the final visceral site, referred to as *late* or *postencystment* immunity.

***Taenia* spp.**

Many other larval taeniid species present a similar immunological pattern, e.g. *T. pisiformis*, *T. saginata*, *T. ovis* and *T. hydatigena*. It might be expected that such a high level of immunity could be passively transferred and the protection of rabbits against *C. pisiformis*, by injection of antibody containing serum from infected rabbits, has been demonstrated; immunity has similarly been transferred by lymph node cells. Activated oncospheres have been used with some success as vaccines against both homologous and heterologous species. Immunity has also been established against. *T. ovis* and *H. taeniaeformis* using larvae grown *in vivo* in filtration membrane diffusion chambers.

Echinococcus granulosus

Most of the work on this species has been carried out on the hydatid stage with relatively little work being done on the adult stage in the carnivore host. Complete immunity to larval infections is probably never developed but in subsequent infections fewer cysts become established.

There is great individual resistance and even in the same liver, for example, both dead and viable cysts may occur. Some degree of 'post-encystment' immunity to *Echinococcus* can develop, however, and can be induced artificially by injection of oncospheres. Host immunity to different strains of *Echinococcus* may prove to be very specific as exemplified by the fact that the 'horse strain' is common in Ireland and yet sheep in that country are not infected. This suggests strong immunological and/or physiological differences between the horse and sheep strains. This hypothesis is supported by the fact that the *in vitro* system in which the sheep strain will grow to maturity fails to support development of the horse strain.

On account of their medical importance, serological tests for the diagnosis of hydatid cysts have been much studied; these include complement-fixation, haemagglutination, bentonite-flocculation and latex-agglutination, immunofluorescence and intradermal tests. The immunology of the adult stages in dogs has been poorly examined. In general there appears to be considerable variation between individual dogs. Thus, in experiments involving 20 experimental dogs and 20 controls, Williams & Esandi (1971) found that a thermolabile homocytotropic skin-sensitizing antibody was detectable in five of the infected dogs but not in the remainder or in the controls. Circulating reagins were not detectable until the seventh week of infection, i.e. just after the worms had become gravid.

Hymenolepis spp.

H. diminuta, *H. nana* and *H. microstoma* are commonly maintained in laboratory rodents and have been most widely studied.

Hymenolepis diminuta

This species exhibits 'premunition,' that is, protection against reinfection as a result of an existing infection (in contrast to 'immunity' which is related to a *previous* one), a phenomenon common amongst cestode species. It has been shown that the size of worms in a secondary infection is inversely proportional to the number of primary worms harboured. When a primary infection is eliminated by an operation, the rate of growth of the worms in the secondary infection was normal. This suggests that 'premunition' in cestode infections may not be due to the action of an immune mechanism but to a crowding effect, which probably has mainly a nutritional basis.

Hymenolepis nana

This species has been widely used as a model for studying immunity to cestodes. It is unusual among cestodes in that cysticercoid may develop directly within the villi of the rodent host. Since the host tissues are thus invaded it can be expected that a high level of immunity will develop and indeed it has been shown experimentally that immunity is detectable as early as nine hours after the initial infection and becomes almost absolute after about 48 hours. In 'immune' mice a few challenge eggs may hatch, penetrate and develop to cysticercoids within the villi, bu none develop to adults. Antibodies may be detected in infected mice serum.

The immune pattern described above refers to an egg challenging dose. If a cysticercoid challenge dose is used instead, some adults do become established although only about one-fifth of a normal infection. Hence, a preceding tissue (cysticercoid) infection provides some degree of lumen protection possibly through the action of antibody in intestinal mucus. Passive immunity has also been demonstrated for *H. nana*, the antiparasitic activity being associated with the IgG (7S) immunoglobulin fraction. That immunity in *H. nana* involves a thymus dependent mechanism has been shown by the suppression of immunity in thymectomized mice, and also by injection, into normal mice, of rabbit antimouse thymocyte serum.

Nematodes

Immunity in nematodes has been extensively studied, most attention having been paid to gastrointestinal nematodes of economic importance (e.g. *Haemonchus contortus*), species pathogenic to man (especially *Tricinella spiralis*) and species easily maintained in laboratory animals (especially *Nippostrongylus brasiliensis* in rats). Various aspects of immunity to nematodes in general have been reviewed by Mulligan (1968); Michel (1968); Soulsby (1966, 1970, 1971); Thorson (1970); Otto (1970); Dobson (1972) and Olson & Izzat (1972). Morphologically, nematodes differ markedly from cestodes and trematodes in possessing a tough, relatively inactive outside surface, in contrast of the 'naked' (and hence strongly immunogenic) cytoplasmic surface of trematodes and cestodes. Thus, nematodes would be relatively unaffected by antibody-antigen precipitations or leucocyte attachment at the surface.

On the other hand, precipitations at the oral or excretory pores would affect metabolism and this appears to be borne out in practice. Only a very limited number of examples of immunity to nematodes are discussed below. A characteristic of many nematode infections is the phenomenon known as 'self-cure' - 'the sudden elimination of a burden of parasites as a result of an immune response on the part of the host'. The term 'self-cure' was first introduced in relation to *Haemonchus contortus* but has been extended to many other species such as *Nippostrongylus brasiliensis*.

Nippostrongylus brasiliensis

The immune reactions against this species have probably been more extensively studied than any other helminth species; the extensive literature has been reviewed by Ogilvie (1969) and Ogilvie & Jones (1971). The value of this species as a model is that a strong, active immunity is generated and this immunity may easily be passively transferred by antiserum so that the study of protective antibodies is feasible. Moreover, high levels of reaginic antibodies are stimulated in rats. The pattern of development of immunity is as follows.

In heavy infections, eggs appear first in the faeces on the sixth day post-infection and reach a peak of production on the tenth day, falling rapidly during the next week or so to zero. Fall in egg production is accompanied by a massive expulsion of worms (i.e. a 'self-cure' reaction) from the intestine but a small number of worms survive the expulsion phase and remain in the intestine for long periods.

A curious feature of these residual worms is that the majority of these are male, the females remaining being largely sterile. With small initial infections (approximately 50 larvae) worm expulsion is more gradual and may extend over 30 days. Rats which have thrown off a primary infection exhibit a marked resistance to a challenge infection, as exhibited by inhibited development of the worms or by complete failure of the secondary infection to become established. Kassai & Aitken (1967) point out that in using *N. brasiliensis* as a model for immunological studies, two measurements have special significance. These are the actual size of the established worm population, i.e. the 'ten-day' take (measured on the eighth and tenth days) and the rate of loss of worms which determines the size of the '20-day' take (measured on days 18-20). By formulating a 'take ratio' calculated as:

$$\frac{20\text{ - day take}}{10\text{ - day take}} \times 100$$

these two values can be related. This should not be confused with the 'percentage take' which is the take expressed as a percentage of the total infecting dose. Kassai & Aitken demonstrated that the adult ability to eliminate worms from a primary infection did not develop in rats until they were about five weeks old; infections originating in four-week-old rats survived for up to ten weeks.

The percentage take was also much higher in three-to four-week-old rats (12-16 per cent) tan in rats aged five weeks or more (1.4-4 per cent). The sex ratio also differs markedly with the age of the rats, being about 50: 50 male: female in young rats but rising to 97: 3 in eight-week-old rats. The mechanism of 'self-cure' in *Nippostrongylus* has been much investigated; there is evidence that it is a two-step process involving (*a*) action of antibodies and (*b*) expulsion. The antibodies are mostly associated with the $7S_{\gamma 1}$-immunoglobulin and it has been suggested that this may interfere with the acetylcholinesterase activity of the worm, although this has not been unequivocally established.

The gut cells of the worms are also apparently damaged by antibody so that normal feeding is disturbed; there is also resorption of sperm and cessation of egg production. The nature of the expulsion step is more fully understood. It appears to be associated with the release of vasoactive compounds from mast cells after interaction on the cell surface of bound reaginic antibodies (which reach their maximum titre

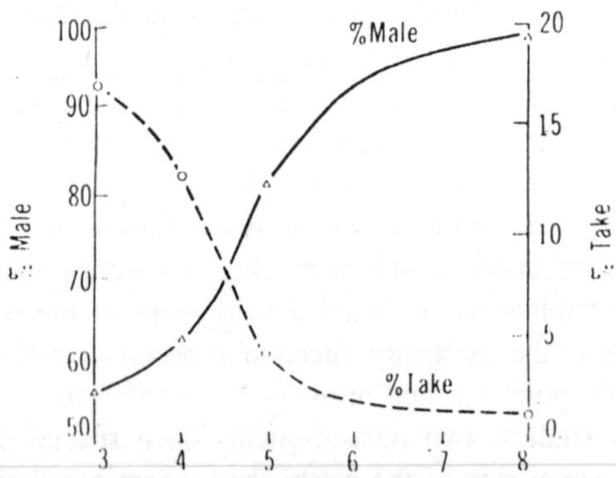

Fig. 11.5. Effect of the age of rats at the time of infection on the 'take' and sex ratio of Nippostrongylus brasiliensis 20 days after infection.

at 20 days post-infection) with worm allergen. It has been suggested that *Nippostrongylus* stimulates the synchronous development of new populations of mast cells and plasma cells which produce immunoglobulins of at least three classes: IgE, IgA and IgG. IgE is associated with the allergen-reaginic antibody system as indicated above.

The pharmacological mediators released by the mast cells are then believed to interact with (*a*) plasma cells containing antibodies of other classes, to release antibody and (*b*) the mucous membrane, to create an intercellular pathway across the mucosa. Such a mechanism could deliver large amounts of locally produced (IgA) and/or systemically produced (IgG) anti-*Nippostrongylus* antibody to sites specifically occupied by the worm, thus effecting a maximal protective response. Most of the work on *N. brasiliensis* has been concerned with single large infections of larvae which terminate abruptly as outlined above.

When, however, 'trickle' infections are given (i.e. small daily infections) no such dramatic expulsion occurs and the worms develop beyond 14 days and, although stunted, are able to remain patent for more than three months. Even rats previously immunized with a high primary infection of *N. brasiliensis* could be infected by low-level 'trickle' infections. Hence it appears that under certain conditions worms can become 'adapted' to the immune response. When these 'adapted' worms were transferred to immune rats—in contrast with normal worms—they established themselves in the intestine and produced viable eggs. The nature of this adaptation is not understood but presents an intriguing problem.

Haemonchus contortus

In this much-studied nematode of sheep 'self-cure' followed by protection was first observed by Stoll (1929). Since then the phenomenon has been extensively studied, as have similar phenomena with other sheep nematodes such as *Trichostrongylus* spp. and *Ostertagia* spp. The relevant literature has been reviewed by Gordon (1967), Mulligan (1968), Michel (1968) and Soulsby (1966). In this species, about two weeks after infecting sheep, egg production begins and continues increasing for about ten weeks; thereafter there is a fall to low egg counts within two to three weeks. Unlike *Nippostrongylus*, the expulsion reaction appears to be precipitated by the intake of a new batch of larvae. This immune response does not occur in lambs under about four months of age, which is comparable to the situation in young rats. Sheep have been vaccinated against *H. contortus* by using third stage larvae partly inactivated by radiation doses of 40 000—60 000 roentgens. These irradiated larvae are still able to undergo a limited migration and retain enough metabolic activity to stimulate a high degree of immunity to reinfection.

Double vaccination is even more effective. Unfortunately, differences in responses of various breeds of sheep have rather invalidated the use of this irradiated vaccine. There is some evidence that the antigen initiating the self-cure reaction is the exsheathing fluid produced at the third ecdysis. Metabolites from third and late fourth-stage larvae cultured *in vitro* and fractionated, have been used with some success as vaccines, and immunodiffusion precipitation reactions have been detected against immune serum and metabolites.

The 'spring rise' phenomenon. As already mentioned that *H. contortus* together with some 30 other species of gastro-intestinal nematodes exhibits a phenomenon referred to as the 'spring rise' or 'postparturient rise' in lactating ewes during early spring. An extensive literature exists on this phenomenon. It is generally accepted that most of this increased egg production is due to maturation of larvae picked up in autumn and arrested in their development throughout the winter, at the early fourth stage. The mechanism of this phenomenon remains obscure but there appear to be two schools of thought. The first believes that the immune status of the host is chiefly responsible and that fluctuation in this leads to the maturation of the worms. The alternative

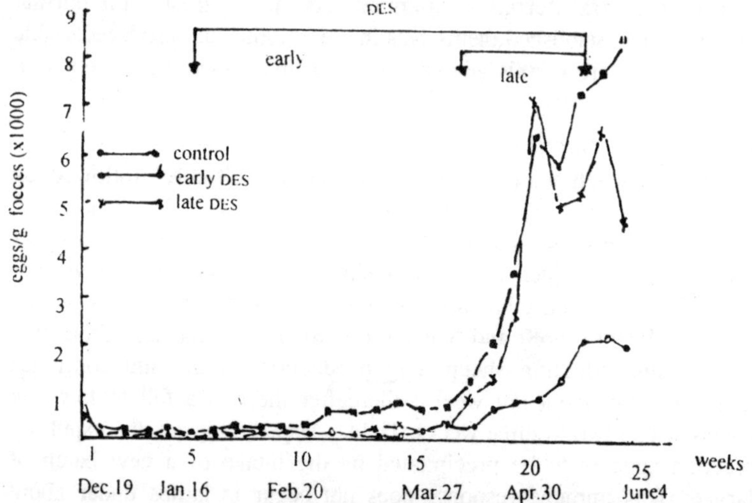

Fig. 11.6. Effect of administration of the hormone diethylstibestrol (DES) to unbred ewes infected with Haemonchus contortus; the onset of the 'spring rise' is not apparently affected, lactation and DES may temporarily suppress the immune capacity of the host.

view is that arrested development of the fourth-stage larvae is induced by environmental conditions acting on third-stage larvae on the pasture during autumn and that during winter arrested development is maintained by a process analogous to diapause in arthropods; development recommences spontaneously when these processes are completed. This view further postulates that these developing larvae trigger a 'self-cure'-like reaction in the sheep with expulsion of worms; it is speculated that this immune response is temporarily suppressed in lactating ewes and is associated with hormone levels. The latter hypothesis is supported by the fact that injections of the hormone diethylstiblesterol raises the egg production but do not advance the date of onset of spring rise.

Dictyocaulus viviparus

The pattern of immunology in this species is somewhat similar to that of *H. contortus*. The use of irradiated larvae as vaccines against the parasitic bronchitis caused by this worm has been successfully produced. Third-stage larvae partially inactivated by exposure to 40,000 roentgens form a suitable vaccine, and calves doubly vaccinated with this vaccine are almost completely resistant to infection when challenged with 10 000 normal larvae. Vaccination is presumably effective against this species because in the natural life cycle the worm population is normally regulated by the protective immunity of the host.

Other trichostrongylids

Immunity to numerous other species of trichostrongylids (e.g. *Ostertagia* spp., *Cooperia* spp., *Trichostrongylus* spp.) has been extensively studied. The relevant literature has been reviewed by Michel (1969) and Fitzsimmons (1969). Instead of using X-radiation, larvae have been attenuated with the cytotoxic agent triethylene melamine (TEM) and these, too, act as satisfactory vaccines.

Trichinella spiralis

This species has been intensively studied as a model of helminth immunity. It has been clearly demonstrated in a number of hosts (e.g. rats, guinea-pigs) that animals which recover from an infection of *T. spiralis* resist further infection. Single infections appear to be rather erratic in inducing immunity but multiple infections appear to be very effective. *Trichinella* antibodies appear in serum within 24 hours post infection and precipitation has been reported to occur as early as the fifth day in the intestine, mouth, anus and excretory pore when worms are placed in the serum recovered. A degree of resistance is also transmitted from immune mothers to their young.

Acquired immunity manifests itself by (*a*) stunting of growth, (*b*) retardation or complete inhibition of development, (*c*) inhibition of reproduction, (*d*) elimination of previously established worms and (*e*) refractoriness to infection. This acquired immunity manifests itself in about 18-22 days after an initial infection of about 1000 larvae. Immunization must be considered in two phases (*a*) parenteral (stimulated by the tissue migration of the larvae), and (*b*) intestinal or enteral (stimulated by older larvae or adults in the intestine). Rather surprisingly, it has been shown that the immunity acquired by mice subjected to an enteral-only infection was at least as strong as that acquired by mice in which the full, biphasic enteralparenteral infection had run its course.

The origin of the functional antigens inducing protection is not known; since newborn larvae appear to be incapable of producing frunctional antigens it has been argued that these only occur in trichinellids which possess a stichosome. There is no agreement regarding the mechanism of immunity or how the enteral phase is able to induce it. Previously it was thought that both humoral and cellular factors operated in immune animals. However, even in the presence of high serum antibody titres hosts were still found to be susceptible. This has led to the hypothesis that elimination of the adult worms is based on delayed hypersensitivity. This would fit in with the experimental evidence that it is the metabolic (excretory and secretory) products which act as functional antigens.

The antigens of *T. spiralis* have been extensively analysed and various serologic tests developed. Larvae irradiated with X-rays or cobalt 60 at doses sufficient to produce sexual sterilization, but not sufficient to prevent growth to adult size serve as reasonably effective vaccines. These do not induce complete immunity, but the number of adults maturing in the intestine, and the number of larvae which reach the muscles, can be measured in tens rather than in thousands.

Other nematodes

The immune reactions to numerous other species of nematodes have been extensively examined; in some cases effective antigens have been examined or vaccines developed. Some of the most actively investigated include: *Ancylostoma caninum*, *Uncinaria stenocephala*, *Litomosoides carinii*, *Ascaris suum*, *Toxocara canis*.

Immunity in Cold-blooded Vertebrates and Invertebrates

Most of the immune reactions dealt with in this chapter have been concerned with warm-blooded hosts. Immune mechanisms do occur, however, in cold-blooded vertebrates and invertebrates although these difficult areas of immunology have been rather neglected. It is beyond the scope of this text to discuss these fields in detail; some relevant reviews are as follows: Cold-blooded vertebrates: Rowlands (1969); fish: Hoar & Randall (1970), Snieszko (1970); invertebrates (in general): Tripp (1969); molluscs: Brooks (1969), Feng (1967), Tripp (1963); insects: Salt (1963), Shapiro (1969), Stephens (1964), Weiser (1969); arthropods (in general): Poinar (1969).

Immune Unresponsivencess: Molecular Mimicry

Perhaps one of the most remarkable aspects of the host-parasite relationship is that parasites are able to survive so long in an 'immunologically hostile' environment such as that provided by (say) the bloodstream of the host. How this is achieved is a matter of great controversy, but - at least in some cases - parasites appear to be able to do this by avoiding stimulation of the host's immune response. This is achieved by presenting to the host a surface (and/or released metabolic products) which, antigenically, so closely resembles that of the hos, that the latter, regards it as 'self' and does not respond to its presence. This concept of antigen sharing by host and parasite has been termed 'molecular mimicry' and there is much dispute as to how this phenomenon develops; the various views are discussed further below. Other mechanisms of avoiding the host's response may also operate and a phenomenon referred to as 'immunological enhancement' in tumour biology may also, in some cases, be involved.

Antigen sharing

The idea that hosts and their parasites might share some antigenic determinants was first put forward by Sprent (1962). Subsequent investigations, especially by Capron and his co-workers have shown that the phenomenon is widespread and that helminths do not only share antigens with their hosts but also with other species within their class, and even with helminths from a different class. Thus, for example, in the trematode *Dicrocoelium dendriticum*, 19 antigens have been identified by immunoelectrophoresis, of which six are shared with *Fasciola hepatica*; the latter shares five antigens with *S. mansoni*. *F. hepatica* also shares four antigens with its intermediate snail host (*Lymnaea truncatula*) and a further six with its definitive cattle host. The occurrence of molluscan antigens in adult trematodes tnabled Capron *et. al.* (1968) to detect the presence of the latter (*Schistosoma mansoni*) in the definitive mammalian host by testing for antimollusc antibodies! A similar pattern of antigen sharing occurs in cestodes and nematodes.

12

Ecology of Parasites

The Turbellaria are in general marine animals; only the Catenulida and Temnocephalida are typically fresh-water groups without evident marine relatives or forebears. The Turbellaria have obviously spread from marine into fresh-water habitats and from thence into humid terrestrial regions. There has also been a considerable amount of return invasion of brackish and even salt water by fresh-water turbellarians, notably among the Catenulida, Macrostomida, triclads, and such rhabdocoel genera as *Dalyellia*, *Castrada*, and *Phaenocora*. The Turbellaria are characteristic and constant members of a variety of habitats. They are typically bottom dwellers. The large forms, as the triclads and polyclads, are generally found on hard bottom, under rocks and among shells and gravel, *etc.*; the small and microscopic members are more apt to occur on sandy and mucky bottoms.

A characteristic assemblage of sand-dwelling rhabdocoels and acoels has been found on north European coasts. These have certain adaptive characteristics in common: long, slender muscular body, long sensory hairs, tail appendages, eyelessness, loss of rhabdites, especial development of food-catching devices, overdevelopment of adhesive organs including girdles of adhesive papillae in a number of species, some reduction of the reproductive system, posterior displacement of the gonopores. The species of such sand communities appear to have a very limited distribution, *i.e.*, those of the north German coast are different from those the west French coast, although ecologically similar. Whether such sand-dwelling turbellarian communities occur along American coasts has not been ascertained but it is highly probable. The benthonic turbellaria are mostly limited to the littoral zone of the

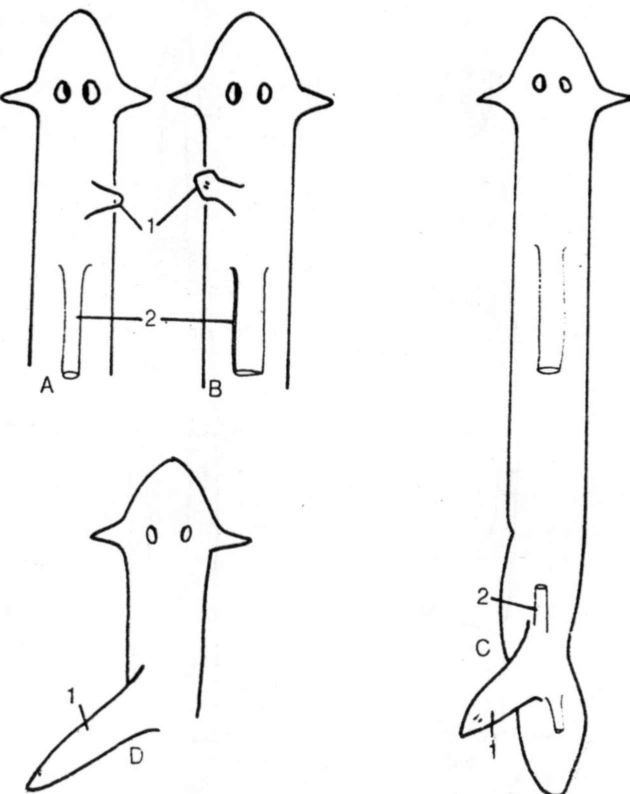

Fig. 12.1. Grafts of planarians. A, B—Types of outgrowth produced by grafts of small pieces from the brain region into the prepharyngeal region of Dugesia. C—Same kind of graft grafted into the postpharyngeal region. D—Postpharyngeal piece grafted into the prepharyngeal region of Dugesia; becomes a tail. 1-regenerate induced by graft; 2-pharynx.

ocean, although some occur at considerable depths; however, there are no abyssal species. Some acoels and polyclads have pelagic habits and show some adaptation of pelagic life by transparency and broadly oval form; many polyclads also have pelagi larvae or juvenile stages. Acoels and polyclads are regular components of the animal association of the floating Sargassum and may have the adaptive brown-and white colour pattern characteristic of Sargassum animals.

Nearly all pelagic Turbellaria are limited to tropical or subtropical waters; the chief exception is the rhabdocoel *alaurina composita* in the North Atlantic of which large number may be found in late summer in the Zuider Zee and adjacent waters (Hofker, 1930). Although the

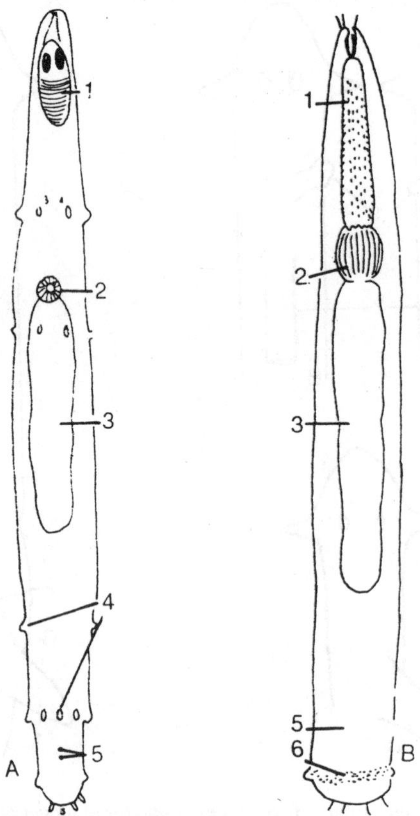

Fig. 12.2. Sand-dwelling types; two kalyptorhynchid rhabdocoels from the Bay of Kiel. A—Acerina remanei, with five bands of adhesive papillae. B—Rhinepera remanei, with a caudal adhesive girdle. 1-proboscis; 2-pharynx; 3-intestine; 4-adhesive papillae; 5-gonopores; 6-adhesive girdle.

ocean is in general a very constant habitat, turbellarians that live along shores in the tidal zone are subject to wide changes of temperature and salinity. The alloeocoel *Monocelis fusca* lives in the tide pools on the Welsh Coast where the nature temperature range is - 1 to 28°C. and can recover from exposure to 44.5°C. (Rees, 1941). This species and the marine triclad *Procerodes ulvae* also great changes in salinity.

The fresh-water Tubellaria are chiefly inhabitants of lentic waters and often have seasonal cycles related to such habitats, especially in the case of planarians and rhabdocoels that inhabit vernal pools. Some tuberllarians are *rheophilous*, *i.e.*, limited to flowing water, such as springs and hill and mountain streams. Outstanding examples of stream planarians are the European *Dugesia gonocephala*, *Polycelis felina* (= *cornuta*) and *Crenobia alpina* which occur along streams with increasing altitude and decreasing temperature in the order named. *Crenobia alpina* requires cold temperature and running water. No alpine planarians are known for North America, but the species *Phagocata morgani*,

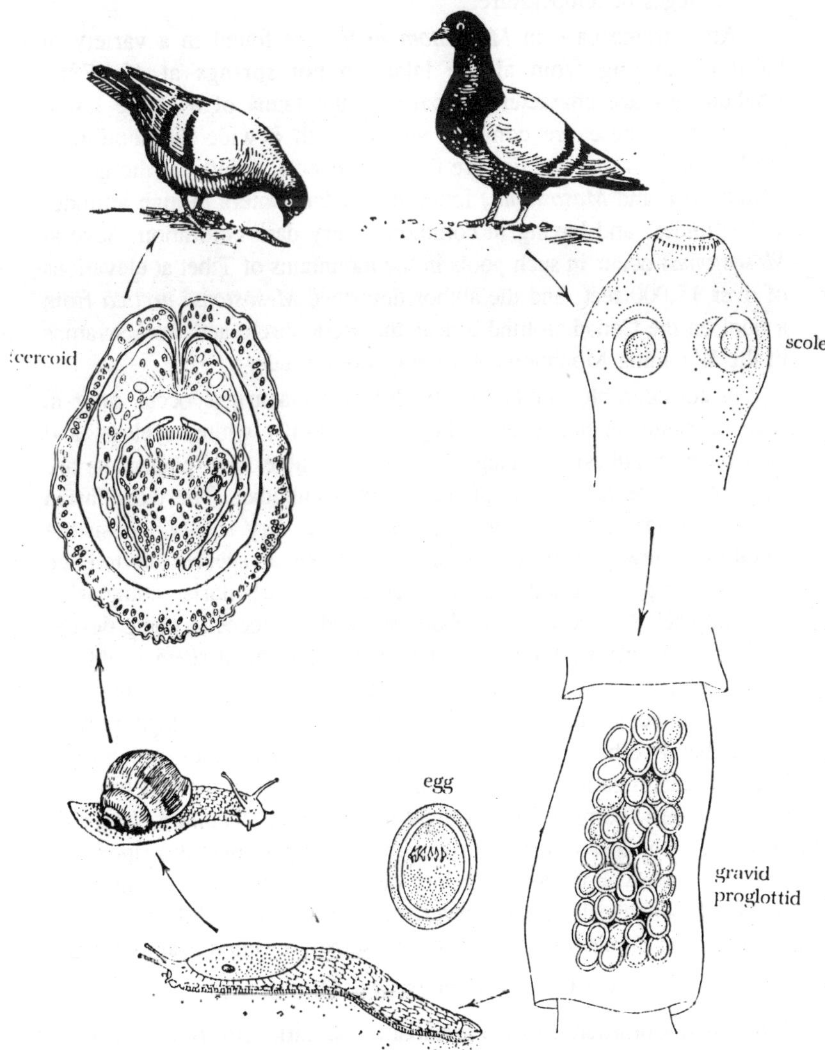

Fig. 12.3. Raillietina bonini, a tapeworm of pigeons.

Appalachians, and *Polycelis coronata*, western mountains, seem limited to flowing waters. Our common *Dugesia dorotocephala* appears to live only in springs or spring-fed waters, probably because of the high calcium content of such habitats. Although some tuberllarians are *stenothermous* (enduring only a limited temperature range) as *Crenobid alpina*, most fresh-water forms are *eurythermous*, *i.e.*, indifferent to wide changes of temperature.

An extreme case in *Microstomum lineare* found in a variety of habitats ranging from alpine lakes to hot springs at 40-47°C. Rhabdocoels are characteristic parts of the fauna of a alpine lakes, but most of these are common species with a wide distribution in fresh waters. More specific are the rhabdocoels, chiefly of the genera *Phaenocora* and *Mesostoma*, found in *shallow* waters in high altitudes and latitudes, and having in common a very dark coloration. Several *Mesostomas* occur in such pools in the mountains of Tibet at elevations of over 15,000 feet, and the author described *Mesostoma arctica* from a pool on the Canadian tundra near the arctic circle and from swamps in the Wyoming Mountains at 10,500-foot altitude.

A considerable number of fresh-water planarians occur only in cave or subterranean waters; they are generally white, eyeless, and provide with adhesive organs. The most notable cave planarians are the North American ones, of the family Kenkiidae. The turbellarian fauna of tropical fresh waters appears to be very scanty, consisting chiefly of a few widely distributed or cosmopolitan species that tolerate a broad range of conditions. The terrestrial Turbellaria, comprising the land triclads, and a few rhabdocoels and alloeocoels, being devoid of protective armor, require high humidity and most seem unable to mature their sex organs except in warm temperatures. Hence they are mostly limited to tropical and subtropical jungles with thigh rainfall where they live by day surrounded by mucus under stones, fallen logs, and leaf accumulations, in earth burrows, or between the leaf bases of tropical plants, coming forth at night to hunt food. *Bipalium*, subjected to desiccation, lost water according to the simple laws of evaporation and could recover if the water loss did not exceed 45 per cent of its weight (Kawaguti, 1932). However, mucous secretion plus aggregation into the most humid spot should furnish some protection against drying.

Relations with other Organisms

Symbiosis, Commensalism, Parasitism, Parasitic Infections, Enemies

Most acoels and some rhabdocoels and alloeocoels harbor symbiotic chlorellae and xanthellae; these occur in the mesenchyme chiefly just inside the subepidermal musculature and appears to be of the same nature and function as in other animals. They are absent in the eggs and newly hatched young, and it is known that they are acquired during juvenile stages by ingestion with food. They pass into the gastrodermis whence they are ingested by free mesenchymal cells that convey them into the mesenchyme (von Haffner, 1925). Much study has been devoted to the chlorellae of the acoel genus *Convoluta*, especially *Convoluta roscoffensis*, which occurs on the channel coast of France in such vast numbers as to colour large areas green. Keeble and Gamble (1907) have shown that the green cells in this worm are flagellates of the chlamydomonad group and that they occur abundantly in the habitat and on the eggs masses of *Convoluta*. These authors state that this acoel feeds while young but after attaining sexual maturity subsists entirely by digesting its chlorellae. A number of fresh-water rhabdocoels of the families Dalyelliidae and Typhloplanidae contain green cells that can be cultivated apart from the worms and appear to be green algae of the genus *Chlorella* or close to this genus. It is notable that the *Dalyellia* species harboring these symbionts are much larger (2 to 5 mm.) than other *Dalyellias*. Sojourn is darkness almost but not entirely eliminates the green cells of *Dalyellia*.

Dalyellia digests senile and degenerating green cells but dies not damage healthy ones. The green and brown symbionts of Turbellaria contain starch, fats, and other lipids and undoubtedly possess some nutritive value for their hosts. Their chief advantages are, however, that they utilize in their metabolism waste products of their hosts (it will be recalled that the Acoela have no excretory system) and that they give off oxygen enabling the host to endure stagnant conditions. Thus the acoel *Amphiscolops langerhansi* will survive enclosure in small airtight bottles of sea water for 2 weeks or more in the light but succumbs in the dark under otherwise similar conditions (Welsh, 1936); in the light, oxygen is added to the water by the photosynthetic activity of the brown cells of the worm.

As already noted, all groups of Turbellaria, except the fresh-water and terrestrial triclads, contain members that inhabit other animals in a relation that appears to be commensal rather than parasitic. Here may be recalled the acoels that live in the interior of echinoderms; the rhabdocoel families Graffillidae and Umagillidae, entocommensals of mollusks, echinoderms, and some other invertebrates; the wholly ectocommensal group Temnocephalida; the marine triclad ectocommensals of *Limulus* and of skates; and the polyclad habit of

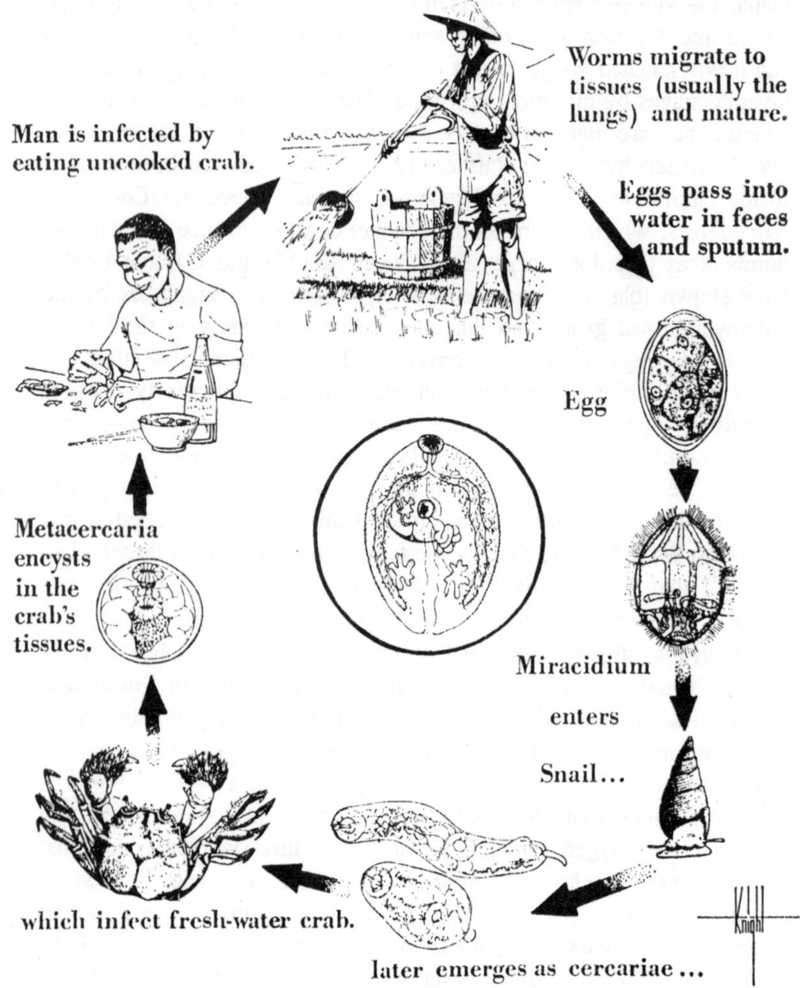

Fig. 12.4. Life cycle of Paragonimus westermani.

dwelling in shells of live snails or in dead snail shells occupied by hermit crabs.

The epizoic or ectocommensal forms are characterized by loss of pigment, cilia (especially dorsally), rhabdites, and often eyes, and especially by the strong development of adhesive areas and organs. The entocommensals, on the other hand, usually lack adhesive organs and are generally completely ciliated although they also tend to lose rhabdites and eyes; their most outstanding feature is the enlarged female system, exemplified by the branching yolk glands, elongated ovaries, *etc.* Such adaptation for increased egg production is a common feature of genuine parasites and no doubt indicates a tendency toward true parasitism among the entocommensal Turbellaria. However, probably the only genuine entoparasites among the Turbellaria are the Fecampiidae and *Oekiocolax*.

The Turbellaria are themselves hosts to commensals and parasites. Euglenoid flagellates of the general *Astasia* and *Euglena* occur in catenulids. A trypanosome-like flagellate, *Cryptobia* (= *Trypanoplasma*) *dendrocoeli*, inhabits the digestive tract and copulatory bursa of *Dendrocoelum lacteum*. Holotrichous ciliates of the mostly free-living genera *Holophyra* and *Ophyroglena* may occur in *Stenostomum* and planarians. Members of the parasitic ciliate group Astomata are also represented, as *Hoplitophyra uncinata* with an anterior hook in the intestinal branches of marine triclads and *Siebolidiellina* (= *Haptophrya*) *planariarum* with an adhesive disk in fresh-water planarians.. *Coleps* and *Trichodina* species often epizoic on fresh-water planarians.

The most common endoparasites of planarians are gregarines, chiefly of the genus *Lankesteria*. Encysted stages of larval trematodes are sometimes found in Turbellaria but probably represent accidental penetrations of wrong hosts. Land triclads seem much subject to nematode infestations. The Turbellaria are seldom eaten by other animals as their surface secretions appear to be distasteful if not actually toxic. Planarians, if taken into the mouth by fish and salamanders, are usually quickly ejected with signs of distaste. Bolen (1938) saw dragonfly and damsel fly nymphs eat planarians. The worst enemies are other turbellarians for the group is highly cannibalistic, promptly attacking weak or injured members or new fission products; the larger species may also eat the smaller ones.

Duration of Life, Cycles, Senescence, Rejuvenescence

The length of life of turbellarians is known chiefly from laboratory cultures. The smaller forms, except those capable of fission, probably live but for a short time, a few weeks or months, usually dying in summer or autumn after laying dormant eggs. Some, however, may survive a year or more by encysting during unfavorable seasons. Species capable of fission can presumably under favorable conditions propagate indefinitely by the asexual method. Thus Nuttycombe and Waters (1938) cultivated *Stenostomum tenuicauda* for 11 years during which there were over 1000 asexual generations without the appearance of sexuality.

They also had male specimens of *S. grande* that reproduced continuously by fission for 4 years. Sekera (1926) kept a *Microstomum* colony going for 3 years. Similar facts are known for asexually reproducing species of fresh-water planarians. Child (1914) maintained *Phagocata velata* in asexual culture for nearly 3 years during which there were 13 cycles of fragmentation, encystment, and hatching of small worms from the cysts.

Individuals of the larger turbellarians survive in laboratory cultures 2 or 3 years and may have several periods of sexual propagation alternating with periods of fission or lack of any reproduction. Something has been said in the preceding pages concerning sexual reproduction in relation to season. This matter has been studied in fresh-water planarians by several workers. In general sexual reproduction is intermittent, alternating with periods of fission or of lack of any reproduction; the periods of capsules lying correspond in nature with external factors, chiefly temperature. In laboratory cultures, sexual reproduction is also intemittent but does to recur with the regularity noted in nature. Kenk, Vandel, and others believe there is an inherent cyclic reproductive rhythm in these planarians, but the author found that the same species used by Kenk (*Dugesia tigrina*) can, when fully grown, be caused at will to develop a copulatory apparatus and lay viable capsules by exposure to low temperature followed by rise of temperature.

In some individuals even 2 or 3 days of exposure to lowered temperature is sufficient to bring on sexuality. Some time must elapse before another sexual period can be induced in the same individual by temperature manipulation. This time is much less than a year, mostly a few months, so that the alleged inherent rhythm is probably only an expression of such a "refractory" period following a period of reproductive activity. After continued laboratory culture or after sexual reproduction in nature, turbellarians often evince signs of senescence such as loss of pigment, sluggish behaviour, unwillingness to feed, and disintegrative processes in the eyes, head, or pharynx. Death by general disintegration frequently follows such changes. In species with high regenerative powers, however, rejuvenescence may be brought abut experimentally or occurs naturally. Thus fission products or animals emerging from cysts are physiologically like young animals.

Through starvation, adult worms may be reduced to the size of very young worms and so far as can be tested have actually become young again. Adult worms may be cut into small pieces, and these after regeneration are also indistinguishable from young worms. The already much mentioned *Phygocata velata* has in temporary pools a natural cycle of senescence and rejuvenescence; the worms grow to mature size, show the above symptoms of senescence, then fragment into small bits, each of which inside a cyst regenerates to a tiny worm that in all respects is identical with young worms hatched from capsules. Apparently this cycle is repeated in nature indefinitely with no loss of vitality. By underfeeding this species so that it does not grow, senescence and fragmentation and prevented, and the worms remain in a state of perpetual youth.

Osmotic Relations

As already noted, both fresh-and salt-water turbellarians may invade brackish habitats. In a fresh-salt-water gradient as the entrance of a river into the sea, there is generally a barren zone, poor in animals, lying between the limits of increasing salinity endurable by the fresh-water animals and of decreasing salinity endurable by the marine forms. To endure changes of salinity animals must have some mechanism for regulating loss or entrance of water and salts. Generally fresh - water forms invading brackish water show reduction in size and greater tissue density as a result of water loss, and the reverse changes occur in marine forms adapting to lowered salt content. The best-studied case is that of the marine triclad *Procerodes vlvae* living on the English Coast at the mouth of streams. This animal is subject in nature to changes lasting several hours from the fresh stream water (salt content, 0.03 g. per liter) to sea water (salt content, 35 g, per liter). It can live indefinitely in all dilutions of sea water to 5 per cent; it swells and loses salts in the greater dilutions but not at all in proportion to the decreased osmotic pressure.

In pure fresh water (*i.e.*, without salts), it swells to nearly twice its volume, loses salts, and survives only 1 to 3 days. The worm can recover after loss of 85 to 90 per cent of its salts. In the natural stream water, swelling and salt loss are much less and survival is better than in pure water. Osmotic pressure appeared not to be an important factor in the effect of the various media tried. It was found that calcium, in which the stream where *Procerodes* lives is rich, is the most effective factor in reducing swelling and salt loss in diluted media. Hence it appears that the presence of calcium in fresh water enables marine animals to invade such waters. The cause of the protective action of calcium is not clear, although probably calcium makes the epidermis resistant to the passage of water and salts. Beadle (1934) studies sections of *Procerodes ulvae* that had been exposed for varying periods to diluted sea water and found that water, passing through the epidermis, at first accumulates in the mesenchyme but later is transferred to the gastrodermis where is accumulates in numerous vacuoles. Through this vacuole formation swelling is reduced, the mesenchyme is restored to normal, and the animal regains more normal appearance and behavior.

In time the epidermis develops resistance to further water intake, although the gastrodermal vacuoles remain as long as the worms are kept in diluted media; *i.e.*, there is not discharge of excess water through the mouth. A number of other littoral turbellarians live under conditions in which they are exposed daily to large changes in environmental salinity. Thus *Monocelis fusca* inhabits tidal pools subject to evaporation and hence concentration of the sea water. The salinity of these pools was found to vary from 26 to 64 parts per thousand; the alloeocoel endured in tests a range from 5.6 to 76.8 parts per thousand but began to show sluggish behavior at 80 parts per thousand. It can, however, form a mucous cyst within which it can survive exposure to salinities of 298 parts per thousand, or over eight times the salinity of sea water (Rees, 1941).

Fresh-water planarians when exposed to solutions hypertonic to natural waters lose water and weight but not in proportion to the increased osmotic pressure, and they show some ability to resist water loss. *Dendorcoelum lacteum* was studied by Gresens (1928) in stream opening into the Danish Wiek where it lives in a salt gradient from pure stream water (salt content, 0.01 to 0.04 per cent) to the Wiek water (0.05 per cent salt). Individuals from the more brackish water proved more resistant to high salinities than stream individuals. Worms could be acclimated by gradual changes so as to live normally and reproduce in 1.5 per cent salinity. Low temperature is an important factor in enabling planarians to resist increased salinity. Some data are available on the salt content of the turbellarian body. According to the work of Pantin (1931), the salt content of *Procerodes vlvae* is about half that of sea water, or about 1.7 per cent. In *Stylochus frontalis*, the salt content is a little over 2 per cent. Hence the osmotic pressure of the body fluids of marine turbellarians is considerably less than that of sea water and the epidermis must be able to resist water loss or salt intake or some to other regulating mechanism must be present.

Dilution of the sea water cause increased oxygen consumption in marine invertebrates, and this represents energy consumed in resisting water intake. The osmotic pressure of fresh-water turbellarians is greater than that of fresh water, and they require some mechanism to resist water intake and salt loss or to eliminate excess water. Murray (1927), seeking a medium suitable for growing planarian cells in tissue culture, found that the balanced salt solutions (*e.g.*, Locke's, salt content, 0.9 per cent) in use for vertebrate tissues are far too concentrated for fresh-water planarians and require one-tenth to one-twelfth or more dilution. This indicates that the somotic pressure of the species used (*Dugesia dorotocephala*) is equal to about a 0.08 per cent solution of sodium chloride. This species also requires a higher ratio of calcium to potassium than is customary in salt solutions made up for vertebrate tissues. The water content of the polyclad *Stylochus frontalis* is 76.2 per cent; that of the land triclad *Bipalium* averages 81.5 per cent, grading from 80.7 per cent in the anterior third to 82.9 per cent in the posterior third (Kawaguti, 1932); and that of a fresh-water planarian, *Dugesia dorotocephala*, is 78 per cent.

Hydrogen-ion Concentration

Turbellarians are in general indifferent to the pH of their environment, enduring without evident effect wide changes in this factor. *Dendrocoelum lacteum* will live indefinitely in a range from pH 4.2 to 9.5; greater acidities or alkalinities are fatal. The toxicity of acids for *Dugesia dorotocephala* depends on the acid used, being less for mineral acids such as hydrochloric and nitric, which kill in 2 to 3 hours at pH 3.0 to 3.4, than for organic acids, especially butyric, which is fatal in the same time at pH 5.0. Similarly alkalinity caused by ammonium hydroxide is more toxic (lethal at pH 9.0) than that caused by sodium hydroxide (lethal at 9.2). In alkalinities and acidities less than those mentioned, *D. dorotocephala* survives indefinitely, although it is to be noted that planarians are able to reduce the alkalinity and acidity of the medium and it is very difficult to maintain a constant pH in solutions in which they are immersed. In general, alkalies accelerate and acids depress the rate of oxygen consumption of planarians, but the acid effect results, not from the pH as such, but from the carbon dioxide set free when natural water is acidified.

LOCOMOTION

The smaller Turbellaria, including juvenile stages of the larger forms, probably move entirely by ciliary action. Although they probably can swim, few of them habitually do so, but for the most part they glide along the substratum or the underside of the surface film. The ciliary waves pass from anterior to posterior, thus driving the animal forward, and appear to be independent of the nervous system. Forms exceeding 2 to 3 mm. in length mostly crawl along a substratum with a characteristic gliding movement. Although the ventral cilia probably play some role in this gliding, if the worm is not too large, this type of locomotion is caused largely or wholly by muscular waves that pass form the anterior end backward either across the whole width (monotaxic) or alternately along the two sides (ditaxic). Triclads have mostly monotaxic, polyclads ditaxic waves. The waves are mostly invisible to the naked eye in planarians, more noticeable in polyclads.

The relative roles played by cilia and by muscular waves in locomotion can be determined by treating worms with lithium chloride which paralyzes the muscles. The first chemical does not stop locomotion in triclads and polyclads, whereas the second does. A number of polyclads swim by means of body undulations; these also may be monotaxic or ditaxic depending on the body shape. Most turbellarians are capable of a more hasty type of locomotion which they employ when stimulated. This consists of large evident muscular contractions and expansions accompanied in those forms with adhesive organs or areas by alternate attachments of these, so that leech-like form of progress results. Usually such an effort lasts but a short time and the usual gliding is soon resumed.

Mucous secretion probably plays an important role in gliding, even in aquatic turbellarians, and is obviously indispensable to terrestrial forms. The slime trail lubricates and smooths out the substratum on which progression occurs. The land triclads and some other Turbellaria also use slime threads in dropping from one level to another or in bridging gaps in a substratum. The role of the nervous system in locomotion has been investigated by several workers. In general, turbellarians from which the cerebral ganglia have been removed tend to cease locomotion and to remain quiescent. They can, however, be stimulated to locomotion by mechanical or chemical means and are able to execute the usual creeping.

Levetzow (1936) found that even quite small postcerebral pieces of *Thysanozoon* and *Eurylepta* are able to crawl. Such experiments indicate that the more simple muscular waves are independent of the brain and are mediated by the general nerve plexi. Most polyclads, however, are unable to swim following decerebration, and some cannot even execute the regular ditaxic crawling. Therefore, the more complicated types of locomotion depends on initiation from the brain, also on transmission along the main nerve cords, since cuts through these inhibit swimming waves posterior to the cuts.

FOOD AND FEEDING BEHAVIOR

The Turbellaria area as a class carnivorous. A few acoels and rhabdocoels are said to eat diatoms and other algae, but the vast majority of turbellarians eat other animals of suitable size or ingest bits of flesh of intact, damaged, or dead larger animals as fish and mollusks. Favorite items of food of the smaller species are rotifiers, copepods, caldocerans, nematodes, annelid worms, etc., and the larger species feed on earthworms, snails, insect larvae, isopods, amphipods, other crustaceans, fish, oysters, and so on. Although the Acoela lack a digestive cavity, they are on the whole strictly carnivorous, ingesting copepods and similar small animals whole and digesting them in their interior cells. Aquatic and land triclads can attack and subdue intact animals of considerable size such as snails, earthworms, and insect larvae, and polyclads often devour bit by bit live sessile, helpless animals such as oysters, barnacles, ascidians, and the like.

Turbellarians often gather, often in numbers, on animal corpses if these are not too decayed. Insects falling into the water constitute a regular food supply for pond turbellarians. Although most turbllarians will eat almost any kind of animal flesh, some appear limited to particular items. Thus *Stylochus frontalis* (= *inimicus*) of the southern Atlantic coast of the United States seems to eat nothing but oysters, slipping between the shell valves and devouring the oyster bit by bit despite shell partitions erected against it, and *Pseudoceros crozieri* apparently feeds exclusively on ascidians. In laboratory cultures, turbellarians can be fed on chopped or crushed earthworms of other annelids, meal worms and other insect, clams, snails, isopods, amphipods, clotted blood, or, most conveniently, bits of beef liver, although some species will not eat the last.

Some species require live prey such as *Daphnia*. In general, turbellarians from lentic waters are most easily kept in the laboratory. Polyclads are usually difficult to maintain under laboratory conditions. Turbellarians may often be captured in numbers by baiting suitable habitats with pieces of beef or dead fish; at short intervals the bait is examined and any worms present shaken off into jars. The presence of intact prey (not emitting juices) is detected either by direct contact with chemoreceptors or by the disturbance created in the water by the prey. The tubellarian then quickly grips the prey with the anterior end, especially the anterior adhesive organ when this is present, and wraps its body about the prey, holding the latter firmly against the substratum. Adhesive margins, disks, etc., are used by the worm to maintain a firm hold on the substratum while subduing the preying it helpless.

Crustaceans and baby fish living in the same habitat with planarians may become coated with the slime left by planarians so as to be unable to move and are then attacked. Some turbellarians use the penis stylet to damage prey. Food emitting juices is sensed in still water at a distance varying with species from 1 or 2 to 10 to 15 cm. or more. A turbellarian, previously gliding aimlessly about, when coming within the radius of diffusing juices exhibits a characteristic behviour. The worm pauses, swings the raised head about, then turns toward the food and goes to it in a direct line. Planarians living in running water detect food juices coming downstream at considerable distances, and if such places are baited with meat or fish the worms can be seen emerging in hordes from under stones and leaves and travelling in processions upstream to the bait.

The European stream planarian *Crenobia alpina* is in fact more or less incapable of directing itself toward to foot in the absence of a current and even if it finds the food may often fails to feed in still water. Having reached the food, planarians test it with the anterior margin, then grip it with the head and move the body along until the pharynx can be brought into play. The protruded pharynx is inserted into the food, and ingestion begins. During feeding the worms keeps some part of its body firmly attached to the substratum. The extirpation experiments of Koehler (1932) and Müller (1936) have shown that the lateral sensory organs of the head-the ciliated pits of *Stenostomum*, the ciliated grooves of *Phaenocora* and *Bothromesostoma*, the auricles including the auricular sense organs of planarians, or the corresponding lateral angles of the head in planarians that lack definite auricles—are the chief or only sensory areas involved in chemical perception of dissolved juices. *Stenostomum* after extirpation of the ciliated pits cannot detect the presence of food or find it, and planarians deprived of their auricles or corresponding regions show great loss in their food reactions. However, fresh-water planarians appear to have some chemo-receptors over the entire body and fatter such operations may in time find food.

The testing of the suitability of food after finding it is vested chiefly in the anterior margin, which also functions to release the feeding reaction of the pharynx. After extirpation of this margin, planarians find the food as readily as before but fail to test it or grip it and often glide away without ingestion. The tip of the pharynx is, however, sensitive to food, and the pharynx may protrude on contact with food in the absence of anterior sensory regions. The pharynx tip can discriminate between normal food and the same food mixed with deleterious chemicals; but, after removal of the both the anterior margin and the pharynx tip, the pharynx loses this discriminatory ability and may protrude into such treated food. The presence of fluid nutrients such as bouillon or blood in the water may induce repeated pharynx protrusions of intact planarians.

According to some workers, headless pieces containing the pharynx or isolated pharynges will not feed when placed in contact with food, but according to other such reactions do occur in some species. According to the available experiment, then, ciliated pits, auricles, auricular sense organs, and corresponding head areas have sensory functions similar to smell in higher animals, and the anterior margin and pharynx tip function like taste. After the prey has been subdued and firmly gripped, ingestion ensues. Turbellaria with simple and doliiform types of pharynx generally swallow their prey whole by means of muscular waves in the distended but not very protrusible pharynx. The rosulate pharynx is not distensible, and hence the contents of the food are sucked in. The triclads also do not swallow their food whole, but the highly protrusible plicate pharynx is inserted into the food or into the prey and the contents sucked in bit by bit through the peristaltic action of the pharynx.

In polyclads the food is usually taken whole into the interior of the voluminous ruffled pharynx. Even those polyclads that have a cylindrical pharynx like that of triclads appear to swallow their prey whole, apparently limiting themselves to small animals. Whether the secretions of the pharyngeal glands have a softening and digestive action on the food cannot be answered categorically. These secretions possibly have such action is some species, but in general they function only to lubricated the food and facilitate its ingestion. Even in those cases where food is help in the pharynx and is seen to undergo digestion, it does not necessarily follows that the enzymes involved are of pharyngeal origin.

Extirpation of the brain (without disturbances of the main chemo-receoptors of the head) may greatly retard or altogether abolish the food-finding and food-ingesting reactions. The result, however, varies with species. The most marked affects of brain removal appear in *Crenobid alpina* which will not ingest food after this operation (Koehler, 1932), but other species may show little alteration of food reactions after decerebration. Whether the surface and pharyngeal secretions of turbellarians have a toxic and paralyzing as well as entangling effect on prey is a question often raised but never satisfactorily answered. In reviewing this question, Arndt and Manteufel (1925) note that large land planarians may produce an unpleasant astringent effect on the human tongue and that a South American land planarian, *Polycladus gayi*, is believed by the natives to be fatal when eaten by horses and cattle. These authors found that water extracts of ground fresh-water planarians were toxic or even fatal when injected into the coelom of small mammals. Later Arndt (1925) located this toxin in the surface secretons of planarians and found that it exercised its effect by stopping the heart.

Mattes (1932) observed that the cercariae of flukes are damages and killed by the surface slime of planarians. However, students of planarians, including the author, in general fail to see any toxic or paralysing action of planarian slime on the usualy prey, such as crustaceans and insect larvae. The nutritional requirements of palanarians are not definitely known, although attempts in this direction have been made by Wulzen, Bahrs, and others.

Vertebrate live, kidney, thymus, adrenal, and gray matter of the brain support excellent growth, whereas little or on growth is obtained with egg white, egg yolk, lean beef and other muscle, thyroid, pancreas, gelatin, beef fat, and the white matter of the brain. Tissues from animals fed on deficient diets also have an adverse effect on planarian growth and may cause a pathological condition in which the worms shrived and turn back. Raw egg white alone or in high proportions in the food is very deleterious, causing disintegrating protrusions on the worms. Refrigeration at 0°C. or heating decreases the growth-promoting power of vertebrate tissues. Greenberg and Schmidt failed to find any favourable effect on planarian growth of adding vitamins and auxins to inadequate foods, although yeast addition was effective. Of all vertebrate tissues, liver has about the best growth-promoting power for planarians, and it contains and ether-soluble heat-labile growth factor of unknown nature.

Digestion, Assimilation, and Food Storage

It was discovered by Metschnikoff in 1866 for *Convoluta* and later verified for rhabdocoels and triclads (Metschnikoff, 1878) that digestion in the Turbellaria is largely intracellular. As already indicated, it is improbable that the pharyngeal secretions have any digestive action, at least in planarians. Wetblad (1922) failed to find any digestive action on protein or fat of extracts of 50 pharynges of *Dendrocoelum lacteum* and 100 pharynges of *Polycelis nigra*. It is also improbable that any digestion occurs in the lumen of the intestine in planarians Willier, Hyman, and Rifenburgh, 1925). The granular clubs, usually regarded as enzymatic cells emitting enzymes into the intestinal lumen, are in the author's opinion protein reserve cells. The details of intracellular digesting are best known for fresh-water triclads where the most complete studies are those of Westblad (1922) and Willier, Hyman, and Rifenburgh (1925).

The latter fed planarians, starved 2 weeks, on beef liver and studied sections of worms fixed at varying intervals after the beginning of feeding. The liver torn into minute bits by the suctorial action of the pharynx fills first the anterior, then the posterior rami of the intestine. Some 45 to 60 minutes were required for complete filling of the intestine in the species used (*Dugesia dorotocephala*), but the phagocytic cells of the gastrodermis begin to engulf the food as soon as this comes in contact with them. These cells swell through intake of water, bulge into the lumen, often fuse into syncytia, and take in food particles in amoeboid manner by putting out pseudopodial processes.

The food particles together with fluid are formed into typical food vacuoles like those of Protozoa. About 8 hours were required for all the food in the lumen to be phagocytized by the phagocytic cells which at this time are found packed with food vacuoles in various stages of digestion. The food in the vacuoles, at first loose and faintly stainable, gradually condenses to a strongly staining (eosinophilous) homogeneous ball shown by chemical tests to be of protein in nature. About 8 hours are required for food vacuoles to reach the homogeneous ball stage. These balls then gradually disappear, presumably as a result of the breakdown of their protein into soluble substances; about 5 days are required for the complete disappearance of the food balls from the phagocytic cells.

In a few phagocytic cells, the food balls do not disappear but subdivide into small spherules, and such cells then become granular clubs. Westblad, using indicator dyes with the food, was unable to detect any change of hydrogen-ion concentration in the food vacuoles so that digestion appears to take place in a neutral medium. He found evidence of proteolytic and lipolytic enzymes but not of diastatic ones; in fact it appears that planarians are unable to digest carbohydrates.

In *D. dorotocephala*, the basal halves of the phagocytic cells are normally packed with fat, which largely disappears during the early stages of intracellular ingestion and digestion. It is also known that, during this period, the oxygen intake is doubled or tripled. It thus appears that the phagocytic activities of the gastrodermis consume extra energy and that this energy is supplied by the oxidation of the fat reserves. During the later stages of digestion fat reappears in the phagocytic cells and increases to its former amount. The indications are that the protein of the food is transformed into fat, without the intervention of carbohydrate stages, and that this fat is stored in the gastrodermis as a reserve. Fat droplets also occur sparingly in the mesenchyme of planarians.

As already noted, planarians cannot digest carbohydrates and no carbohydrate reserve (glycogen) has been detected in most species studied, although von Brand (1936) found 3 per cent by weight in *Planaria torva* in autumn, less in winter. The available researches indicate, the, that in fresh-water planarians practically the entire processes of digestion, assimilation, and storage of food occur within the phagocytic cells of the intestine. Nothing is known of the manner in which the food reserves are transported to and utilized by other body tissues. In those Turbellaria that swallow their food whole one infers that considerable digestion must take place in the lumen before phagocytosis can occur.

Metschnikoff (1878) and Westblad (1922) claim that digestion is wholly extracellular in those rhabdocoels having a ciliated gastrodermis (Catenulida, Macrostomida). Westblad saw protein and fat particles undergo digestion in the intestinal lumen of *Stenostomum* while starch grains remained unaltered. However, Jacek (1916), also using *Stenostomum*, found that digestion is wholly intracellular and that protein particles (egg white) and oil droplets are digested in vacuoles in the gastrodermis; during protein digestion the vacuolar fluid changed from an acid to an alkaline reaction and protein was converted into and stored as fat. Starch grains were ejected unaltered. Some observations of the author confirm the occurrence of intracellular digestion in *Macrostomum*. It may be assumed that in rhabdocoels small enough particles are ingested at once by the gastrodermis whereas larger ones must first undergo some extracellular digestion.

Extensive storage of glycogen in temnocephalids was reported by Fernando (1945). Nothing is known of digestive processes in polyclads, which, it may be recalled, also have a ciliated gastrodermis. Defecation has been witnessed by several observers. The worm first takes in water through the pharynx, then by vigorous muscular contractions expels indigestible remnants. Those polyclads that possess them use the anal pores as well as the mouth for defecation.

Starvation

The fresh-water planarians can endure prolonged starvation, from 6 to 14 months, depending on species. During this time they may reduce to 1/10 to 1/13 the original length and 1/300 of the original volume. This reduction results from tissue degeneration and does not involve any alteration of cell size. The nervous system undergoes no alteration so that greatly starved animals have disproportionately large heads. Epidermis and musculature are also practically unaltered, but the mesenchyme degenerates and disappears for the most part. The digestive system retains its form, and there appears to be on reduction in the number of diverticula, but the volume of the intestine is greatly diminished through degeneration of large numbers of the phagocytic cells. The gastrodermis loses food inclusion and reduces to a flattened syncytial state.

In the later stages of starvation the fat stored in the phagocytic cells begins to be utilized but Willier, Human and Rifenburgh (1925) found that the surviving gastrodermis still contained considerable fat after 3 months' starvation in *D. dorotocephala*. However, the complete breakdown of most of the gastrodermis must furnish a large amount of fat and protein to the planarian. The greatest degeneration in starving worms is seen in the reproductive system. The copulatory apparatus, sex ducts, and yolk glands entirely disappear, and the gonads reduce to small clusters of cells. The oxygen consumption of planarians per unit weight declines during early starvation but rises continuously later, and indication that starvation in these worms acts as a rejuvenating factor. The ability of other Turbellaria to endure prolonged starvation is not well known, but the smaller forms are probably able to survive for only short periods in the absence of food.

Respiration

The Turbellaria are devoid of any special respiratory structures; the exchange of respiratory gases takes places by simple diffusion through the body surface. So far as known, respiration is of the ordinary aerobic type, comprising the intake of free oxygen and the emission of carbon dioxide. Exact measurements of respiratory rate have been made chiefly on fresh-water planarians. The rate of oxygen consumption of *Dugesia dorotocephala* is about 0.2 to 0.3 *cc.* per gram per hour in adult worms; the rate per unit weight higher the smaller (younger) the worms, and is also increased by regeneration, during the early stages of intracellular digestion (see above), and during later stages of starvation. The oxygen consumed is also affected by the osmotic pressure of the medium as well as by the kind of salts in the medium. The rate of oxygen consumption of *Dugesia* species is constant at different oxygen concentrations in the water until this concentration falls below about one-third saturation. The cutting of planarians into pieces accelerates the rate of oxygen consumption, and this increase is greater the smaller the pieces, *i.e.*, if planarians are cut into six or eight pieces the rate of oxygen take of the combined pieces so greater than that of intact worms.

This stimulatory effect of cutting declines with time and disappears within 24 hours. After allowance has been made for this effect, an axial series of pieces of *D. dorotocephala* shows an anterio-posterior decline in rate of oxygen intake, another expression of the axial gradation already mentioned several times. Some experiments on the carbon dioxide production of the polyclad *Stylochus ijimai* are also available. Here, too, the rate of carbon dioxide production was found the decline with increasing size and to show an axial gradation in the form of a U, *i.e.*, a decrease from the anterior end toward the middle and an increase again toward the posterior end. Unfortunately sexual worms were used, and as the reproductive system is not evently distributed throughout the worm, a factor incapable of evaluation was present.

The oxygen consumption of planarians is depressed in the presence of potassium cyanide. It can be depressed as much as 90 per cent without injury to the worms. This indicates that the oxidative mechanism of planarians is chiefly of the cyanide-sensitive type, that it belongs to the "Warburg-Keilin" system, involving an iron-containing oxidase, a cytochrome complex, and a dehydrogenase. The dehydrogenase enzyme attacks the substance furnishing the energy, oxidizing it, and reducing the cytochrome complex which in turn becomes reoxidized by reducing the oxidase enzyme. The latter is restored by means of the free oxygen in the air or water, and so the system continues indefinitely as long as free oxygen is available. Cyanide acts by preventing the oxidase from taking up free oxygen. The cyanide-stable respiration, which constitutes only a small part of the respiration of planarians, operates by way of enzymes other than oxidases.

Excretion

Nothing whatever is known of the form in which turbellarians eliminate their nitrogenous wastes. From the fact cited above that protein food is stored chiefly as fat in planarians, it seems evident that a process of deaminization must occur and hence that ammonia or some related substance must be given off. Several attempts have been made by use of vital dyes to determine what tissue is involved in elimination. Such experiments rest on analogy with conditions in vertebrates where it is known that vital dyes introduced into the body are treated like waste products and ejected by way of the kidneys. When the Acoela (which have no protenephridial system) are exposed to vital dyes, the dyes collect in the central or digesting mesenchyme and are ejected through the mouth.

In triclads, also, the gastrodermis serves as the chief means of elimination of vital dyes, the dyes granules collect in mesenchymal vacuoles which are then passed to the intestine. In a number of kalyptorhynchid and typholoplanid rahabdocoels, also in *Stenostomum*, the dye collects in certain portions of the protonephridia, mostly the more proximal portions, and in the paranephrocytes which also pass it into the protonephridia so that in these forms some evidence is available of an excretory function of the protonephridia, aided by paranephrocytes. No evidence has been seen of any passage of substances through the flame bulbs in any turbellarian.

It seems probable that in Turbellaria the necessity of eliminating nitrogenous wastes is avoided by converting them into insoluble granules ("concrements") that are held permanently in the mesenchyme and in pigment cells. Regulation of water content is in all probability one of the chief functions of the protonephridial system and the only function of the flame bulbs. This view is supported by the fact that the protonephridia are much better developed in fresh-water than in marine Turbellaria. Atrophy of the protonephridium of *Stenostomum* is followed by edema and general degenerating. New flame bulbs were formed by the evagination of capillaries in response to the accumulation of water in the worm. In a comparison of specimens of *Gyratrix hermaphroditus* from fresh, brackish, and salt water, Kromhout (1943) noted a gradual reduction in the protonephridial system with increasing salinity of the medium. Probably influx of water into the turbellarian body evokes activity of the flame bulbs, resulting in a current of water along the protenephridial tubule.

Chemotaxis

This was already discussed in connection with finding of food, which is in effect a chemotactic reaction. Experiments have also been performed on the behavior of planarians to chemical substances in capillary tubes placed in the same dish with them. Pearl (1903) tested in this manner various concentrations of several acids, alkalies, and salts, and found that in general a positive response is given to weak concentrations of all the substances tried, and a negative one to stronger concentrations. A positive reaction is the same as that toward food, *i.e.*, the planarian on perceiving the diffusing chemical turns towards and then proceeds directly to the capillary. In the negative response, the animal turns away from the capillary. Weak acids proved especially attractive, and planarians would often grip the capillary and extend the pharynx into its orifice.

Koehlar (1932) also employed the capillary method, using snail blood or crushed planarians, and found highly positive responses in some species, which can be caused to follow such as capillary around. Reactive species will also take a median course between double capillaries. Planarians with one auricle removed may show "circus movement" in a dish of diluted snail blood, circling toward the intact side. By adding India ink to snail blood diffusing from a capillary, it can be seen that response occurs when the carbon particles have passed along the auricular sense groove.

Reactions to Contact; Thigmotaxis

The reactions of fresh-water planarians to mechanical stimuli have been investigated in detail by Pearl (1903). To a light touch on the side of the head, the worm may give a positive response, pausing, turning the raised head toward the stimulus, and moving to it. A similar stimulus along the body sides causes usually a small local contraction and may evoke the positive turning if not too far posterior. A positive reaction is never exhibited by postpharyngeal regions. Stronger tactile stimulation of the anterior end evokes the typical negative response, a turning of the head away from the stimulus and a gliding off in the new direction.

Very strong or repeated prodding of the head causes jerking back by means of longitudinal contraction and turning through a larger angle. Moderate to strong mechanical stimulation of the sides of the body or the posterior region evokes local contraction and the hurried type of locomotion for a short time, followed by the usually gliding. If a planarian is cut in two, the anterior piece reacts as to strong posterior stimulation, *i.e.*, moves by the hurried method for a short time, then returns to the usual gliding. The posterior piece generally crawls backward immediately after the cut and continues for some time to give this response to mechanical stimulation of the cut surface. To such stimulus elsewhere it gives either the negative turning away or the hurried crawling. A positive reaction to contact is not obtained in the absence of a brain.

The Turbellaria are markedly thigmotactic; the dorsal surface is negatively, the ventral surface positively thigmotactic. This is another way of saying that flatworms strive to keep their ventral surface in contact with a substratum, regardless of gravity, and their dorsal surface freely exposed to the medium. Pearl (1903) notes that if a light object is gently places on the dorsal surface of a resting or moving planarian, the worm will promptly move away from under the object. Planarians will not come to rest in a situation where both surfaces are in contact. When placed on their backs, planarians, polyclads, and probably all flattened Turbellaria give a characteristic *righting reaction*. The worm twists into a spiral in such a way that the ventral surface of the head finally comes in contact with a substratum. The head then glides away and the spiral unwinds as the rest of the body follows.

Planarians dropped through water (planarians cannot swim) give the righting reaction, usually several times, as they fall. Pieces of planarians also give the righting reaction although more slowly than when the head is present. Brainless pieces of polyclads also right themselves, but the reaction differs from the normal in that any part of the piece, not always the anterior end as in the case of intact worms, may attach first. However, this may also occur in pieces of planarians. The entire body of planarians is sensitive to contact, but in general tangoreceptivity is highest on the auricles or sides of the head and decreases in an anteroposterior direction. In the absence of the brain, the auricles are no longer more sensitive than other parts of the body margin.

Phototaxis

Although some Turbellaria are positive to light, especially pelagic types, larval and juvenile stages of polyclads, acoels, and rhabdocoels that contain holophytic symbionts, and the rhabdocoels of Lake Baikal, the members of the class as a whole avoid light and unless other factors are operative seek the darkest areas of their environment. Many studies have been made on the response of turbellarians, chiefly fresh-water planarians, to light. Exposure to light increases the activity of planarians and causes resting individuals to start locomotion; the rate of locomotion may also be more rapid in the light. This stimulating actions of nondirective light has been termed *photokinesis*. After a time the effect decreases, and the worms slow down and finally come to rest in the least illuminated region of the container, if such exists.

The avoidance of light by planarians is an example of a phobotactic or phobic response, that is, the darkened area is found by random movements, a negative reaction when passing into more lighted areas and a lack of reaction when passing into more darkened regions. The most careful test of the phobic nature of light avoidance in planarians is that of Ullyott (1936). In a light gradient in which all possibility of directed rays from reflections, etc., was avoided, the planarians eventually came to rest in the darker part of the gradient. This phobic response to light is given equally well by specimens from which the eyes have been removed; these also show no change in rate of locomotion. Decapitated specimens also avoid light, although they move more slowly.

In most species light is perceived by the entire surface, although in some the posterior region shows no reaction to light. Both dorsal and ventral surfaces are sensitive of light, although planarians tend to keep their dorsal surfaces toward the light. Planarians perceive changes in light intensity and may give evidence of such perception by hesitation, head, swingings, and change of path when encountering a new intensity. Various experiments that have been performed on the possibility of color perception by planarians are really tests of intensity discrimination according to the more critical papers. On contrasting backgrounds of black and white or dark and light colors, planarians come to rest on the darker ground and in general on colors of longer wave lengths. Such reactions are probably merely avoidance of light of higher intensity. A very exact phobic reaction was reported by Lemke (1935) for *Dugesia lugubris* on a background of alternate black and white circles; the planarian follows the black circles, turning back at the boundary with the white circles. Eyeless specimens lose the reaction.

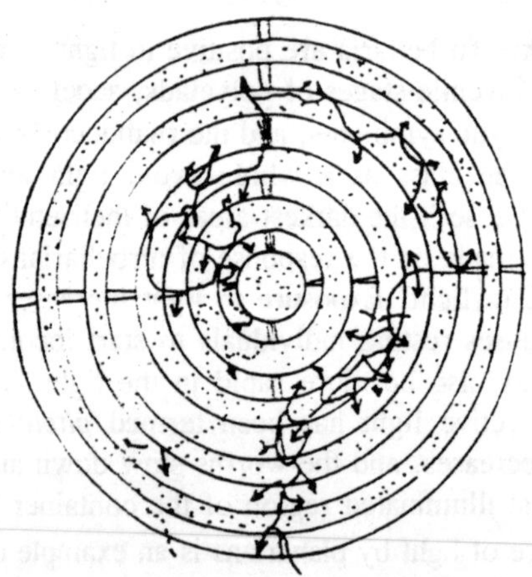

Fig. 12.5. Reaction to light.

Other species may not give such exact response to intensity boundaries and may in fact exhibit no change of behaviour to considerable changes of intensity. To directed light, planarians give a fairly precise negative response, turning away from the light and proceeding in a direction approximately opposite to the incidence of the light rays. The path of planarian can therefore be directed by means of light rays. This phototopic reaction is mediated primarily by the eyes, for in worms from which the eyes have been removed (without injury to the brain) the response is much less definite and precise, although still elicitable. If a spot of light is shone upon the eyes or head the negative turning away results, but if it is shone on middle or posterior regions there is not such response. Hence the topic reaction is mediated by the anterior end only, whereas the phobic response to light is mediate by the entire body in most species, with sensitivity decreasing in an anter-oposterior direction.

The sense organs responsible for the phobic reaction have not been identified, but the eyes are evidently the chief receptors for the topic response. According to the generally accepted analysis of Taliaferro (1920), directed light affects only those retinal cells whose long axes are parallel to the light rays. Because of the curvature of the planarian eye, only a few such cells would be affected in any given position of the planarian. The pigment cup also acts to shade retinal cells except those parallel to the light rays. This mechanism accounts for the highly directive effect of light on planarians, *i.e.*, their ability to perceive rather exactly the direction of light, and it also makes it rather probable that planarians can see movement, although not objects. The decrease in intensity of light as a moving object interrupts light rays would affect a succession of retinal cells.

Probably perception of movement is the function of the numerous eyes of land planarians and polyclads since a moving decrease of intensity would affect successive eyes rather than successive retinal cells of the same eye. According to Kawaguti (1932) the rear part of *Bipalium* is insensitive to light because it lacks eyespots. Planarians can perceive ultraviolet radiations; they react negatively to them but move slowly than in visible light. The behavior is the same in normal and eyeless specimens. Short exposures to ultraviolet and to direct sunlight are fatal to planarians. White species appear to be more sensitive to sunlight than colored ones. Some data are available on the behaviour to light of cave planarians that live in perpetual darkness. *Sphalloplana percoeca* from Kentucky caves showed no avoidance of light or seeking of dark areas but would writhe on exposure to direct sunlight and might be killed by less than 2 minutes of such exposure. *Phagocata subterrannea* from an Indian cave, however, was negative to light but less exact and more given to wandering than epigean species. Whether this different behaviour is correlated with the fact that the first species is eyeless and the second had eyes cannot be stated.

Rheotaxis

In general fresh-water turbellarians inhabiting still waters fail to react to water currents. Strong currents usually elicit only a "clamping" reaction, contraction and increased adhesion to the substratum. Weak currents may, however, call forth a positive rheotactic response, a turning of the anterior end towards the current and placing of the body in line with it. This has been reported for planarians and the rhabdocoel *Bothromestoms*. In the latter animal, extirpation experiments indicated that the eight groups of sensory bristles along the lateral body grooves are rheoceptors. The reaction of still-water species to *weak* currents is probably part of the food-catching mechanism, since turbellarians are able to detect the presence of intact prey by water disturbances created by them. Pond planarians may develop a rheotactic response after being kept for some time in a rather strong water current. Planarians that habitually live in flowing water usually exhibit a positive rheotaxis, for obviously without such a reaction they could not maintain themselves in their habitat. Thus, *Dugesia dorotocephala*, a spring-dwelling species, was found to show positive rheotaxis (travelling upstream) in nature and in the laboratory; but the positive reaction could be altered to a negative one (travelling downstream) by change of water or lowering the temperature.

The outstanding example of a positively rheotactic planarian is *Crenobia alpina*, inhabitant of rapid alpine streams. As already noted, this animal is more or less incapable of finding food in the absence of a water current. It is highly positive to currents and will follow a tube from which a water current is issuing. In case of two currents of equal velocity opposite or at right angles to each other, it will take a resultant path between them, and of two currents of unequal velocity it will select the faster one. The rheotactic reaction is given by all parts of the body, although most readily and to weaker currents by the head. Loss of the entire body margin does not eliminate the reaction so that rheoreceoptors must be distributed over the entire body surface being, however, most numerous on the head. After being kept for long periods (6 months to a year) in still water this species retains its rheotactic response undiminished. Beauchamp (1933) worked with a British stream planarian[2] in which the rheotactic response showed a definite seasonal cycle. In winter when temperatures were low and the reproductive system developing, the animal was positively rheotactic and accumulated at the head of hill streams; after the breeding season, negavite rheotaxis ensued and the worms travelled downstream and even into the lake into which the streams flow.

Thermotaxis and Temperature Relations

The majority of the Turbellaria and undoubtedly eurythermous, enduring wide ranges of temperature, but some species appear to be somewhat stenothermous. The behavior to temperature changes is considerably dependent on the temperature at which the worms have been living prior to the tests. Available data are limited to fresh-water planarians. For local application of temperature change, a narrow hairpin-shaped tube is used through which water of the desired temperature is circulated. Planarians react to temperature change of 2 to 3°C. from that at which they have been living. To moderate increase applied to the anterior end a positive reaction may be given at first; the worm turns towards the tube and may test it with the anterior end, before turning away. Usually a negative reaction is eventually given. If the tube is very hot, a negative reaction may occur at once.

Reactions to lowered temperatures are similar but less definite, and often no reaction is given. Localized heat applied to posterior regions results in local contraction, more rapid gliding away, or the hurried type of crawling. The lower limit of reaction to increased temperature was found by Koehler to be 24°C. for the anterior end of *Crenobia alpine*, 32 for its posterior part, 28-30°C and 40, respectively, for *Dugesia lugubris*. The ventral surface of planarians is also sensitive to temperature, so that thermal perception is evidently distributed over the whole body. When placed in a temperature gradient, *D. dorotocephala* which had been living at 22°C. aggregated in a zone ranging from 17 to 26; *Phagocata gracilis* aggregated at 0 to 10°C., whether taken directly from its natural habitat at 9.5°C. or kept for a month the laboratory at 20 to 22°C. The thermal death point of the former species is about 42°C, of the latter 30°C.

The most interesting behaviour to temperature is that of *Crenobia alpina*. This animal when presented with two streams of different temperature selects the cooler one and follows this about, being able to distinguish a difference of 3°C. Planarians possess a high capacity for acclimation to temperature, being able when kept at high temperately (27 to 30°C.) to lower their rate of metabolic activity below and when kept at low temperature (8 to 10°C.) to raise to above that expected for such temperatures.

Geotaxis

The Turbellaria in general show perception of gravity for when the oxygen supply of the medium decreases below requirements they come to the surface, *i.e.*, exhibit negative geotaxis. This is not a response to an oxygen or other gradient for its is given by worms in a closed vessel completely filled with water of uniform constitution and turned upside down. Unfed planarians in water of good oxygen content in the dark tend to seek the bottom of the vessel, but after feeding they tend to come to the surface probably because of the increased oxygen demand during intracellular digestion. The outstanding example of geotactic response among Turbellaria is seen in the acoel *Convoluta roscoffensis*. This animal is negatively geotactic when undisturbed, positively so when disturbed. On the sandy tidal zone on the channel coast of France, where it lives in countless numbers, it comes to the surface of the sand at low tide, and disappears into the sand when the tide returns or an any other disturbance. This rhythmic geotactic behaviour with reference to the tides is retained for about a week by *Convoluta* when kept in the laboratory in vessels of still water. It is independent of day and night. That the tidal response is really a geotactic one was proved by Fraenkel (1929) who showed that when *Convoluta* is simultaneously exposed to centrifugal force and gravitation it takes a resultant path between the two forces.

Although Turbellaria without statocysts react to gravity, it appears that the statocyst is of some importance in the geotactic reactions of the Acoela. The anterior tip of *Convoluta*, which contains the statocyst, reacts normally to gravity, but the rest of the body has lost the geotactic response. In regeneration experiments on Acoela it has been noticed that pieces without the statocyst have difficulty in righting themselves and in general react poorly to stimuli.

Galvanotaxis

The reaction of turbellarians to the direct constant electric current has been ascertained by a number of workers. All forms tested are cathodic, *i.e.*, they turn their anterior ends towards the cathode and may swim or crawl towards this pole. *Stenostomum* in an electric current behaves exactly like ciliate Protozoa and its cilia also beat with reference to anode and cathode as in ciliates. *Convoluta* is highly cathodic. Fresh -water planarians turn the anterior toward the cathode and may place the body in line with the current and proceed toward the cathode or the body may curve into a C-, U, or W-shape with ends and often ventral surface facing the cathode. This cathodic orientation of posterior as well as anterior end may be attributed to the presence in such species of a fission region or some region of especial activity, such as a adhesive area, in the posterior part of the animal. Decapitated planarians show the same response as entire ones, and middle pieces move anterior end first to the cathode or backward to the anode.

The land planarian *Bipalium* is likewise cathodic, moving by loops to the cathode, and pieces obtained by cutting the body in thirds also loop to the cathode. The one polyclad tested behaved like planarians, turning toward the cathode and usually also curving its body into a U-shape toward this pole. The various explanations of galvanotaxis that have been advanced are based chiefly on the idea of compulsory effect of the current directly on the ciliary or neuromuscular mechanism. This type of explanation appears unsatisfactory to the author, for it fails to explain reversals that occur under different physiological conditions. It can hardly be supposed that the neuromuscular mechanism also undergoes reversal during temporary changes of conditions. The fact that individuals facing the anode at the time the current is made also turn and proceed to the cathode is further difficult to explain on any such theory. According to the author's theory of galvanotaxis also expressed independently by Robertson (1927), the flat-worm takes that position with reference to the electric current which best corresponds with its own electrical currents.

For at least the lower animals, as has been shown by numerous experiments have inherent electrical currents related to their anteroposterior axiation. These currents presumably originate from different rates of chemical activity along this axis. In general, the anterior or head end is electropositive to lower levels, and hence it orients so as to face the cathode. The rear end of flatworms through the presence of a fission plane or other active region may also be electropositive with reference to the middle; hence the C or U attitude taken by such species the current.

Desiccation

The Turbellaria have no surface protection against evaporation and hence require an aquatic or humid environment. Some species secrete a mucous cyst or covering when exposed to desiccation and some land planarians can endure relatively dry environments as the South American pampas by habitually dwelling under cattle dung. In general, however, exposure to desiccation results in death. Fresh-water planarians when subjected to drying curl up so as to expose as little surface as possible and from time to time attempt to escape by extending the head in seeking movements or giving backward crawling waves but are unable to progress on a dry surface. When placed on a wet spot in a dry environment they give a phobic reaction whenever the head extends outside the wet spot.

The land planarian *Bipalium* was found by Kawaguti (1932) to lose water according to the simple laws of evaporation in relation to the humidity and temperature of the environment. Water was lost even in an atmosphere of 100 per cent humidity. Recovery occurred if the water loss did not exceed 45 per cent of the fresh weight.

Mass Protection

Following the discovery by Bohn and Drzewina (1920) that a number of *Convoluta* are more resistant to fresh water or much diluted sea water than single specimens in the same (limited) volume of fluid, much work was done by these authors and also by Allee and his students on such mass protection in various turbellarians (and other animals) against adverse conditions and agents, such as toxic chemicals, ultraviolet radiation, salinity changes. In general, resistance to such agents and conditions is much greater when there are a number individuals as contrasted with single or a few individuals in the same quantity of medium.

The protection is not caused primarily by the fact that, in the case of a toxic chemical, for instance, there is less of the chemical per worm in the mass experiment, but is does result from the greater quantity of animal in a limited amount of medium. The French authors postulate the secretion under these conditions of a special protective substance of which of course more would be emitted by the larger number of animals, but the experiments of Allee indicate rather that the normal secretions and exudates of the animals furnish the protective materials. Protective action is furnished by disintegrated animals or the water in which the same or other animals had lived. In the case of turbellarians the mucous and other glandular secretions emitted from the surface are probably the chief means of protection. Such secretions, exudates, or disintegrated remains neutralize toxic chemicals and, in some unknown way, probably not by adding salts to the medium, also protect marine turbellarians against fresh water.

LEARNING

The Turbellaria are capable of associative memory. *Stenostomum* can be trained to turn back into the dark at a light-dark boundary if punished by an electric shock when entering the light area. As *Stenostomum* is negative to light only a very weak light to which it does not react can be employed. The reverse experiment, training it to avoid the dark and remain in the light, requires longer training and gives a less definite result. Retention is poor. Similar experiments have been performed with the fresh-water planarian *Dugesia gonocephala* using shaking and electric shock as punishments, and rough versus smooth substratum, weak light versus darkness, and vertical versus horizontal surfaces as contrasting conditions.

In general, the planarian could be taught by repeated punishment to avoid that condition to which it normally gives a negative response, as light and rough substratum *i.e.*, to turn back at the boundary in the absence of punishment, but the reverse attempt, to teach it to turn back into conditions that it normally avoids, gave poor or no results. The experiments of Soest and Dilk came from the same laboratory as the experiments with the formation of associations in Protozoa reported in P–C. Much more convincing is the work of Hovey (1929) on the formation of an association in a leptoplanid polyclad. These worms were subjected for hours to alternating periods of light lasting 5 minutes and darkness lasting 30 minutes.

Light induces movement, darkness a quiescent state as is general in Turbellaria. Whenever a worm started to move on exposure to light it was stopped by being touched on the anterior margin. The number of touches required to cause the worms to cease movement in the light decreased notably with repeated trials although no worm learned to remain entirely motionless when exposed to light. Possibility of motor fatigue, injury to the anterior margin, or adaptation to light was ruled out. Following extirpation of the cerebral ganglia, worms did not learn to move less in the light, and repeated hindrance to such movement had not effect. Evidently the brain is involved in associative learning in polyclads.

13

LIVING ACTIVITIES IN TREMATODES

The physiology of trematodes has been much less studied than that of cestodes or nematodes, probably on account of the difficulty of obtaining suitable experimental material. Before considering the physiology of trematodes, it is as well to review certain features typical of the group which directly or indirectly impose certain limitations or requirements on their physiology.

1. Their life cycles are complex (excepting the Monogenea) and always involve a molluscan intermediate host, sometimes a second intermediate host and, more rarely, a third intermediate host. The physico-chemical conditions and the nutritional levels of the different environments encountered may thus vary profoundly, so that striking changes in the metabolism of various stages may be expected.
2. All trematodes possess at least one sucker, which means that they are brought into intimate contact with host tissues. The definitive habitat must thus present a suitable surface to which attachment can be made and on which feeding can take place.
3. Without exception, they possess a well-developed alimentary canal usually with a muscular pharynx. This kind of system is especially suitable for ingesting semiliquid or viscous food such as intestinal contents, mucus, blood and bile, and with the exception of a few progenetic forms distribution of adults is limited to habitats, chiefly in vertebrates, where such food materials are available.
4. Their body covering (tegument) bears a striking resemblance to that of cestodes and, as in that group, is absorptive in nature; nutrition may thus be taken in both through the gut and through the tegument.

5. Their egg capsules are generally delicate structures, incapable of withstanding desiccation, and normally require water for development.

Chemical Composition

As is stressed elsewhere the chemical analysis of parasitic organisms seldom provides significant information unless it is related to the nutritional state of the host, and the degree of maturity of the parasite. Thus the carbohydrate content of an intestinal parasite may vary within considerable limits, depending on whether the host has

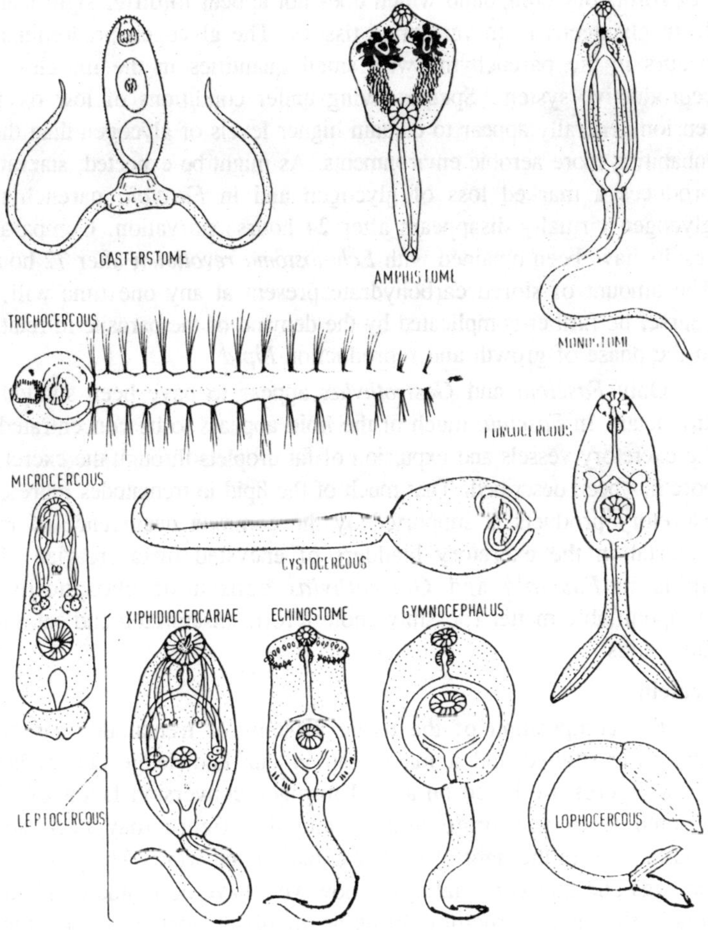

Fig. 13.1. Various types of cercariae.

been actively feeding or starving before autopsy. This point is illustrated particularly by experiments with *Fasciola*. Chemical analysis data are usually expressed as a percentage of the dry weight. The dry/fresh weight ratio may vary appreciably with the maturity of the fluke. Thus in the case of *Echinostoma revolutum*, experimentally developed in chicks, the dry weight after 10 days was approximately 10 per cent and after 30 days 20 per cent.

Carbohydrate

As in other parasitic helminths, the main polysaccharide found in trematodes is glycogen, and that obtained from *Fasciola* is a highly dextrorotatory compound which does not appear to differ significantly from glycogens from vertebrate tissues. The glycogen predominantly occurs in the parenchyma with small quantities in the muscles and reproductive system. Species living under conditions of low oxygen tension generally appear to contain higher levels of glycogen than those inhabiting more aerobic environments. As might be expected, starvation produces a marked loss of glycogen and in *Fasciola* parenchymal glycogen virtually disappears after 24 hours' starvation. Comparable results have been obtained with *Echinostoma revolutum* after 72 hours. The amount of stored carbohydrate present at any one time will, of course, be further complicated by the demand of the parasite in relation to the phase of growth and reproduction *Lipid*.

Only *Fasciola* and *Gastrothylax* appear to have been studied in any detail. In *Fasciola* much of the lipid appears to be concentrated in the excretory vessels and expulsion of fat droplets through the excretory pore has been described. That much of the lipid in trematodes represents excretory products is supported by the common occurrence of fatty material in the excretory bladders of encysted metacercariae. The lipids in *Fasciola* and *Gastrothylax* consist of phospholipids, unsaponifiable matter (possibly cholesterol), unsaturated and saturated fatty acids.

Protein

The composition of the tissue proteins of trematodes has been very little studied and a major gap in our knowledge occurs here. These figures are based on a total nitrogen conversion factor of 6.25 adopted from vertebrate tissues, and this figure may need some modification in the light of further experimental work. The proteins of the eggs and metacercarial cysts, however, have been studied in some detail; the main structural proteins involved appear to be chiefly sclerotin and keratin.

Water Relationships

The osmotic relationships of trematodes with their environments have been little studied. This is surprising, for their complex life cycles may involve in turn passage through fresh or sea water (by the miracidium and the cercaria), an intermediate molluscan host, and often a second intermediate host (usually an arthropod or a vertebrate) before reaching the definitive host. A study of the osmotic changes encountered and how they are overcome might produce some interesting results. Both miracidia and cercariae of *Fasciola hepatica* face considerable osmotic changes on passing to or from tissue ($\Delta = 0.58°C$ for cattle) to fresh ($\Delta = 0°C$) or sea water ($\Delta = 2.2°C$), or to snail tissues ($\Delta = 0.22°C$ for *Lymnaea stagnalis*).

In a miracidium, except for the simple flame-cell system, there is no special device for osmotic control. Rather surprisingly, in *Fasciola* large changes in osmotic pressure (0-1.0 per cent NaCI) produce no change in the flame cell activity of the miracidia. The few studies carried out on adult trematodes suggest that they are poor regulators and relatively permeable to water; data is available for *Fasciola gigantica*, *Gastrothylax crumenifer*, *Haematoloechus medioplexus*.

Nutrition

The nutrition of a number of species has been examined by Halton (1967a) who, together with Jennings (1968), have given excellent reviews of this problem.

Nature of Food Materials

The habitats of trematodes are so diverse that it is not surprising to find that the food materials utilized cover a wide range and include blood, bile, mucus, intestinal contents. The question is further complicated by the fact that some species can migrate through the body and are exposed to different kinds of food materials during this process. Probably the most controversial species studied has been *Fasciola*.

According to Dawes (1963*a*, *b*, *c*) this species is essentially a tissue feeder utilizing epithelium and connective tissue and only incidentally takes in some blood. Other workers, notably Halton (1967*a*) have produced evidence that *Fasciola* feeds predominantly on blood but that tissue and mucus is also ingested. The relative amounts of blood and tissue taken in may be a matter of degree, depending on the site and activity of the organism at any time. The nature of the food materials is further complicated by the fact that absorption can take

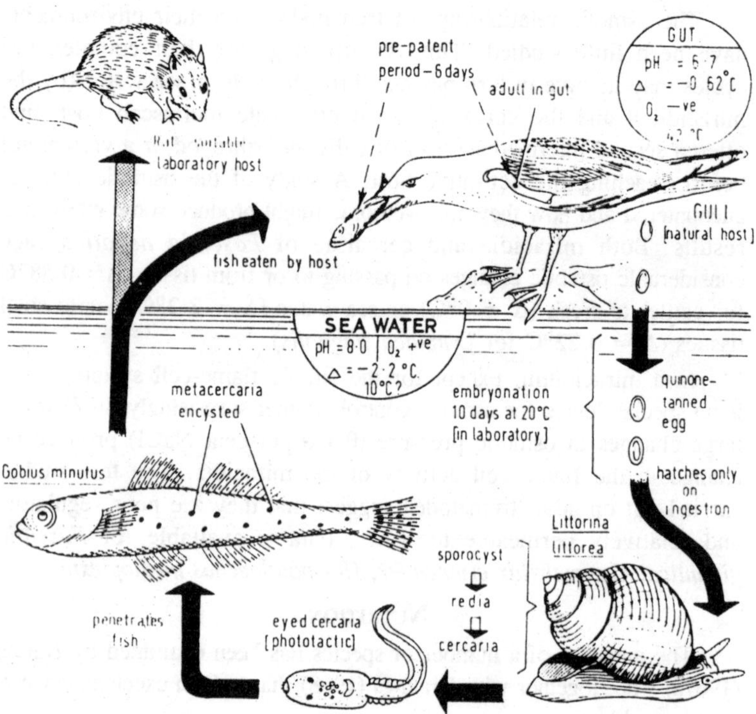

Fig. 13.2. Cryptocotyle lingua—life cycle and some physiological factors relating to it.

place through the tegument. Thus schistosomes, which live in the hepatic portal system, probably utilize soluble food materials such as amino acids and monosaccharides via the tegument, in addition to blood.

In schistosomes and in some other blood-feeding trematodes a blackishbrown pigment, derived from ingested blood is found; although the exact nature of this pigment has not been studied, there is some evidence that it is identical in composition with the malarial pigment. The nutrition of larval trematodes has been largely neglected. There is evidence that the rediae of *Parorchis acanthus* and *Cryptocotyle lingua* can also absorb nutrients, such as glucose, through the body surface as well as ingesting particulate matter in the gut.

Digestion

Understanding of the process of digestion in trematodes is complicated by presence of unusual morphological features, the nature of which are not properly understood. Thus, E.M. studies have shown the presence of protoplasmic extensions in the form of lamellate processes, ridges or 'microvilli' or combinations or modifications of these in the gut. Species studied include: *Schistosoma mansoni*; *Schistosomatium douthitti*; *Gorgodera amplicava* and *Haematoloechus medioplexus*. Various hypotheses have been put forward regarding the function of these processes. Thus it has been suggested that (*a*) at least some digestion occurs in vacuoles formed by these processes, (*b*) absorption of macromolecules occurs across the membranes of the lamellae, (*c*) the processes provide a porous surface on which membrane digestion could operate.

An extensive examination of nine species by Halton (1967a) has revealed that digestion in Digenea appears to be predominantly extracellular. Enzymes make the gut contents soluble, the products are then absorbed and the process of digestion is completed in the gastrodermis. Attempts to demonstrate digestive enzymes have not been very conclusive. In homogenates of *F. hepatica* and *H. cylindracea* a protease, a dipeptidase, a lipase, and acid and alkaline phosphatases have been detected. Few of these have been detected histochemically but problems of technique may be involved. There is, however, evidence of an esterase in the gut. From *S. mansoni,* a globin-digesting enzyme with a pH optimum of 3.9 has been extracted. Direct evidence that haemoglobin is ingested and digested by schistosomes has been obtained by labelling reticulocytes with tritiated L-leucine and showing its later incorporation in the worm tissues.

In *Schistosoma rodhaini*, histochemical methods for the exopeptidase, leucine aminopeptidase, failed to demonstrate this enzyme in the gut although it was present in the tegument. This further confirms the important metabolic role played by the tegument. One additional factor operates in trematode digestion, namely the fact that certain adhesive organs serve not only as organs of attachment but also as extracorporeal organs of digestion and absorption. Thus the tegumental lining of the suckers and associated regions of *Haplometra cylindracea* has been shown to give strong reactions for carboxylic esterases suggesting that extracorporeal digestion takes place in this region. It is in the strigeoid trematodes, however, that extracorporeal digestion appears best developed. This process has been extensively studied by Erasmus and his co-workers.

The three bird species, *Cyathocotyle bushiensis*, *Apatemon gracilis minor* and *Diplostomum phoxini* have been particularly studied. In these

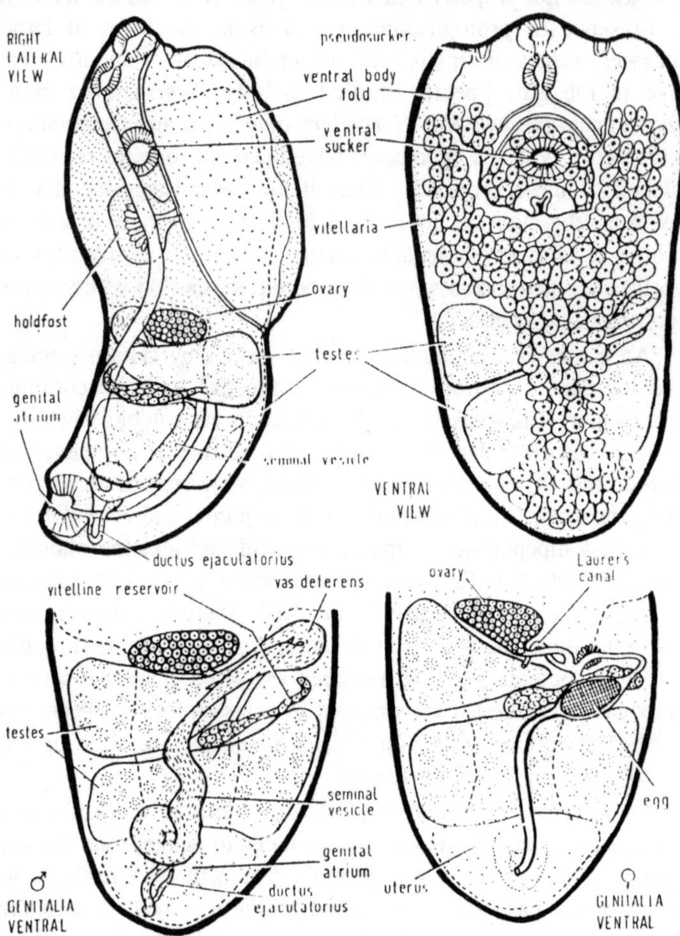

Fig. 13.3. Diplostomum phoxini—general morphology.

species three main host-parasite interfaces are found: (*a*) the general tegument, (*b*) the lappets and (*c*) the adhesive organ lobes. Well-developed glands are associated with the lappets in some species. There is now abundant evidence that both the lappets and the adhesive organs release histolytic enzymes which digest host tissues. This results in the production of finely fragmented material which may then be ingested. In *A. gracilis* and *C. bushiensis* the oral sucker does not appear to be permanently attached and the organism is free to move and digest material.

In *A. gracilis* the host-parasite interface is so intimate that it is 'impossible to distinguish precisely the extent of the parasite tegument.' The adhesive organ is not only proteolytic in function but also appears to be 'placental' and serves an important role in the absorption of materials. Apart from the strigeids, little work appears to have been carried out on the host-parasite interface and this is an area in which much work remains to be done.

Respiration

Respiration metabolism of trematodes has been reviewed by Vernberg (1968).

Eggs

The eggs of most trematodes embryonate in an aerobic environment but the extent to which oxygen is necessary for embryonation is not known. In *Fasciola*, the oxygen consumption is 0.3 mm^3 /mg (d.w.) but rises to five times this figure after about 12 days' development at 25°C. In several species - *S. mansoni*, *S. japonicum* and *Paragonimus ohirai* - miracidia show an 'ageing effect' and consume more oxygen after five hours than after one hour; differences between strains do, however, occur. In *Schistosoma* but not *Paragonimus* respiration was stimulated by glucose, a result suggesting the presence of larger glycogen reserves in the latter species.

Larvae; Rediae and Sporocysts

Work on larval respiration has been reviewed by Vernberg (1963) and Smyth (1966). Little work has been carried out on rediae and sporocysts; these in general utilize oxygen if it is offered to them. In the starving sporocysts of *Cercaria dichotoma*, oxygen consumption appears to be related to the number of contained cercariae.

Cercariae

As might be expected of predominantly free-living organisms, cercariae have a marked aerobic metabolism and cannot survive long under anaerobic conditions. The cercariae of *S. mansoni*, for example, become inactive without oxygen after one hour and some 75 per cent die within four hours. Some cercaria (e.g. *Himasthla quissetensis*) are 'conformers' i.e. their respiration decreases in direct proportion to the decrease in oxygen tension. Other species, however (e.g. *Zoogonus rubellus*), are conformers at some range of oxygen tensions only.

Adults

As in the case of most other helminth parasites it is extremely difficult to determine the significance of these figures. The fact that trematodes utilize oxygen when it is offered to them *in vitro* has little bearing on the *in vivo* situation unless the level of oxygen available *in vivo* is also known. The influence of different substrates, such as those occurring in the Krebs or Embden-Meyerhof pathways, has been studied in a number of trematodes and there is some evidence of the operation of part or all of these pathways. The influence of a number of recognized inhibitors of mammalian metabolism on trematode respiration has been examined.

The effects of various inhibitors on the oxygen consumption of *Fasciola* in the presence of succinate or NADH. The succinoxidase present in *Fasciola* appears to be relatively insensitive to amytal, requiring 10-100 times the concentration normally required to inhibit the corresponding mammalian enzyme. An even more striking result was obtained with Antimycin A for which a concentration of 10^{-5} was required - about 1000 times that required by mammalian systems. In contrast to mammalian systems (in which the electron transport system can be inhibited by 10^{-3} cyanide), appeared to *stimulate* oxygen uptake in *Fasciola*.

A similar stimulation of respiration by cyanide has been reported for *Paramphistomum cervi* and some cestodes and nematodes. Thus, although *Fasciola* appears to possess enzymes similar to those of the mammalian chain - NADH-cytochrome *c* oxidoreductase, succinate-cytochrome *c* oxidoreductase, NADH oxidase and cytochrome *c*-oxygen oxidoreductase - it is apparent that major differences in the electron transport chain of these two systems occur. Cytochrome *c* does not appear to be present in *Fasciola* although it has been reported in *Echinostoma revolutum*. In *Schistosoma mansoni*, however, cyanide does have an inhibitory effect on oxygen uptake which may indicate the presence of a cytochrome system. Respiratory pigments with absorption spectra approaching those of mammalian haemoglobin have been described in several species of trematodes but there is no evidence that this substance acts as a respiratory pigment.

Metabolism

Carbohydrate Metabolism

Like most other parasitic helminths, trematodes have a pronounced carbohydrate metabolism. This is evidenced by the substantial amount of stored carbohydrate, the extensive use of endogenous carbohydrate, the high rate of uptake of exogenous glucose and production of fatty acids as metabolic by-products. Under starvation conditions, endogenous glycogen is rapidly used but its reserves are built up again when exogenous sources become available. Adult trematodes can make use of hexoses, such as glucose and mannose. Experiments with gut-ligatured *Fasciola* have demonstrated that absorption can take place not only via the gut but also through the tegument; this result is in keeping with the metabolic activity implied from ultrastructure studies. Although much further remains to be done in this area, the few studies carried out - largely on *Schistosoma mansoni*, *Dicrocoelium dentdriticum* and *Fasciola hepatica*-indicate that trematodes generally follow a Embden-Meyerhof pathway of phosphorylating glycolysis.

The end-product of carbohydrate metabolism in *Schistosoma* is lactic acid; up to 91 per cent of the glucose utilized being converted to this acid. In contrast, in *Fasciola*, up to 95 per cent of the carbohydrate is converted to the volatile fatty acids, acetic and propionic acids. The proportion of these acids produced is a function of the oxygen tension, ranging from 2.5:1 in nitrogen/carbon dioxide to approximately 1:1 in oxygen. There is also evidence of a complete Krebs cycle in some trematodes and for a partial cycle in others, but the terminal oxidation appears to be deficient.

In *Fasciola*, however, the Krebs cycle appears to be of relatively minor importance in metabolism, but the presence of the enzymes isocitrate lyase and malate synthetase suggests that the glyoxylate cycle is functioning in this species. There is evidence, too, that the pentose-phosphate pathway operates in this species. In the majority of helminths, the period immediately following anaerobiosis is characterized by an increase in the rate of respiration. This is often referred to as the repayment of an 'oxygen debt.' In trematodes, however, there appears to be no oxygen debt, and the fatty acids are almost entirely excreted without further metabolizing.

Protein Metabolism

It has already been pointed out that trematodes in general can ingest and digest complex proteins. An efficient protein metabolism must occur, for in some species astonishing numbers of eggs can be produced; *Fasciolophsis buski*, for example, produces up to 25 000 daily. The potential protein synthesis of larval trematodes is even greater - a singel miracidium of *S. mansoni* may produce more than 200 000 cercariae. It is interesting to note that an aminopeptidase has been detected in the rediae and cercariae of *Philophthalmus gralli*. This could explain the lytic action of these larvae in the snail tissue.

There is very little direct evidence on the amino acid requirements of adults as investigation of this depends largely on *in vitro* studies and these, in turn, depend on knowledge of metabolism so that somewhat of a dilemma exists. As in the case of carbohydrates, amino acids may be absorbed through the tegument and, as in cestodes may even 'leak' back through this structure. Experiments on the uptake of *S. mansoni* appear to substantiate this view.

Transamination

A number of transaminases have been detected especially in *Fasciola* and *Schistosoma* (Kurelec & Ehrlich, 1963; Huang *et al.*, 1962). In *S. japonicum*, glutamicpyruvic and glutamic-oxaloacetic transaminases have been demonstrated, and in *Fasciola* additionally, a pyruvic acid/alanine transaminase has been found. Much more, however, remains to be done in this field.

End-products of metabolism

Ammonia and uric acid appear to form the nitrogenous compounds most commonly excreted, although urea has also been reported. The materials excreted may, however, be affected by the host diet. Thus, *Fasciola* appears to be largely a protein feeder and excretes some ten times the amount of ammonia excreted by the cestode *Hymenolepis diminuta*, which utilizes largely carbohydrate. *In vitro*, the addition of glucose to media in which *Fasciola* is cultured greatly decreases the amount of ammonia excreted but increases the lactic acid released.

Lipid Metabolism

The lipid metabolism has been very little studied, although the lipid analysis of several forms is known. In *Schistosoma mansoni*, the major lipids are cholesterol, triglyceride, phosphatidylcholine, phosphatidylethanolamine and phosphatidylinositol. Like cestodes this species has lost its ability to synthesize *de novo* sterols and fatty acids but when supplied with dietary sources it has retained the capacity to synthesize all its complex lipids. During starvation, the fat content rises, presumably as the result of accumulated fatty acids as by-products of carbohydrate metabolism. Lipases have been detected in a number of species, often in the lappets and adhesive organs of strigeids. As could be expected, acetylcholinesterase and pseudo-cholin-esterase have been detected in a number of species, because histochemical methods for these are well developed. It was first shown in Monogenea that such methods provide elegant techniques for the demonstration of nervous tissue in whole mount preparations. These methods are readily applied to most species, such as *Schistosoma* spp. and will undoubtedly be of considerable value in further studies on the trematode nervous system.

14

LIVING ACTIVITIES IN CESTODES

Cestodes may be considered to be biological models of exceptional physiological interest on account of (*a*) the unusual habitat in which they live, (*b*) their basic morphology, which is characterized by the lack of an alimentary canal, and (*c*) their pattern of strobilar growth. Expanding these points further, it can be seen that:

(*a*) The growth of cestodes is unusual in that it results in the continuous production of *embryonic* tissue (*i.e.* the strobila) which differentiates into a strring of sexual individuals which eventually producef fertile eggs. Certain species (*E. granulosus*, *Taenia serialis*, and probably *T. crassiceps*) are capable of dedifferentiation and, under certain conditions, cystic forms can give rise to either strobilar or cystic forms depending on which stimulus has been applied.

(*b*) Cestodes live in an environment whose physicochemical properties and nutritional level may vary in relation to the feeding pattern and nutritional condition of the host; some species (e.g. *Hymenolepis diminuta*) undergo diurnal migration patterns in relation to the host feeding pattern.

(*c*) Cestodes lack a dlgestive tract and a defined circulatory system; their outer body covering (the tegument, is essentially a 'naked' protoplasmic surface through which all substances must both enter and leave the body; *active* rather than passive transport appears to be generally involved in these processes. The occurrence of this naked surface implies that cestodes would be strongly antigenic.

(*d*) It has generally been assumed that, for nutritional purposes, cestodes utilize small molecules derived from the digested food sources of the host. The situation appears to be more complex, however, for

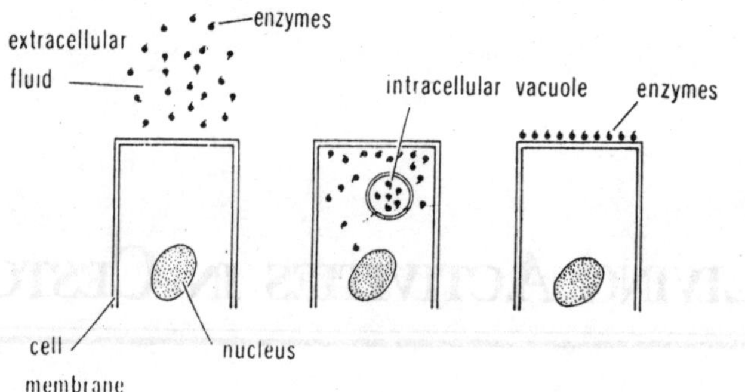

Fig. 14.1. Comparison of membrane (contact) digestion with extracellular and intracellular digestion.

it is now known that cestodes are capable of fixing carbon dioxide and that this process serves as a valuable source of carbon atoms. Furthermore, there is some evidence that cestodes may be capable of utilizing larger molecules, such as protein, either by direct uptake or by a process known as membrane (= contact) digestion this hypothesis is, however, by no means proved. In the latter process membrane-bound enzymes are believed to be capable of digesting large molecules with which they come in contact. Pinocyytosis has not, however, been demonstrated in cestodes, although several attempts have been made to do so.

(*e*) Possibly in relation to the process of membrane digestion, mentioned above, it is likely that in some cestodes (*e.g.*, *Echinococcus granulosus*) the scolex acts as an organ of nutrition (*i.e.*, has a placental-like function) as well as of attachment.

The physiology of this group of parasitic helminths has been more extensively studied than any other and a large literature exists. Various aspects of cestode physiology have been reviewed by Read & Simmons (1963), von Brand (1966) and Smyth (1969). The majority of experimental work has been carried out on those species which are comparatively easily maintained in the laboratory, *e.g.*, *Hymenolepis* spp; *Taenia* spp. and *Echinococcus granulosus* amongst the Cyclophyllidea and *Spirometra mansonoides*, *Diphyllobothrium* spp., *Schistocephalus solidus* and *Ligula intestinalis* amongst the Pseudophyllidea. With rare exceptions, the physiology of the remaining cestode orders has been little investigated and this area offers much scope for future work.

CHEMICAL COMPOSITION

As emphasized elsewhere, data on the chemical analysis of parasitic worms are of little value unless qualifying information is provided on the nutritional state of the host at the time of autopsy. The carbohydrate content especially is liable to fluctuate, and as this affects the dry weight, changes in other constituents, such as protein when expressed as a percentage, may be more apparent than real. The lipid content is also directly related to that of the host. Moreover, the chemical composition may vary wity the 'strain' of both parasite and host. It can also vary in specimens from different regions of the intestine. The data must also be accepted with caution on technical grounds, because some of the older analytical methods have been shown, by more modern workers, to be unreliable. Some of the problems inherent in analysis have been reviewed by von Brand (1960) and Hopkins (1960).

Major Chemical Constituents

The proportions of the main tissue constituents - protein, lipid and carbohydrate - show a somewhat different pattern from most other invertebrates in that the carbohydrate content tends to be high and the protein relatively low. Larval cestodes are particularly rich in glycogen which may reach astounding levels (>50 per cent in the plerocercoids of *Ligula* or *Diphyllobothrium*).

Calcareous Corpuscles

Cestodes (and trematodes) often contain enormous numbers of curious bodies termed *calcareous corpuscles*, made up of an organic base together with inorganic material. They vary much in size; in some species being very large - 16 - 32 μm (*Spirometra mansonoides*, *Echinococcus granulosus*), in most species, they may be as small as 12 μm. The organic material contains RNA, DNA, proteins, polysaccharides and alkaline phosphatase. The inorganic materials are mainly calcium, magnesium, phosphorus and carbon dioxide with traces of other metals, but these constituents, especially phosphate, can vary considerably in relation to metabolic conditions and to origin.

Ultrastructural studies have shown that in *T. taeniaeformis*, the corpuscles are formed intracellularly and a single corpuscle is formed in one cell, which is apparently destroyed in the process. The role of calcareous corpuscles is not clear but it has been speculated that they act as major reserves for (*a*) phosphates and for (*b*) other organic ions together with carbon dioxide. These materials may be called on suddenly - phosphates for phosphorylation, the ions to act as enzyme catalysts and the carbon dioxide for CO_2-fixation - such as when a larva enters a host intestine and immediately requires a large amount of energy for its establishment process; such processes frequently involve muscular attachment.

Carbohydrates

As with trematodes and nematodes the main carbohydrate reserve in cestodes is 'glycogen', which is a typical energy reserve of helminths inhabiting biotopes with a low oxygen tension. The properties of cestode 'glycogens' have not been well characterize and their physical characteristics vary somewhat with their methods of extraction. A few workers have distinguished between *lyo-glycogen* and *desmo*-glycogen (precipitated by TCA) but this distinction has not been widely made. Glycogen is distributed largely in the arenchyma and muscles. The data shows that the total glycogen content of cestodes may vary in the range 6-48 per cent of the dry weight.

Larval cestodes generally show a higher and more constant glycogen content than the corresponding adults; this may reflect the more stable intermediate host environment - usually the coelomic cavity or tissues. The glycogen content can show marked variation within any species in relation to the nutrition of the host and the degree of maturity of the strobila (orpart of it). Thus, in *Raillietina cesticillus*, after 20 hours of host starvation, the glycogen level falls from 4.6 to 0.25 per cent (wet weight). Again, the glycogen, protein and lipid content of *Hymenolepis diminuta* varies very greatly with the age of the worm and the degree of maturity. It is thus clear that the glycogen content of a strobila, or a particular part of it, may depend on a number of factors, only some of which are understood. Any quoted figures for the carbohydrate content should thus be accepted with caution, unless the previous metabolic history is known, and the region of the strobila defined.

Protein

Like the glycogen content, the figures are meaningful only against the age, degree of maturation and previous metabolic history of the worm. Cestodes are unusual in the animal kingdom in that in many species the protein content is less than the sum of the carbohydrate and fat. This appears to be related to the high level of glycogen reserves in some species. Figures for protein content are based on multiplying the total nitrogen content by 6.25. It must be emphasized, however, that this figure is based on average analysis of *mammalian* tissues and may not hold for invertebrate tissues generally.

There is evidence that some species of cestodes contain significant quantities of non-protein nitrogen so that early analyses for 'protein' in cestodes should be accepted with some reservation. Fractionation of tissue proteins has been carried out in a few species, and in some, protein has been found to be conjugated with substances such as bile acids, glycogen or cerbrosides. The chief structual proteins in cestodes are the sulphur-containing keratin (hooks and embryophores) and sclerotin (egg-capsules); the latter has already been considered.

Analysis (both in the scolex and in the oncosphere) of the keratin in hooks and embryophore of several cyclophyllid species suggests that its structure and composition do not differ substantially from that of vertebrate keratin. An exceptionally detailed examination of the physical and chemical characteristics of the keratin of the hooks in the oncosphere of *H. microstoma* and strobilocercus of *H. taeniaeformis* has been made by Dvorak (1969*a*, *b*). The amino acids in the tissues of a number of species have been examined but, in general, no unusual information has emerged.

Lipid

The lipid content of cestodes clearly varies considerably from species to species. Moreover, the lipid content may vary considerably even in the same species grown in the same host species fed on different diets. This is related to the fact that host intestinal fatty acids and sterols are *directly* (*i.e.*, without further digestion) absorbed by cestodes and the qualitative composition of cestode lipids generally follows that of the host. The lipid content can also vary with the age of the proglottid. Not many detailed fractionations of cestode lipids have been made.

Some older methods of analysis for some lipid components have been replaced by methods, such as gas chromatography for fatty acids, which are now acepted as being more reliable; earlier data may thus need re-evaluation. All the usual lipid fractions have been identified in cestodes - phospholipids, glycolipids, fatty acids, glycerol and unsaponifiable material. The unsaponifiable material, which is largely cholesterol, may account for more than 20 per cent of the total lipids. Perhaps the most detailed lipid analyses are those for *H. diminuta*.

Carbohydrate Metabolism

From a series of studies carried out on both cyclophyllidean and pseudophyllidean cestodes it is possible to build up a general picture of the carbohydrate metabolism. These experiments fall into two groups: (*a*) *in vivo* experiments: carried out by feeding the host on experimental diets and observing the effect on the chemical composition of a worm; (*b*) *in vitro* experiments: concerned largely with the kind and quantity of carbohydrate utilized and the end-products of carbohydrate metabolism. The carbohydrate metabolism dominates other aspects of metabolism and this aspect of cestode physiology has been extensively studied. Only a broad review will be presented here.

In vivo carbohydrate utilization

It has long been recognized that carbohydrate must be present in the host diet for the normal establishment and growth of cestodes. If rats already infected with *H. diminuta* are placed on carbohydrate-deficient diets or one deficient in sucrose, there is a decrease in the number and size of strobila and the numbers of gravid and mature proglottids. A number of other experiments have been carried out on the effect of restricting the carbohydrate in the host diet to other substances such as glucose, galactose, maltose and dextrin. Care must be taken with such experiments, however, or diets with different calorific intakes may result.

Evidence from experiments, in which the carbohydrate content of the diet was varied calorifically, appear to indicate clearly that the factor limiting growth of cestodes *in vivo* is the actual availability of glucose in the intestinal lumen. With this species, taking glucose as a standard, dry weight of worms from rats fed on a maltose diet showed a decrease of 3.9 per cent, dextrin a decrease of 11.8 per cent and sucrose a decrease of 22.8 per cent. In these experiments galactose failed to support worm growth although some limited growth has been obtained by other workers. These results are in keeping with the sequence of events after the carbohydrates, mentioned above, are ingested by a mammal, *i.e.*, both sucrose and dextrin require further enzymatic degradation before being available (as glucose) for worm growth. Starch, which also produces glucose on hydrolysis in the intestine, supports cestode growth *in vivo*.

In vitro carbohydrate utilization

Extensive experimental work on the *in vitro* utilization of carbohydrates by cestodes has been carried out. There is increasing evidence that the conditions under which some of these early experiments were carried out were not sufficiently close to the *in vivo* situation to be reliable. Hence, some early work may need re-evaluation. For example, the PCO_2 level can have a profound effect on the utilization of glucose and synthesis of glycogen. This is probably related to CO_2-fixation and its role in the production of fumarate used for the reoxidation of NADH under anaerobic conditions. The presence or absence of ions (such as Na^+) can also influence uptake of glucose. Most species studied have been found to utilize glucose and many use galatose (contrast *in vivo*); a few species can utilize maltose and one species (*Cittotaenia*), sucrose. The mechanism of uptake for all kinds of carbohdrates is not known in detail but experiments using ^{14}C have already shown that the uptake of glucose is by active transport and not by simple diffusion. Such a mechanism would permit the uptake of glucose from low levels in the gut.

Utilization of glucose in many species results in a build up of glycogen in the parenchyma. In larval *Taenia taeniaeformis*, however, significant glycogen synthesis only took place under aerobic conditions when *both* glucose *and* glycerol were present in the media. When either of these were present alone some sparing of tissue glycogen utilization took place, but no glycogen synthesis occurred.

The significance of the interaction between the metabolisms of glucose and glycerol is not yet clear but there is evidence that the α-glycerophosphate (Bücher) cycle is involved. Glyceol is known to stimulate respiration in other species, e.g. *Mesocestoides corti*. It is interesting to note also that glucose uptake and glycogenesis are stimulated *in vitro* by the presence of insulin in the medium; hence the endocrine balance of the host may have an influence on the carbohydrate metabolism of a cestode.

Intermediary Carbohydrate Metabolism

Most of the work on this has been carried out on the genera *Hymenolepis*, *Echinococcus*, *Taenia* and *Monezia*. It generally can be said that the three main metabolic pathways best known in animal tissues, also occur in cestodes although often in a modified form. These are: (*a*) Embden-Meyerhof pathway (glycolysis) (*b*) Krebs cycle or citric acid cycle (oxidative decarboxylation). (*c*) oxidative phosphorylation (the electron transport system: the cytochrome system). Lesser well-known pathways, the pentose-phosphate pathway (the hexose-mono-phosphate shunt) and the a-glycerophosphate cycle also occur in a few species. It is beyond the scope of this book to discuss these pathways in detail, but the extent to which they are utilized by cestodes is summarized, briefly, below.

Embden-Meyerhof pathway

A number of studies-especially on *H. diminuta* and *E. granulosus*-have demonstrated most, but not all, of the enzymes of this pathway. Thus, myokinase, phosphorylase, phosphohexomutase, hexokinase, aldolase, phosphoglyceraldehyde dehydrogenase and lactic dehydro-genase have been detected. Although all the enzymes of the Embden-Meyerhof sequence have not been identified, the results strongly suggest that this pathway is followed in these species and is probably the general pattern in cestodes.

Krebs (citric acid) cycle

This cycle is essentially the metabolic centre at which the carbohydrate, fat and protein metabolisms make contact and are enabled to exchange intermediate compounds. In most cestodes, evidence for the occurrence of the enzymes involved in the Krebs cycle is incomplete. Although the existence of many enzymes of the cycle, such as succinic dehydrogenase, is well established, it yet remains to be demonstrated (with a few exceptions) that the cycle is actually operative in the intact organism (*i.e.*, not just in homogenates). A species in which the Krebs cycle has been studied in great detail is *Echinococus granulosus* in which it has been shown that a number of Krebs cycle intermediates were oxidized not only in homogenates but also in intact protoscoleces. There seems no doubt that in *Echinococcus* a Krebs cycle, in whole or in part, operates. Only very incomplete evidfence is available for the existence of the cycle in other genera such as *Taenia*, *Moniezia* and *Hymenolepis*.

Pentose phosphate pathway

Although the Embden-Meyerhof pathway is the main conversion route for carbohydrates in some species of cestodes a small proportion of carbohydrate may be utilized via the pentose phosphate pathway. In *Echinococcus*, although 60 per cent of the glucose utilized is metabolized via the Embden-Meyerhof pathway, some 20 per cent may be utilized by the pentose phosphate pathways. Evidence for the pentose phosphate pathway in other species is very incomplete.

CO_2-fixation in cestodes

Although CO_2-fixation was not thought to be common in animals, experimental work provides increaseing evidence that the process is more common in parasitic organisms than formerly thought, and occurs extensively in cestodes. In terms of the host-parasite relationship it is clearly economical for an intestinal parasite to utilize carbon dioxide because its use has two advantages, (*a*) carbon dioxide is a waste-product of the host metabolism and thus freely available and (*b*) it penetrates cells very readily and thus can be rapidly utilized by an intestinal parasite.

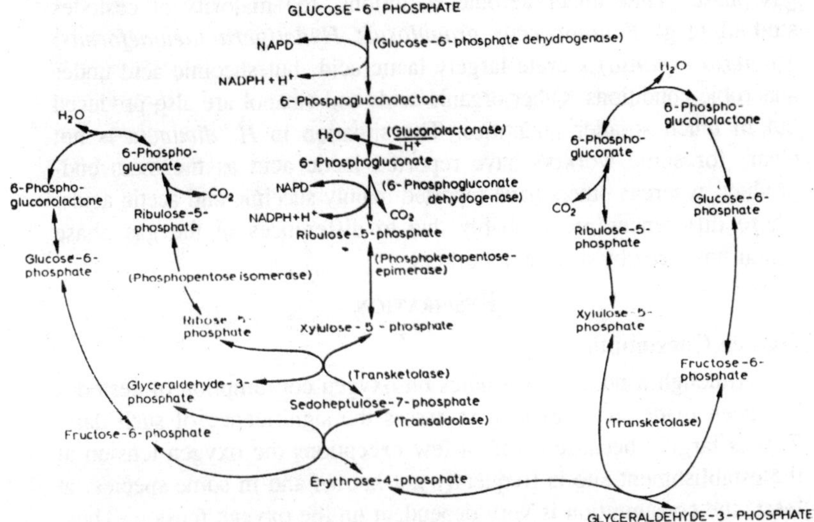

Fig. 14.2. Pentose phosphate pathway.

The most active CO_2-fixing enzyme appears to be phosphoenol-pyruvic acid carboxylase according to the equation:

Phosphenol pyruvic acid (PEP) + CO_2 + guanosine diphosphate (GDP) = oxaloacetic acid (OAA) + guanosine triphosphate (GTP)

The main metabolic end-product in *Echinococcus* is succinic acid. This is often indicative that CO_2-fixation is taking place and following a pathway from pyruvic to succinic acid. Note that in some nematodes (*e.g., Ascaris*) the succinic acid is further decarboxylated and reduced to other organic acids.

It is known that fumarate can act as a hydrogen acceptor for the reoxidation of NADH, a substance involved in glycogenesis and glycolysis. Thus the fumarate is essentially substituting for oxygen in terminal respiration in an anaerobic environment. This would explain the fact that in certain cestodes (e.g. *H. diminuta*) the presence of carbon dioxide has a marked effect on the synthesis of glycogen and utilization of glucose (Fairbairn *et al.*, 1961). This stimulating effect of carbon dioxide suggests that early work on carbohydrate metabolism carried out in the absence of carbon dioxide is open to re-evaluation.

End-products of carbohydrate metabolism

In considering this question, it is important to note that the nature of the end-products is greatly influenced by the PCO_2 and PO_2 of the gas phase. Thus under aerobic conditions, the majority of cestodes studied (e.g. *Echinococcus granulosus*; *Hydatigera taeniaeformis*; *Moniezia expansa*) secrete largely lactic acid, but succinic acid under anaerobic conditions. Other organic acids and ethanol are also produced but in much smaller quantities. The situation in *H. diminuta* is not clear, for some workers have reported lactic acid as the main end-product, whereas others have reported mainly succinic and acetic acids. These differences are probably due to differences in the gas phase (often unstated) by different workers.

RESPIRATION

Oxygen Consumption

Although a number of studies on oxygen consumption of cestodes has been made, it is difficult to assess the significance of such data. This is largely because - with a few exceptions the oxygen tension at the establishment site is frequently not known and in some species, at least, the consumption is very dependent on the oxygen tension. Thus, in *Echinococcus*, at tensions >5 per cent oxygen, consumption is directly proportional to the tension, yet below the 5 per cent level the consumption ceases completely. Again, in larval *Mesocestoides corti*, the respiration rate declined at a tension of 2 per cent oxygen. Two general conclusions can be drawn from the data available.

(*a*) all species of cestodes examined have been found to utilize oxyten *in vitro* when available.

(*b*) consumption of glucose in the absence of oxygen (*i.e.*, fermentation) proceeds at almost the same rate under anaerobic conditions as under atmospheric oxygen.

As pointed out above, the fact that oxygen is consumed *in vitro* does not imply that oxygen is utilized *in vivo*, unless we hae evidence that a similar level of oxygen is available *in vivo*. Several species, such as *H. nana*, are capable of growing to maturity in the complete absence of oxygen. Again, a careful study of the development *in vitro* of *H. diminuta* by Roberts & Mong (1969) found that this species appeared to be 'completely facultative with respect to oxygen presence or absence, under the conditions of the experiments.' This ability to develop under aerobic or anaerobic conditions would clearly be of considerable adaptive value.

Terminal respiration: electron transport

Although a substantial amount of work has been carried out on electron transport of cestodes, very little unequivocal information is at present available. This is related largely to the difficulties of obtaining cestode tissues in sufficient quantities and of making mitochrondrial preparations of sufficient purity. Progress in this field has been reviewed by Bryant (1970) and Smyth (1969). It is beyond the scope of this book to deal with this matter in great detail and only a brief account will be given here. Probably two pathways operate: (*a*) the conventional one utilizing oxygen and the cytochrome system and (*b*) the pathway utilizing fumarate derived from carbon dioxide as indicated above. A 'classical' cytochrome system of the mammalian type involving cytochrome oxidase can only operate efficiently at minimum oxygen tensions of the order of 5 mm. The very few observations made on the PO_2 of the gut suggest that at least on the surface of the mucosa tensions as high as 30 mm may be encountered.

In the crypts of Lieberkühn, in which some species (e.g. *Echinococcus*) become embedded, tensions may be considerably higher, it is of interest, therefore, to note that this species appears to require oxygen for evagination. There is evidence from cyanide inhibition of aerobic respiration and from spectra studies, that components of a 'classical' cytochrome system may be present in at least some species of cestodes. Thus, direct or indirect evidence suggests that a cytochrome system occurs in *D. latum*, *Echinococcus granulosus*, *Triaenophorus lucii*, *H. diminuta*, *Hydatigera taeniaeformis*, *Taenia hydatigena* and

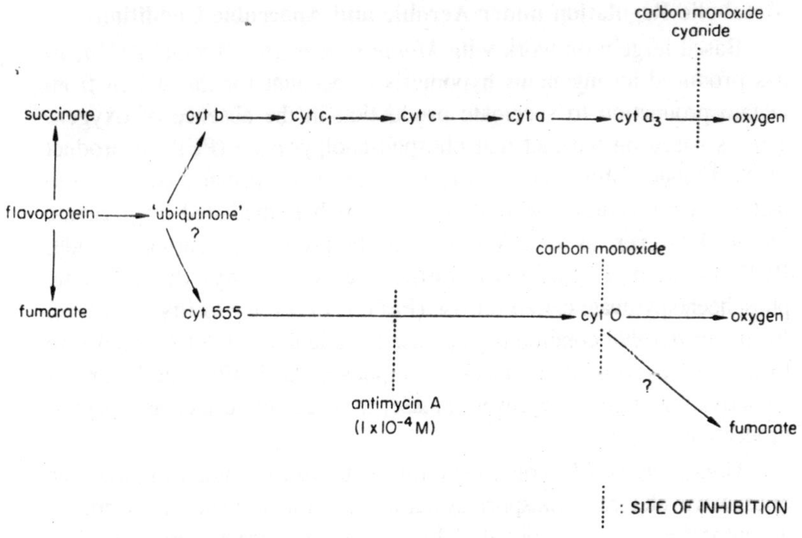

Fig. 14.3. Pathway of electron transport in Taenia hydatigena.

Moniezia expansa. The most complete study appears to have been carried out on *Moniezia expansa* and *T. hydatigena* in which the respiratory chain components contained two terminal oxidases one of the cytochrome *a* type and one containing cytochrome.

In *Moniezia*, cytochromes *b, c* and aa_3 have been positively identified and the experimental evidence indicates that they do participate in electron transport. This, together with the fact that cristate mitochondria have been isolated from this species indicates that *Moniezia* can undergo oxidative phosphorylation. The fact that an aerobic system can operate does not exclude anaerobic metabolic processes, because, as discussed earlier, in the absence of oxygen fumarate can act as an electron acceptor; in this situation succinic dehydrogenase acts as a fumarate reductase.

The extent to which aerobic or anaerobic pathways are followed will clearly depend on the availability of oxygen in the particular intestinal site in question. A possible mechanism controlling these alternative pathways is described below. In *Moniezia*, hydrogen peroxide appears *in vitro*, apparently as a result of the action of cytochrome o, but it is questionable if this substance ever appears *in vivo*. The occurrence of a peroxidase in the tegument of *Moniezia* and *Hymenolepis* may suggest that this enzyme functions to remove peroxide - a substance which is highly toxic.

Metabolic Regulation under Aerobic and Anaerobic Conditions

Based largely on work with *Moniezia expansa*, Bryant (1972*a*, *b*) has produced an ingenious hypothesis to account for the switch from lactate production to succinate production in the absence of oxygen. This is based on the fact that phospho-enolpyruvate (PEP), a product of the Embden-Meyerhof glycolysis pathway, is normally converted to lactate in a reaction, the first step of which is mediated by pyruvate kinase. However, as already shown, in the presence of carbon dioxide, PEP is also capable of being acted on by the enzyme phosphoenolpyruvate carboxykinase (PEPC). According to Bryant (1972*a*, *b*), under *aerobic* conditions, increased availabiltiy of ATP would lead to increased levels of fructose-1, 6-diphosphate (FDP). The latter has an activating effect on pyruvate kinase, thus leading to increased lactate production.

Under *anaerobic* conditions i.e. with oxygen not available, an alternative electron transport system, with fumarate as the terminal acceptor becomes functional. This alternative system produces less ATP, which in turn leads to the lowering of concentration of FDP,

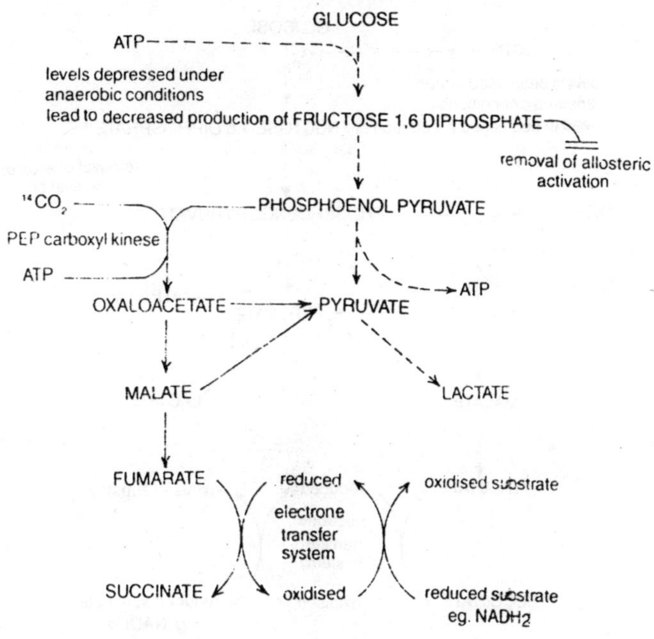

Fig. 14.4. Hypothesis to account for succinate and lactate production in Moniezia expansa: the possible aerobic pathway resulting in lactate production.

thus removing the activation effect on the pyruvate kinase. The result is that the enzyme PEPC can now compete more favourably for PEP and increased succinate production takes place. Moreover, malate inhibits the action of FDP, so that activity along one pathway would inhibit the activity along another. This ability to switch pathways, has two advantages. Firstly, the oxidation of glucose can continue in the absence of oxygen. Secondly, the NADH generated during the oxidation of glucose to PEP provides electrons for the reduction of fumarate to succinate instead of pyruvate to lactate. The occurrence of these alternative pathways thus makes cestodes exceptionally adaptable to environmental conditions. The efficiency of this adaptation in *Echinococcus* is evident from the fact that consumption of glycogen is almost the same under anaerobic as aerobic conditions.

Protein Metabolism

It is evident from the remarkable growth rate of some species that rapid protein synthesis takes place. It must be borne in mind,

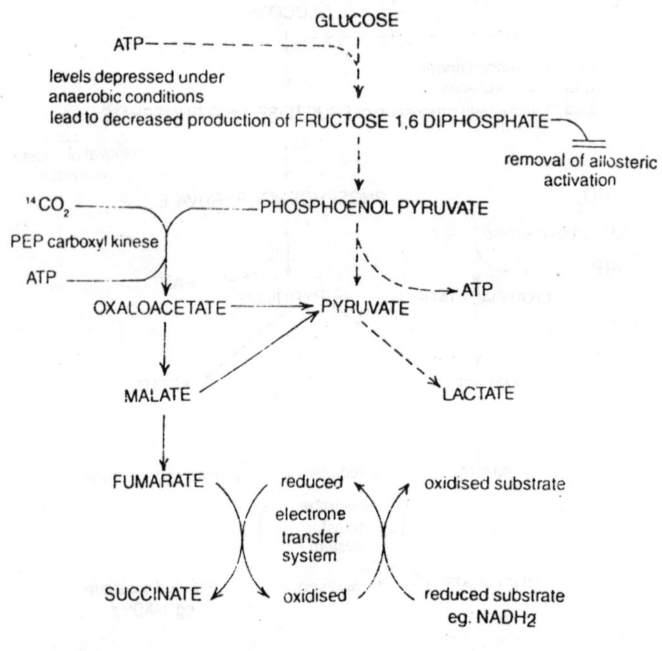

Fig. 14.5. Hypothesis to account for succinate and lactate production in Moniezia expansa: the possible anaerobic pathway resulting in succinate production.

however, that cestodes are essentially made up of a string of embryos and hence different parts of a strobila may not only be synthesizing at a different rate, they may have different nutritional demands depending on the degree of maturation of the proglottids in question. Our concept of the way in which protein or its breakdown products are taken up by an absorptive surface, such as the intestinal mucosa, has undergone drastic revision within recent years. It is now known that not only are small molecules, such as amiino acids, absorbed, but also larger ones such as dipeptides or even whole protein molecules may be ingested.

The same processes may take place in the cestode tegument, although there is, as yet, no unequivocal evidence that pinocytosis - which would be necessary for the ingestion of large molecules - actually occurs in cestodes. As discussed earlier the processes of absorption and/or digestion in cestodes are further complicated by the possibility that a process known as 'membrane' (= contact) digestion can occur. This concept developed by Ugolev (1965; 1968), envisages digestive enzymes *bound* in or on the plasma membrane covering the surface in question.

There is some evidence from *in vitro* studies on *Echinococcus* that the scolex in this species can digest proteins at the host/parasite interface by this proces. There is also some indirect evidence, from experiments with *Hymenolepis* spp. and *Moniezia expansa* that starch hydrolysis in these species takes place by membrane digestion. The occurrence of proteolytic enzymes in cestodes has not been much investigated, but they have been reported in *Taenia*, *Diphyllobothrium* and *Echinococcus*.

Amino acid uptake

The uptake of amino acids by cestodes has been extensively studied and it is now known to take place largely by active transport, although a diffusion component may also be involved. This may generally be recognized experimentally by the fact that the reciprocal of the velocity of uptake (I/V) against the reciprocal of the concentration (I/S) (a Lineweaver-Burk plot) is a straight line. The evidence for this is based on: (*a*) cestodes can accumulate amino acids aginst a gradient, (*b*) the uptake of one amino acid can inhibit the uptake of another (i.e. there is 'competition' for the carrier locus in the amino acid transport system). Early work has been reviewed by Smyth (1969) and Read & Simmons (1963); later representative studies have been those carried out on *Taenia crassiceps*; *Hymenolepis citelli* and *H. diminuta*.

In *H. diminuta*, a species which exhibits a diurnal migration pattern the uptake of methionine was found to differ in the anterior and the posterior parts of the worm. This can be accounted for by the fact that the level of dietary methionine in the anterior region reaches a peak 1.3 hours after feeding, whereas the level in the posterior region of the gut remains relatively stable. The latter fact is related to a homeostatic mechanism which maintains the molar ratios (but not necessarily the concentrations) relatively constant.

The mechanism whereby this is achieved has already been discussed. The amino acid 'pools' (i.e. the free amino acids in the tissues) in cestodes of elasmobranch fish have been examined in detail by Simmons (1969). The pools showed remarkable consistency in some cases. For example, *Lacistorhynchus tenuis* and *Calliobothrium verticillatum*, which are parasites of the elasmobranch *Mustelus canis*, have almost identical amino acid pools. This suggests that the amino acid pool of the host intestine and the ability of the parasite to adapt to it, plays an important role in host specificity.

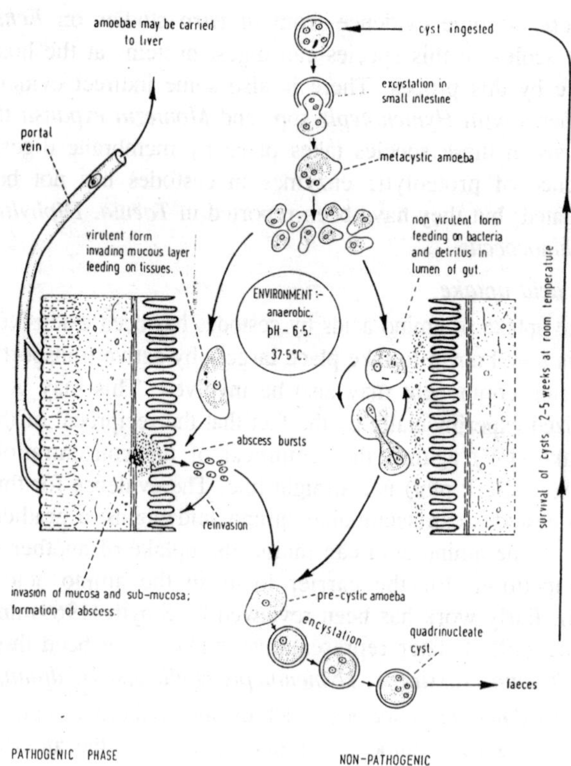

Fig. 14.6. Uptake of proline by Hymenolepis diminuta. A—after 2 minutes' incubation. B—after 1 minute incubation.

Intermediary metabolism

Little is known of the precise amino acid requirements of cestodes, i.e. which amino acids are 'essential' and cannot be synthesized by the host. *In vitro* studies should ultimately provide the answer to this question. The chief mechanism of amino acid formation is by transmination and the a-ketoglutaric acid/glutamic acid and/or the pyruvic acid/alanine systems have been demonstrated in *Hymenolepis* spp., *Anoplocephala magna*, and *Raillietina cesticillus*. In *Hymenolepis*, some 17 amino acids failed to act as amino donors; this suggests that this species is largely metabolically dependent on the host for these materials, which would, of course, nomally be freely available in the intestine. It has already been pointed out above that the amino acid pool in some cestode species in the same host have almost identical amino acid pools. Cestodes have also been shwon to take up purines and pyrimidines either by active transport or by facilitated diffustion there appear to be at least two binding sites on the purine-pyrimidine permease.

End-products of nitrogen metabolism

The intermediary metabolism of cestodes has been little studied. Although the nitrogenous compounds excroeted are known for some species, the metabolic pathways whereby these products are produced are poorly known. The end-products most commonly reported are urea, uric acid and ammonia but numerous other compounds have been found. The production of urea points to the occurrence of an ornithine cycle and two important enzymes of this cycel ornithine transcarbamylase and arginase have been detected. Despite the presence of these enzymes, a *functional* ornithine cycle could not be detected in *M. expansa*, *D. caninum*, *T. pisiformis* and *E. granulosus*. A number of cestodes have also been shown to excrete amino acids (at least *in vitro*) but it is not known whether or not this represent 'abnormal' leakage from the tissues. The same phenomenon occurs in trematodes and nematodes.

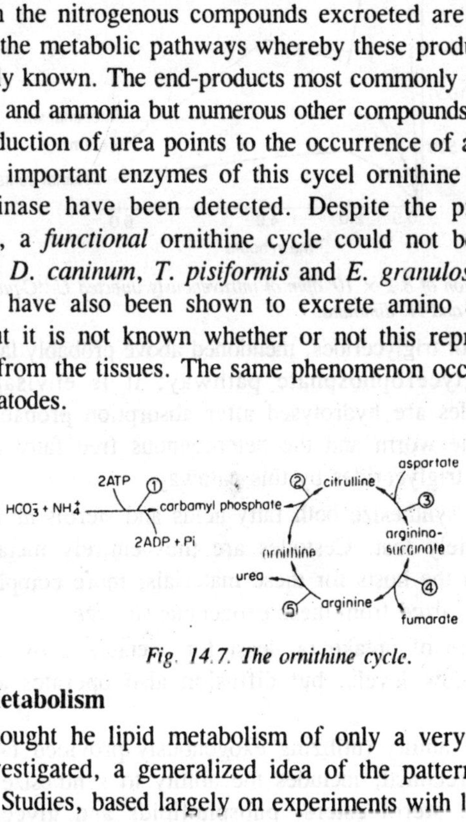

Fig. 14.7. The ornithine cycle.

Lipid Metabolism

Althought he lipid metabolism of only a very few species has been investigated, a generalized idea of the pattern is beginning to emerge. Studies, based largely on experiments with labelled materials, carried out on *H. diminuta Spirometra mansonoides* and *R. cesticillus* have led to the following general conclusions about the lipid metabolism of cestodes:

(*a*) the sterol and fatty acids of the parasite and host are thus (not surprisingly!) qualitatively almost identical; they are not, however, quantitatively identical and some control mechanisms in lipid synthesis must exist in cestodes.

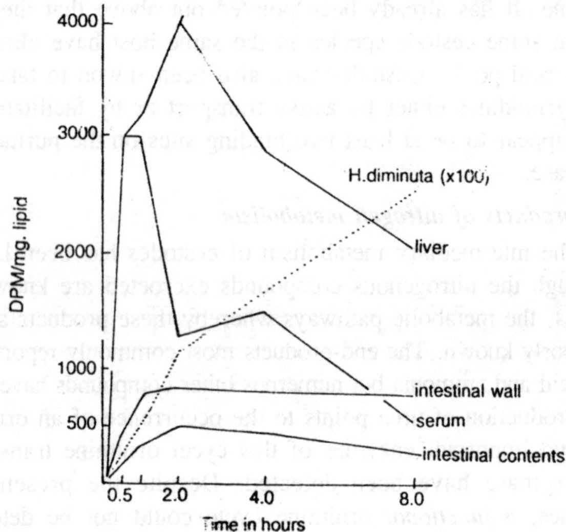

Fig. 14.8. Incorporation of 8.2 × 10^6 dpm of intravenously injected U-^{14}C-palmitate into rat tissues and H. diminuta.

(*b*) the synthesis of triglycerides, mentioned above probably takes place via the α-glycerophosphate pathway; it is envisaged that monoglycerides are hydrolysed after absorption probably at the surface of the worm and the heterogenous free fatty acids are converted to triglycerides by this pathway.

(*c*) the ability to synthesize both fatty acids and sterols *de novo* has been completely lost. Cestodes are thus entirely metabolically dependent on the hosts for these materials; more complex lipids may be synthesized from these exogenous sources.

(*d*) the mechanism of uptake (at least for acetate) is by meditated transport at low levels, but diffusion also operates at higher concentrations.

(*e*) The synthetic ability (utilizing exogenously provided fatty acids and sterols) retained, includes the ability to synthesize its own triglycerides, sterol esters, phospholipids and glycerolipids. Cholesterol, however, cannot be synthesized and must be absorbed from the host. Since labelled atoms from ^{14}-C-glucose and ^{32}p can be incorporated with these molecules it is evident that the synthetic ability retained includes that to synthesize the non-fatty portion of lipids.

As emphasized above, most of the conclusions have been based on work on *H. diminuta* and *S. mansonoides*, and it would be interesting

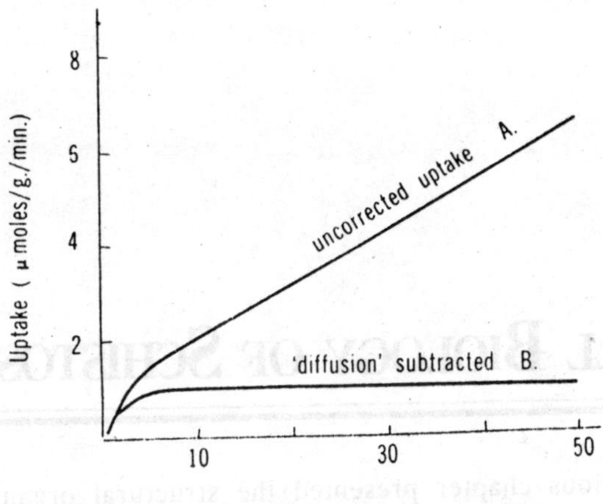

Fig. 14.9. Uptake of ¹⁴C-acetate by H. diminuta, as a function of acetate concentration.

to extend similar investigations to other species. From the above results, it is evident that the lipid content of any cestode will depend largely on the lipid content of the host diet. This has been strikingly demonstrated in *R. cesticillus*.

In worms from chickens fed on a natural balanced diet, the lipid content was found to be 20.1-21.6 per cent. In worms from chickens fed on a low-fat diet the lipid content fell to 10.1 per cent compared to 29.6 per cent in worms from chickens fed on a similar per cent diet with added corn oil. It is well known that the uptake of lipids is assisted by the presence of surface-active agents and it is interesting to speculate if the well-known ability of bile or bile salts (such as sodium taurocholate) to induce evagination of the scolex in cestodes could be due to its presence assisting the uptake of a lipid or sterol by the worm. This absorbed material could then act as a hormone or a trigger substance which could initiate evagination. It should also be noted, however, that bile itself is an important source of lipids and in bileless hosts (experimentally prepared) the uptake of fatty acids such as palmitate is greatly reduced.

15

Cell Biology of Schistosomes

The previous chapter presented the structural organization of schistosomes, showing that the surface of the parasite facing its host was always a syncytium in either the mammalian host or the snail. Here we extend the discussion into the functional properties of the synctium, particularly its surface membranes. The structural, molecular, and functional properties of the schistosome surface are critical in explaining the parasite's long survival in the human host, because the surface is a significant potential target of host defence. Unlike some protozoa that hide within host cells, schistosomes are exposed to the circulating components of the immune system. To survive, the parasite surface presumably possesses mechanisms for evading immune attack. We wish to explain how the parasite can be attacked and killed by some cells, such as eosinophils, but evade that attack of other cells.

The chapter will focus on *Schistosome mansoni* because most of the experimental work has been done on this species. Presumably, the paradigm constructed for *S. mansoni* will be applicable to the other species of schistosomes that infect humans. We will consider initially knowledge of the structure and composition of the surface membranes and then review functional studies performed on the membranes themselves. The chapter will conclude with a discussion of the interactions between various human blood cells and the parasite.

Structure of the Parasite Surface

During its life span in its mammalian host, the parasite is covered by a synctium called the tegument. In all stages, the tegument is connected to the cell bodies that are located deep within the organism below the large circumferential bands of muscle that are responsible for such gross movements of the whole parasite as contraction, elongation, and turning. The cell bodies contain the basic machinery for protein synthesis, namely, nucleus, endoplasmic reticulum, and Golgi apparatus. The surface of the tegument has a varied topology consisting largely of protruding spines, which contain crystalline actin, and invaginating pits. The distribution of the spines and pits is altered during larval development. During the first 4 days after transformation from cercariae to schistosomula, the spines are lost over much of the body, there is more pitting at the surface, and the surface area increases by 30 per cent. In adults, different regions of the parasite have characteristic patterns of folding, pitting, or spines.

Despite these large variations of structure, the ultrastructure of the surface membranes that face the host tissues or bloodstream remains relatively similar from the completion of cercarial transformation through adulthood. Transmission electron microscopy reveals that the parasite surface consists of adjacent lipid bilayers that appear as two apposed membranes by transmission electron microscopy. These membranes are anisotropic, however, as shown by freeze-fracture technique. In newly transformed larvae, the inner of the tegumental membranes, that is, the one that is closed to the cytoplasm of the tegument, contains many intramembrane particles and resembles the plasma membrane of most mammalian cells.

Intramembrane particles are associated with membrane proteins, and membranes devoid of these particles usually contain little or no protein. The outer membrane which faces the host in vivo or culture medium in vitro has few intramembrane particles and resembles a simple protein-poor lipid bilayer. Interestingly, the outer membranes of worms isolated from the lungs of mice about a week after infection contain intramembrane particles, whereas outer membranes of worms cultured in vitro for the same period do not. The parasite may acquire these particles or proteins from its host. However, the host may induce the parasite to synthesize intramembrane particles in its outer membrane.

Electron microscopy has also suggested that the outer membrane contains cholesterol, since the membrane structure is perturbed by digitonin and the polyene antibiotic filipin, both of which characteristically alter cholesterol-containing membranes. These structural observations suggest that the schistosomula are covered by two different membranes: an inner one that probably can function in ion transport much like mammalian cell plasma membranes and an outer one that is simply a lipid bilayer composed of sterol and phospholipid. The schistosome surface the bears a strong resemblance to gram-negative bacteria, which also have a double membrane system.

Surface Membranes

When the plasma membrane of most mammalian cells is exposed to a molecule that binds specifically to a receptor on the membrane, for example, low-density lipoproteins or transferrin, or to reagents that bind to specific chemical groups on the membrane, for example, lectins or polycations, the bound molecules are usually internalized by the process of endocytosis and transported to internal compartments within the cell, such as phagosomes or lysosomes. Some mammalian cells, notably neoplastic cells, also can shed bound ligands into the culture medium.

Schistosomes bind antibodies and complement at various stages of their life cycle in the mammalian host but most extensively shortly after transformation. However, these multivalent ligands bound to the surface are not endocytosed by the parasite. Furthermore, low-density lipoproteins, which also bind to the tegument, and fluid-phase markers such as horseradish peroxidase are also not endocytosed by the tegument. Thus, the tegumental membranes appear incapable of endocytosis and are very different from typical normal mammalian cell plasma membranes. Although complement and antibody bind and are not endocytosed, they do not remain on the parasite surface but are slowly lost. This loss has been explored by applying the lectin concanavalin A (Con A) to schistosomula.

The lectin binds to a unique population of molecules that are present for 24-48 hours after transformation but are not present on cercariae or schistosomula older than 48 hours. Thus, the Con A - binding molecules are lost from the surface independent of the binding of ligand, and the loss therefore is physiological and not induced by Con A. Furthermore, the kinetics of the loss of Con A binding is the same as that of the loss of bound rat and human antibody and bound complement, suggesting that the mechanism of loss is not confined to this lectin; a more general mechanism of membrane turnover is probably at work. The kinetics of the loss of these three molecules-antibody, complement, and Con A-follows that of a decay curve and indicates that the loss of bound molecules is random event.

In addition, Con A can be bound to the parasite and recovered in the culture medium over time. The recovered Con A is the same molecular weight as the Con A originally bound. These studies suggest the parasite is sloughing or shedding the bound ligand and attached membrane component into the culture medium. In mammalian cells, multivalent ligands such as Con A or antibody bound to the plasma membrane at 0°C are capped when the temperature is raised, so the ligand becomes localized to a region of the cell surface. The studies with Con A also show that there is no capping of the bound ligand on the parasite surface, suggesting that the contractile proteins of the tegument do not functionally interact with the outer membrane. As suggested by the kinetics of loss of Con A that follows a decay curve, bound Con A is randomly distributed on the parasite surface. However, after the addition of hemocyanin that is rich in mannose and binds to the Con A, clustering of the hemocyanin and hence the lectin is observed. This clustering occurs after the parasites have been killed with azide, demonstrating that contractile proteins are irrelevant in cluster formation.

Instead, the hamocyanin itself induces aggregation of the Con A in the membrane. These experiments also demonstrate movement in the plane of the membrane by the moieties to which Con A is bound. The loss of bound Con A and its recovery in the culture medium suggest that the outer membrane or portions of it are being sloughed into the medium. This phypothesis can be tested by labelling the parasite surface and examining the culture medium over time to see if labeled membrane components appear there. Several general labeling schemes are available, particularly iodination and labeling oxidized sugars on glycoporteins and glycolipids with tritium.

The sugar-labeling techniques have the advantage of being two-step procedures, permitting the nonspecific binding of the radiolalbel to be determined in the absence of oxidation. Further, the oxidation step can be performed with an enzyme, such as galactose oxidase, which limits the specific labeling to the parasite surface, since the parasite does not endocytose. Experiments of this type as well as subsequent experiments employing monoclonal antibodies show that many surface glycoproteins and glycolipids are being shed into the culture medium with a half-time of approximately 12-14 hours. Given that the surface area of a schistosomulum is 20,000 square micrometers (μm^2), this figure represents losses of approximately 800 μm^2/hour, an astronomical rate by mammalian cell standards. Furthermore, the rate of synthesis must be double that figure, since the surface area is increasing by 100 per cent per day. Studies of this type are usually not carried out on adult schistosomes, largely because such studies would be complicated by the presence in adults of a functioning gut, which can ingest sloughed proteins. Although the surface membrane proteins and lipids are clearly being sloughed into the culture medium, some experiment do not fit this model and there are serious methodological limitations in all experiments of this type. In particular, studies with monoclonal antibodies have suggested that some proteins may not be sloughed and may persist for long periods of time on the surface.

More generally, the interpretation of the turnover experiments is clouded by the fact that schistosomes are covered by two membranes and that the distribution of labeled or antigenic membrane components within the double-membrane system is usually unknown. In particular, it is unclear whether only components of the outer membrane are labeled or those of the inner membrane are also labeled. Definitive experimentation in schistosomes is complicated by difficulties in obtaining large numbers of organisms and ultimately in determining the composition and distribution of the components of the two membranes.

Chemistry of Surface Membranes

Many laboratories are now actively investigating the composition of the surface membranes, but a definitive statement cannot yet be made about which molecules constitute the two membranes. Investigations have employed isolation of the membranes, surface labeling, selective extraction, and analysis of biosynthetically labeled materials shed into the medium. Taken together, they yield a general picture. The outer membrane is primarily composed of phospholipids. (Note that phospholipids are made up of a glycerol backbone attached to two fatty acids, called acyl chains, and phosphate. The phosphate is linked to another molecule, such as choline or serine, called the head group.) The major phospholipids in the outer membrane are phosphatidylcholine and phosphatidylethanolamine, with small amounts of phosphatidylserine, phosphatidylinositol, sphingomyelin, and lysophosphatidylcholine. Cholesterol is also present as well as glycolipids, some of which are antigenic.

There are a variety of antigenic and nonatigenic proteins, most of which are glycoproteins as well. The inner membrane probably contains the various enzymes known to be associated with the tegumental membranes, namely acetycholinesterase, Na-Ca ATPase, and alkaline phosphates as well as phospholipids and cholesterol. Schistosomes do not synthesize cholesterol but instead use host cholesterol; they do not modify host acyl chain length during phospholipid biosynthesis but perform a limited number of head-group modifications. Presumably, the lipid components are assembled into membranes in the cell bodies and transported to the tegument in multilamellar bodies. There they fuse with the base of the pits and insert the new membrane into the parasite surface.

Interactions with Cells

Schistosomes dwell in the bloodstream from about 4 days after they penetrate through the skin until they die years later. Being in the circulation exposes them to cells and plasma components that have the potential to destroy the parasites. On the other hand, one of the ways the parasite can protect itself from immune attack is to acquire host antigens from serum or blood cells themselves. In this way, the parasite mimics the host, is not recognized as foreign, and in a sense becomes the wolf in sheep's clothing. The attack of cells and parasite defense can be studies in vitro by mixing purified leukocytes, schistosomula, and infected patient's serum that contains antibodies against the parasite. Let us consider four blood cells: eosinophils, neutrophils, monocytes, and erythrocytes. Eosinophils adhere to the parasite surface and degranulate within minutes.

The parasites die about 18 hours later when most of the surface is covered by material discharged from the granules. Since a single eosinophil discharge over an area of about 80 square micrometers (μm^2) and since about 1800 μm^2 of the parasite are covered by discharge material after 12 hours, several hundred eosinophils have to discharge onto each parasite. The successful attack of the eosinophils should be regarded as a mass onslaught directed broadly against the entire surface of the parasite that is carried out over the course of a half a day or more.

Eosinophil degranulation is mediated through antibody bound to the parasite surface. The cells bind to the worm by means of an Fc receptor in the plasma membrane binding to the Fc portion of the IgG on the worm. The granules move through the cytoplasm to the zone where the plasma membrane is bound to the antibody. The granule membrane then fuses with the plasma membrane, thereby releasing the granule contents at the site of eosinophil-parasite contact. During the attack, eosinophils die and lyse on the parasite surface. Cells can be seen with electron lucent cytoplasm and granules that are still intact and undischarged. Presumably the cells died before they can exocytose all of their granules. Furthermore, fragments of the plasma membranes of eosinophils are seen covering degranulated material on the parasite's surface, perhaps assisting in the attack by confining exocytosed material to the target and preventing diffusion of material from the surface. Whether the eosinophils are killed by the parasite or by material released by their own or other eosinophil granules remains to be determined. Unlike eosinophils, neutrophils do not kill schistosomula in the presence of antischistosomal antibodies alone, but they will kill if complement is also present. Although neutrophils adhere to the worm in the presence of antibody alone, they do not degranulate, which probably explains their failure to kill the worms.

The lack of degranulation is in turn explained by experiments examining how neutrophils process multivalent ligands such as antibodies and lectins on the parasite surface. Fluorochrome-conjugated antischistosomal antibodies or lectins promote cell-parasite adherence and can be observed simultaneously by fluorescence microscopy. Immediately after labeling, the ligands are evenly distributed on the parasite. Soon after neutrophils are added, dark areas appear on the parasite and fluorochrome can be identified in the lysosomes of cells. Neutrophils endocytose antibody or lectin from the parasite surface. Thus, unlike the interaction of the Fc portion of the antibody with the Fc receptor on the eosinophil, which promotes fusion of the granule membrane with the bound membrane at the cell surface, the neutrophil membrane receptor ligand complex is internalized and fuses with the granule membrane in the cell.

Parasite antigens are also endocytosed along with the antidody, as demonstrated by autoradiography. Removal of both antigen and antibody from the parasite by neutrophils reduces the density of recognition signals available to eosinophils. About 10 per cent of adherent neutrophils fuse their plasma membranes to the outer of the two tegumental membranes of the worm when worms are first coated with antibody or lectin. A hybrid membrane is derived from the plasma membrane of the neutrophil and the outer tegumental membrane of the parasite is formed as a result. No antibody or lectin is detectable by electron microscopic autoradiography in the fusions, suggesting that the ligand is removed by endocytosis before fusion occurs. These fusions are probably the largest (up to 10 μm) to be described between biological membranes. They may be a mechanism for acquiring host membrane components, particularly integral membrane proteins, such as histocompatibility antigens, which are thought to defend the parasite against being recognized as foreign by the host immune system.

The fusions are not reversible and ultimately result in lysis of the fused neutrophil. After this lysis, the fused membrane appears to mix with the normal worm outer membrane. The third major effector cells in human blood is the monocytes. Human monocytes are not as effective as eosinophils in killing schistosomes. Monocytes that have been cultured for several days are called monocyte-derived macrophages because of a certain resemblance to tissue macrophages. Quiescent monocyte-derived macrophages do not effectively kill tumor cells or schistosomula that have been prein-cubated with antibody. However, after exposure to interferon-g (a mediator that activates the host cells), the monocyte-derived macrophages acquire the capacity to kill tumor cells; their ability to kill schistosomula is inconsistent. Human monocyte-derived macrophages fuse with the outer tegumental membrane.

Murine macrophages, in contrast to human monocyte-derived macrophages, consistently kill schistosomes and apparently are major host defense effector cells in murine schistosomiasis infections. Parasite interactions with erythrocytes are potentially important, particularly since adult worms can acquire human ABH blood group antigens in vitro. After coculture of schistosomula and erythrocytes, an electron-dense plaque is seen between the parasite outer membrane and the plasma membrane of the attached cell. Membrane proteins and glycolipids are not transferred to the parasites from adherent cells. However, carbocyanine dyes, which are anchored in the red-cell membrane by acyl chains, are transferred rapidly in 30 minutes to 3 hours. These studies also demonstrated that erythrocytes lyse on the parasite surface and that the lysed membrane fragments remain attached to the parasite. This mode of host antigen acquisition and the resultant masking of parasite antigens differs from the process of host antigen acquisition in which thc antigens are presumably inserted into or absorbed onto the outer tegumental membrane.

Additional modes of host antigen acquisition could exist. Since adult worms ingest red blood cells, they also might be able to transport intact glycolipid antigens of erythrocytes from the their gut to the tegumental surface. Furthermore, glycolipid antigens can be acquired from serum. Each cell type examined-eosinophils, neutrophils, monocytes, and erythrocytes—is lysed on the surface of the parasite. With the first three cell types, studying the mechanism of lysis is difficult because the cells themselves have cytolytic molecules. It therefore is not clear whether lysed effector cells represent a form of suicide or homocide, with the parasites as the murderer. Erythrocytes, on the other hand, do not have such autocytolytic capacities. Erythrocyte membranes bound to parasites have been studied by fluorescence photobleaching recovery (FPR) methods to further address the role of the parasite in host cell lysis. In this technique, a fluorochrome is inserted into the red cell membrane. For example, either glycophorin or band 3 can be selectively fluorescinated or the phospholipid analogue, fluorescein phosphatidylethanolamine (Fl-PE), can be partitioned into the membrane. After the labeled erythrocyte adheres to the worm, a laser is used to bleach the fluorochrome in a small spot on the surface of the cell.

If the molecules bearing the flurochromes are able to diffuse in the plane of the membrane, then the fluorescence will reappear at the bleached spot. Two measurements of recovery can be obtained: the mobile fraction, or percentage of labeled molecules that are able to move, and the diffusion coefficient, or rate of movement of the fluorochrome into the bleached area. In such experiments, both glycophorin and band 3 are totally immobilized in lysed red cell membranes adherent to the worm (e.g., they do not move into the bleached spot). (The phospholipid probe, Fl-PE, is partially immobilized (50 per cent mobile fraction). These results obtained with red-cell membranes adherent to worms contrast with what is observed in studies of intact red-cell suspensions, in which the mobile fraction of glycophorin and band 3 is about 60 per cent and that of Pl-PE is 100 per cent. Red-cell membrane ghosts produced by hypotonic lysis exhibit enhanced membrane movement, with mobile fraction of the proteins nearly 100 per cent.

Relative immobilization of the proteins observed on the parasite - associated erythrocyte membrane can be reproduced by cross-linking the protein in intact erythrocytes with lectins, polylysine, or diamide. Under these conditions there is no decrease in Fl-PE mobile fraction, however. The only treatment of labeled erythrocytes that caused red cells lysis and alterations of the FPR measurements of all three test probes in a manner similar to those found in red cells adherent to the worm is the addition of exogenous lysophosphatidylcholine (lyso-PC). Interestingly, lyso-PC is also produced by worms metabolically labelled with palmitate or choline. Furthermore, the parasite releases lyso-PC into the culture medium in quantities sufficient to account for the FPR measurements. Together, these observations and the results of photobleaching experiments strongly suggest that the parasite is capable of lying erythrocytes attached to its surface by the generation and release of lyso-PC. Perhaps the lysis of effector cells (eosinophils, neutrophils, and macrophages) is also due to the action of lyso-PC.

Coda

The schistosome appears to operate according to three general principles: (1) I'll take anything I can get; (2) If you want it back, you can have it; (3) If you get too close, I'll bite. The parasite outer membrane appears to be deceptively simple, containing mainly phospholipids and cholesterol derived from the host as well as adsorbed host proteins and lipids. The true parasite molecules, proteins and antigenic glycolipids, are probably only a small fraction of the outer membrane. The outer membrane can be shed rapidly because it is energetically cheap to make. It is also poorly antigenic because so much of its is derived from the host. The presence of lyso-PC in the outer membrane or the ability to generate this molecule explains many of the membrane properties.

Lyso-PC is a detergent very similar to sodium dodecyl sulfate. When its critical micelle concentration is reached in the membrane, proteins are solubilized in lyso- micelles or mixed micelles formed from lyso-PC and the other phospholipids and cholesterol. This property of lyso-PC explains the shedding of both membrane proteins and ligands from the parasite. Endocytosis by neutrophils of ligands such as antibody and lectins and the membrane components to which they are bound can be seen as an extension of this process. Lyso-PC is a fusigen, which accounts for the fusion of neutrophils and monocytes with the outer membrane. Finally, as suggested by its name, lyso-PC is a powerful lytic agent and is most likely responsible for the lysis of various cells on the parasite surface.

The success of eosinophils in attacking the parasite probably owes to the high levels of lysophospholipase on their surface and in their granules. This enzyme cleaves lyso-PC into a fatty acid and glyceryl phosphorylcholine, thereby inactivating it. Many important questions about the function and molecular architecture of the tegumental membranes remain unanswered. In particular, it is not clear whether the outer membrane has high steady-stage concentrations of lyso-PC or generates it enzymatically in a controlled manner to in response to stimuli. It is also not clear how the inner membrane is separated from the outer membrane, how the membranes maintain their identity, and how the inner membrane functions as a true biological membrane when covered by an apparently intact separate lipid bilayer. Finally, the biogenesis of these membranes needs further study to determine if the models based on morphological data-namely, assembly and transport in multilamellar bodies—are correct.

16

Nematod Control

The main objective of the control of phytonematodes is to improve the growth, quality and yield of crop plants. Phytonematode control is achieved mainly through management of their populations because even with the best anti-nematode agricultural chemicals hundred per cent elimination of their numbers is not possible and the remaining population multiplies rapidly to injurious levels in the presence of a suitable host. Therefore, their population management involves attempts to keep the populations low at a safe level where they do not cause economic loss. This can be achieved by a single method or by a combination (integration) of several methods. The latter approach is always preferable. Whatever the nematode control programme is used, it must be profitable to the grower. It is important to evaluate the cost:benefit ratio of each or combination of methods to make the approach effective.

Regulatory Measures

Regulatory control involves certain measure, self-imposed or enforced by state law, which prevent entry of any serious pest or disease agent into a geographic area, to which it is exotic, from outside that area, within or outside the country. These regulatory measures fall under the category of "quarantine regulations." A country undergoes certain expenses and inconveniences to enforce these regulations so that it can save on continued expenses and losses that could occur if the unwanted crop enemy got itself established. Some quarantine measures have been highly successful while others have not. The term 'quarantine' is derived from the latin word *quarantum* meaning forty. It was originally applied to the period of detention of ships arriving from countries where epidemic fevers occurred. The passengers and crew of the ship were compelled to remain isolated on board giving enough time to permit latent cases of disease to develop. When compared to human quarantine plant quarantine is of recent origin. The first plant quarantine legislation in the world was enacted in the year 1877 in Indonesia. India adopted plant quarantine measure in 1914 by introducing the Destructive Insects and Pests Act.

Necessity of Regulatory Measures

Easy transportation of food materials (plants, seeds, etc.) and necessity of such materials in many areas of the world has created the problem of easy transport and entry of plant parasites also. In many countries, including India, intensive developmental research work to increase crop productivity is going on. This involves import of germ-plasm of important crops, especially from areas of their origin. In many instances the parasites of such crops are also established in these areas. Associated with these exotic introductions has been the danger of entry of some of the most serious plant parasitic nematodes so far not known in India.

Interception of several destructive nematodes such as *Heterodera schachtii*, *H. goettingiana* and *Ditylenchus dipsaci* from imported planting material were recently reported from this country. These nematodes are not known to occur in India. The presence of sugar-beet cyst nematodes (*H. schachtii*) in inert seen material such as dried root bits, plant debris, and in accompanying soil clods suggests that there is every possibility of these cyst forming nematodes gaining entry into the country along with imported seed, provided the crop is grown widely. This strongly emphasizes the importance of examining introduced seen or plant material for nematode infestation in order to prevent or at least restrict the introduction of serious nematode pests into the country. Thus, plant quarantine should function as the first line of defence in plant protection programmes.

The modern concept of plant quarantine is that it should function as a filter but not as a barrier. There are some essential requirements for implementations of quarantine measures. The regulations should clearly specify methods and procedures for inspection, treatments and certification, etc. The quarantine stations should be equipped adequately to carry out the duties. Treatment schedules should be clear so that the exporting country and the inspector at the point of entry are able to follow them with reasonable resources.

Quarantine Methods and their Application

Pre-entry precautions

Pre-entry precautions are to be ensured by the exporting country as well as by the importing country. The latter should avoid, as far as possible, importing plant materials from countries where dangerous nematodes exist but are not existing in the importing country. On their part, the exporting country should ensure that the material being exported has been produced in properly inspected fields and proper laboratory testing and treatments have been provided before export was made.

Post-entry control

These involve inspection of the material at the first point of entry to detect and refuse delivery of anything which may show infestation. The inspection may involve laboratory testing, treatment if necessary and also quarantine of the material, especially plants, for specific duration in isolated area. The necessity for mandatory inspection at the port of entry, in spite of the fact that consignments are accompanied by phytosanitary certificates from the exporting country, can be seen by the following example: About 112 metric tons of potato tubers were imported from Scotland, Ireland and West Germany through Madras Seaport in 1965 and 1968. The consignments were accompanied by the mandatory phytosanitary certificates and were additionally certified to be free from soil as well as potato golden nematode cysts. The examination of about 80 to 90 kg of soil accompanying the consignment, although import of soil is prohibited, yielded 16 cysts of the golden nematode and one cyst of sugar-beet cyst nematode. This necessitated destruction of the entire consignment by dumping into the sea.

International cooperation in plant quarantine

Full benefit of plant quarantine can be achieved only when countries situated in the same geographic area, having similar agricultural interests, take common action in preventing the introduction of pests and diseases. Realizing the need for cooperation in enforcing quarantine laws, governments coordinate their activities under the guidance of F.A.O. Eight international plant quarantine commissions were set up both and without F.A.O. participation.

The main objectives of these commissions were to prevent the introduction and spread of destructive pests and diseases in their respective areas through appropriate plant protection measures with mutual cooperation among countries in their zone. Asia and Pacific

Protection Commission is one of these commissions. There are 22 countries, including India, in this commission. F.A.O. plays a major role in organizing and supervizing as well as disseminating information regarding pest and disease incidences, regulations regarding prevention of introduction etc.

Administration and organization of plant protection in India

In India the first law, an act called Destructive Insects and Pests Act, 1914 (Act 11 of 1914), was passed on third February 1914. Under this act various notifications have been issued from time to time prohibiting or restricting the import of certain plants, plant materials, insects and fungi, by air or sea. These regulations are called Foreign Quarantine Regulations. Regulations have been imposed prohibiting transport by rail, road, sample post etc. of various types of plants and propagating materials from infested areas in a state or Union Territory. These are called Domestic Quarantine Regulations and draw their authority from Agricultural Pests and Disease Acts of the respective States. The effectiveness of quarantines depends on complete information regarding the biology of the nematode (life history, dispersal methods, environmental relations, and host range), present distribution in the concerned area, probability of introduction in the absence of quarantine and probability of establishment and relative importance.

There should be periodic review of the measures being followed and if and when necessary, modifications should be made in the existing regulations. Objective evaluation of effectiveness of quarantine measures is not possible since there is no check for comparison. However, the long-term benefits compensate for inconveniences and cost incurred by the government and the growers. Under special circumstances the government may pay compensation to the growers for the financial loss suffered by them in the enforcement of the regulations.

Cultural and Land Management Practices

The aim of these practices, based on the principle of attrition, is to reduce nematode populations to a low level before a suitable crop is grown satisfactorily. Usually the cultural operations are carried out in conjunction with normal crop production practices, at little or no extra cost, and are aimed at a particular nematode pest. The art of weakening the parasite to the point of exhaustion by constant harassment, breaking down or wearing down from repeated attacks or constant diminution of the parasite is known as attrition and is an important principle of plant disease control.

The nematodes are gradually weakened to the point that they are incapable of causing invasion and multiplying or spreading. Plant parasitic nematodes are obligate parasites and do not feed, reproduce or multiply in the absence of their host plants. When the infested soils are kept free from host plants, reproduction and multiplication of nematodes is limited.

There are several other practices which causes premature death of nematodes. If these practices are followed for several seasons the populations are starved due to lack of food and killed. These methods are practices can be briefly listed as below:

1. Prevention of spread and multiplication by destruction of host weeds, selection of nematode free, healthy planting material and field sanitation.
2. Growing antagonistic crops and cover crops.
3. Land management practices such as flooding, fallowing, deep summer ploughing, and drying of soil.
4. Crop rotations.

Crop Rotation

Crop rotation, sequence or arrangement of crops grown in a particular, is an ancient agronomic practice used by farmers throughout the world to avoid crop failure due to soil problems. It continues to be one of the most problem component of any cropping system to combat the problem of soil-borne plant pathogens including nematodes. Mention has been made earlier that more than 100 years ago Julius Kuehn had recommended crop rotation as the best method of control of the sugar-beet cyst nematode, *Heterodera schachtii*. Extensive work on the effects of crop rotations on phytonematodes has been done in other countries but in India such studies are limited. The value of crop rotations in the control of root knot nematodes, *Meloidogyne* spp. cyst nematodes, *Heterodera* spp. and migratory ecto and endo-parasitic nematodes is well established.

There are specific beneficial crops which, when they follow a susceptible crop, eliminate or reduce the population of phytonematodes; for example, barley following cotton or groundnut following tomato in *M. incognita* infested soil, wheat following rice in *Pratylenchus indicus* infested soil, or beet following cereals in *P. penetrans* infested soil. Control of phytonematodes by crop rotation is based on the fact that being obligate parasites these nematodes can feed and multiply on their host plants. They must have a suitable living host to complete their life cycle. Better the host for feeding larger will be the number of generations of the nematodes in one season.

By introducing non-host crop plants or resistant/immune cultivars between the susceptible crops or their cultivars, there is significant decrease in the number of generations and, finally especially in long rotations, the population of the nematode declines to insignificant or innocuous level. Field trails carried out by Chawla and Prasad and Mukhopadhyay and Prasad have shown that when a fallow land, with insignificant numbers of phytonematodes, was put under one crop (wheat) the total numbers of Tylenchids (including *Tylenchorhynchus*, *Hoplolaimus*, *Pratylenchus*, and *Helicotylenchus*) almost doubled. Taking fallow population as a unit, the Tylenchid population build up in one crop (wheat), double cropping (maize-wheat) triple, cropping (mung bean-maize-wheat), and four cropping (mung bean-maize-potato-wheat or mung bean-maize-*toria*-wheat) was 16.5, 27.1, 29.8 and 10.2, respectively. In Punjab the field trials conducted by Sharma and associated to determine the effect of *kharif* cropping sequences (groundnut, sesamum, soybean, tomato) and four rabi corps (wheat, mustard, chick-pea, tomato) in different combinations on the population of *M. incognita* have shown substantial increase in populations in tomato monoculture and in sequences having tomato in combination with chick-pea and soybean. The population was reduced considerably under groundnut-mustard rotations.

Monoculture of tomato, brinjal, okra, chilli, sponge gourd or rotations in all combinations with these crops brought about manifold increase in the population of *M. incognita*. A decrease in the nematode population was obtained when marigold, spinach and bottle gourd were grown or when the field was left fallow. In the U.S.A. the barley root knot nematode, *M. nassi* was effectively controlled by growing a non-host, potato, for two years. For controlling the potato cyst nematode there is a law in the Netherlands prohibiting growing growing potatoes or other crops susceptible to the nematode more than once in three years. A four to five year rotation is usually sufficient to ensure that the population left after harvest of susceptible potato has fallen to a safe level before the next susceptible crop is grown. Populations decline about 30 per cent annually in the absence of the host crop. If long rotations of five to seven years are followed loss in yield can be minimized. The cereal cyst nematode, *Heterodera avenae*, is highly host specific, being restricted to wheat and barley among the common cereals in India. There is, therefore, good scope for crop rotation as a method of control for this nematode.

In the presence of the host crop the nematode multiples five to six times. When a non-host is sown in the infested soil there is about 60 per cent decline in nematode populations. The average decrease in population after lucerne has been found to range from 45 to 65 per cent. In the U.K., the damage by this nematode to oats, the preferred host there, is considerably reduced when oats are avoided for two years in the infested soil. Thus, if non-host crops are grown for two to three years there is good possibility of controlling this pest of wheat and barley. Crop rotations for four-year duration with such non-hosts as chick-pea, mustard and carrot have shown 50 to 75 per cent decline in *H. avenae*.

Mustards have special advantage being a good source of cash returns and also because the mustard roots have adverse effect on cyst nematodes. The mechanism of control phytonematodes with crop rotation has not been studied in detail. The general belief is the populations of these obligate parasites in the absence of the host and reduced because they are unable to multiply and produce new generations. However, there could be other possibilities such as changes in soil physico-chemical and biotic environment caused by different types of crops, chemically different crop residues of the rotation crops releasing decomposition products harmful for the nematodes, and better nutrition and root growth of the plants.

There are many reports of the decline being attributed to parasites and predators of the phytonematodes in specific rotations. This biological control is discussed later. Roots of some plants such as marigold and some cucurbits produce toxic substances harmful to nematodes as discussed later. In spite of the fact that crop rotation is without doubt an effective method of disease control, it has certain limitations, especially in countries where land is in short supply and choice of crops is limited. A nematode having a wide host range will not favourably respond to crop rotation within the narrow choice of crops unless the whole cropping system is changed. Root knot nematodes of vegetable crops (*M. javanica*, *M. incognita*) is an example. These nematodes have very wide host range and attack almost every vegetable crop in India. Vegetable-cereal rotation effectively reduces the incidence of root knot. However, vegetable cultivation in India is a specialized cropping system and normally the vegetable growers will not put the land under cereals. In situations where, for economic reasons there are difficulties in changing the cropping system, cultivar rotations, instead of crop rotation can be followed.

In cultivar rotation, the crop remains the same but rotation includes resistant or immune cultivars of the crop(s) in place of non-host crops. In such rotations susceptible cultivars may be included occasionally to prevent emergence of new races or biotypes of the nematodes. Even in the case of host specific phytonematodes such as *H. avenae*, farmers may be reluctant to practice long rotations of three to four years in areas where this nematode is widespread because the choice of alternative crops is limited. It is not only choosing a crop but also the consideration of cash return from such crops that matters. Farmers always prefer to follow prefer to follow a rotation in which each crop yields a reasonably good cash return. The success of crop rotation, thus depends on a variety of factors which such as selection of crops in the sequence, relative susceptibility of the crops, identity of the nematode(s) predominantly present, races of the nematode present, host range, initial inoculum density, length of life cycle and other aspects of population dynamics. The length of rotation is related to the magnitude of initial population level and to the rate of population decline during the presence of the rotation crops.

Fallows

Fallowing is the practice of keeping the land free of all vegetation for a specified period by occasional ploughing of soil, especially during not summer months. Fallowing may have to be adopted out of sheer necessity, such as because of limited water supply or failure of a crop soon after planting and no time left for replanting or planting of a substitute crop. It can also be adopted by choice in order to get specific benefits. In fallowing out of the necessity the field should be kept free from weeds and repeated turning of the soil should be done to expose the nematodes to adverse elements of weather.

Fallowing by choice is largely an economic consideration, taking into account revenue from crops that could be grown during the fallow period, alternative methods of reduced nematode populations in the soil and benefits of the fallow on the succeeding crops. Even where fallowing is known to give definite advantages in disease control the farmers have not accepted it because of economic consideration. Fallows are of different types such as dry fallows, wet fallow and flood fallow.

Dry fallowing for nematode control is generally restricted to extensive farming where other methods of disinfestation of soil are not economical. It is combined with specific crop rotations, soil amendments, and fertilizer and tillage practices. Clean fallow deprives nematodes of their food source and exposes those in surface soil layers to extremes of weather, especially heat and desiccation. While most non-cyst forming nematodes are considerably reduced by this treatment, even cyst forming nematodes such as *H. avenae* which are highly susceptible to desiccation are also adversely affected. Root knot nematodes may be killed by drying soil for two days in 2.5 cm depth, for four days in 5 cm depth and for 32 days in 25 cm depth.

The nematodes present in deeper layers can also be killed by giving frequent turning of soil. Considerable reduction in the population of cereal cyst nematode can be achieved by ploughing the fields during summer. It is, therefore, a good practice to keep the fields free from vegetation during hot summer months (May-June), give one early irrigation and then two to three ploughings to turn and expose the soil to hot sun and desiccation. There are some objections to the practice of including a fallow in crop rotation or in a cropping system. From an economic viewpoint, in countries where due to pressure on land maximum crop production per unit area is the aim, keeping the land free from crops may not be acceptable. However, during May-June when water scarcity makes crop cultivation expensive fallowing can be practised to advantage. From an agronomic viewpoint, frequent turnings of the soil without vegetation cover causes depletion of nutrients. This may necessitate use of cover crops.

Cover Crops, Trap Crops, Antagonistic or Enemy Crops

Cover crops usually have not much economic value and are grown to provide vegetation cover to the land and for fodder etc. Such crops are highly resistant to phytonematodes. Growing cover crops in nematode infested soil has been advocated in some countries to reduce nematode populations. Cover crops can be harvested as fodder crop or used as green manure. *Crotalaria* spp. have been used for green manure and reduce root knot nematode damage. Trap crops are plant highly susceptible to a nematode parasite of the major crop and are planted so as to get infested by the nematode at a particular stage of growth of the plant and both the crop and nematode can complete its life cycle. This reduces the primary inoculum for the susceptible major crop that follows. When trap crops are grown in root knot or cyst nematode infested soil, the second stage juveniles enter the plants but before they reach adulthood to lay eggs the plants are destroyed. Oats have been used in England as a trap crop for cereal cyst nematode.

The crop is harvested for fodder before the nematode produces the next generation of cysts. *Solanum nigrum* stimulates hatching of eggs of *G. rostochiensis*. The second stage larvae enter the roots as on normal host but cyst development is poor. Enemy or antagonistic plants produce some toxic compound that destroys the nematodes in soil. Some grasses, varieties of mustard, marigold, species of *Crotalaria*, and asparagus have been listed as enemy plants.

The relationship between marigold (*Tagetes* species) and a number of root infecting nematodes was studied by Oostenbrink and his associates. They found that by growing marigold, populations of *Pratylencus* could be reduced by 90 per cent. Later, it was found that these plants reduced populations of *Paratylenchus*, *Tylenchorhynchus*, and *Rotylenchus* also. The effect is due to root exudates of marigold that contain terthienyl compounds toxic to nematode. Prolonged stand of the plant in the soil is essential before sufficient concentration of the toxic compound is reached in the soil. Since marigold is not a commercial crop in most places its cultivation for at least for months in the field may not be feasible. The effects of potato root exudates, essential for hatching of cyst nematode, are neutralized by growing mustards along with potato. This prevents the larval release from cysts.

Later, it was found that mustard oils contain allyl isothiocyanate toxic to nematodes. Ellenby reported that mustard oils reduced the severity of cyst namatode infestation and increased the yields of potato. Populations of *Trichodorus christiei* are not supported by *Asparagus officinalis*. The fleshy roots of the plant contain a nematicidal agent, asparaguisic acid being one component. The compound also suppresses *Pratylenchus curvatus*. Root exudates of *Brassica nigra* and *B. hirta* neutralize the effects of potato root exudates on cysts. The root exudates of certain grasses contain pyrocatechol that is toxic to nematodes. While the effectiveness of these antagonistic plants against several phytonematodes has been demonstrated in laboratory and glasshouse tests their practical utility on a field scale is not certain.

Flooding

Submergence of land under water can be a part of fallow (food fallow) in which the crop-free land is submerged in water for a specific period or it can be a part of crop culture such as wet rice culture in which the soil is flooded with water for most of the duration of the crop. While the former practice can be followed where the availability of water is not a problem and the land is well levelled, the latter is a normal practice with the crop grown during rainy season when only occasional watering of the field may be required. Flooding of the land can be employed as a means of reducing certain phytonematode populations. It was first reported for control of root knot nematodes of

Florida, the U.S.A. by Watson. Many reports have appeared on the effectiveness of this method in nematode control. Periods from seven to 40 days are recommended to get decreased infestation of root knot nematodes. Compared to continues submergence, alternate flooding and drying increases the destruction of root knot nematodes.

Thomas and Stoner had reported reduction of root knot infestation in vegetable crops taken after wet land rice culture. The mechanism(s) involved in the control of phytonematodes by flooding are not clearly understood. It is presumed that flooding eliminates the desired host plants and the nematodes die from starvation. Exhaustion of second stage juveniles at the expense of their body reserves of energy is hastened by increased mobility in water. High soil moisture hastens the decay of nematode galls exposing the eggs and larvae to adverse conditions. Although there is lack of aeration and free oxygen in flooded soils free oxygen is not a limiting factor. Swarup and Pillai have reported that there is no marked difference in the total number of juveniles hatched when eggs were kept submerged or under free oxygen conditions.

Experimental evidence presented by Hollis and Johnson had shown that non-sterile flooded soils exhibited more reduction in nematode populations than sterilized flooded soils. This points to the role of microbial activity in the anaerobic conditions created by flooding or submergence. Anaerobic bacteria such as *Clostridium* sp. are known to release highly nematotoxic substances during decomposition of organic matter in flooded soils. Nematicidal concentrations of fatty acids are also formed within four days in corn meal amended saturated soils. The role of hydrogen sulphide in wet rice soils has also been pointed out by Rodriquez-Kabana and associates and Rhoades. In laboratory tests *Tylenchorhychus martini* was killed by hydrogen sulphide in five to ten days at concentrations similar to those found in flooded soils.

Organic Amendments and Fertilizers

The value of decomposition of organic matter in soil in the reduction of nematode damage was first demonstrated by Linford and his associates in 1938. Since then a large number of research reports have confirmed that addition of a variety of decomposable organic matter to nematode infested soil results in the reduction of population of several phytonematode species.

Organic manures

A brief review of the effects of manuring on nematode populations in soil was given by Oostenbrink. Addition of farm-yard manure and compost reduces the infestation level of *G. rostochiensis* in potato roots. These manures hampered the development of the nematode in plant roots where they developed more slowly than in plants grown in unmanured soil. There seemed to be some type of biochemical resistance imparted by the manure to the plants. Deep ploughing during hot summer months followed by application of farm-yerd manure at planting spots increases the yield of tomato in root knot infested soil. Reduction in the citrus nematode populations by application of steer manure and chicken manure has also been reported. Organic nitrogen in sewage sludge has been found more effective then ammonium nitrate for reducing populations of *Belonolaimus longicaudatus* and *Hypsoperine graminis* in turf grass.

Green manuring

Green manuring is a conventional practice in which green plants are ploughed into the soil and allowed to decompose to provide nutrients to the crop grown subsequently. The decomposition of organic matter brings about important physical, chemical and biotic changes in the soil. Linford and associates had demonstrated that the incidence of root knot could be significantly reduced by incorporating chopped pineapple leaves into the soil. In India, chopped leaves of *Pongamia glabra*, *Crotalaria juncea*, *Azadirachta indica*, *Melia azadirachta* and *Sesbania aculeata* to infested soil have reduced root knot incidence.

Cellulosic soil amendments

Excellent success in control of phytonematodes has been achieved with the use of cellulose and sawdust. Miller and Edgington reported of chopped paper and white pine sawdust at the rate of 2 g per 500 g soil. Similar control of *Heterodera tabacum* with a 1 per cent amendment of paper and white pine sawdust was observed by Miller and Wihrheim. Addition of fertilizers accentuated the effect of cellulose. Singh and Sitaramaiah, in extensive studies, have shown that sawdust amendment of soil supplemented with nitrogen in the form of urea gave a highly significant control of root knot caused by *M. javanica* in okra and tomato. Similar reduction in the activity of citrus nematodes by soil amendment with cellulose is also reported. Chitin has been found a very effective cellulose material against many nematodes.

Mature or dry crop residues

Application of mature crop residues such as straw is not a conventional manuring practice. However, substantial quantities of dry crop debris are annually ploughed into the soil as a regular agronomic practice. The amount of such residues depends on the type of crops included in the rotation being followed. The effect of decomposition of such residues in suppression of nematodes in soil has been studied by many. Johnson found that every type of crop residue introduced into the plots infested with root knot nematodes reduced the number of galls per plant. Maximum reduction was obtained when the residue was allowed to decompose for 30 days before the host was planted. Later, Johnson and associates confirmed that when mature dry residues of lespedeza, alfalfa, oats and flax were chopped to about 3 mm particles and incorporated into field plots infested with *M. incognita* the incidence of root knot of tomato was significantly reduced. A 10-ton/acre dose was more effective than 5-ton/acre dose.

Longer durations allowed for decomposition gave better results than shorter durations. Subsequently, Johnson found that application of 1 per cent (w/w) oat straw with a fertilizer mixture was effective in controlling *M. incognita*. Similar observations were reported by Miller and Wihrheim. Morgan and Collins got a very high degree of control of *Pratylenchus penetrans*, as good as D-D fumigation, by application of rye straw into soil in summer fallow. Other such residues are cotton waste, cocoa bean hull, timothy hay, lucerne hay, rice straw etc. which have suppressed such phytonematodes as *Tylenchulus semipenetrans*, and *Belonolaimus longicaudatus*.

Meals and oils-cakes

Fertilization of soil with meals and oils-cakes has been a practice in parts of the world. In Florida farmers believed that the application of tungnut meal to soil reduced losses from root knot. Lear reported significant in populations of phytonematodes when large quantities of castor bean pomace were mixed with soil. Populations of the lesion nematode are reduced by soil amendment with soybean and corn meal. However, soybean meal has no effect in sterilized soil. Use of oil cakes of margosa, castor bean, and many other oil bearing plants were extensively studied by Singh and Sitaramaiah who found significant reduction of root knot of tomato and okra in field trials. Earlier, Singh had, in 1964, demonstrated in pot experiments almost complete suppression of root galls of tomato (*M. javanica*) with oil cake of *Pongamia glabra* (*karanj*). The results of use of sawdust and oil cakes in nematode control was subsequently confirmed in studies conducted in India and many other countries.

Oils and sugars

Vegetable oils and carbohydrates such as sugars and sugar related products like molasses are not normal soil amendment materials but have been shown to reduce phytonematode populations in soil. Adverse effect of black mustard oil that contains allyl isothiocyanate on the potato cyst nematode was reported by Ellenby as early as 1945. In his field trials 0.1 cwt/acre oil increased potato yields by 100 per cent and decreased the emergence of larvae from cysts.

Spraying of 1.5 per cent water dispersible fish oil on soil surface is reported to reduce populations of *Radopholus similis* in citrus roots. Corn oil, cotton seed oil, groundnut oil and soybean oil when applied to soil at the rate of 1 per cent (w/w) reduce populations of lesion nematodes, corn oil being the most effective. Osmotic destruction of nematodes in the presence of high concentrations of sugar in soil and in solutions has been reported by many workers. As little as 1000 ppm of dextrose is nematicidal. Sugarcane bagasse and molasses with 0.2 per cent sucrose content give 50 per cent larval mortality in *M. javanica*. Most workers have used such substances in very large dosages which are not practicable on the field scale.

Mechanisms Involved in Nematode Suppression by Organic Amendments

Organic amendments using sawdust, dry or green crop residues, oil-cakes and other easily decomposable materials have proved highly effective against nematodes in subtropical and tropical regions. Intense microbial activity and quick decomposition of the added materials is hastened in tropical climate. The effects of such soil amendments vary from soil to soil and region to region. The mechanisms involved are not clear and require detailed studies. The available literature suggests the following mechanisms.

Parasitism and Predation of Phytonematodes by Nematophagous Fungi, Bacteria and soil Microfauna

The intense microbial activity during decomposition of organic matter improves the chances of increase in numbers of microflora and fauna that can destroy phytonematodes. This is a form of have been reviewed by Mankau, Sayre, Singh and Singh and Sitaramaiah. Till the sixties the attempts on biological control of nematodes were mostly observational records of parasites and predators attacking nematodes. After 1960 detailed studies on the nature and ecology of these enemies of nematodes were intensified.

Nematode Destroying Fungi

Most investigators believe that nematode destroying fungi or nematophagous fungi are potentially most useful biocontrol agents. These fungi are found in all types of environments and soils. Their ecological study may provide means to encourage development of their populations to the level where they can become successful destroyers of the nematodes. Most of the nematophagous fungi belong to the order Moniliales of Fungi-Imperfecti but some are from Zoopagales in Phycomycetes. These fungi can be divided into two groups: (a) the trap forming fungi and (b) the non-trap forming endozoic parasites.

Nematode trapping fungi

These fungi develop special structures from their hyphal branches in the presence of nematodes to trap or catch the latter. Most of them are ordinary saprophytes than can be cultured on synthetic media where they do not form traps. But when nematodes are added to the medium trap formation is initiated. Death of the nematode occurs due to mechanical damage by the trap and also due to release of certain nematotoxic secretions. Nutritional conditions are the primary factors affecting their growth and trap formation. The active morphogenic substance believed to induce trap formation of *Arthrobotrys* is known as "nemin". The predaceous activity of nematode trapping fungi is often independent of nematode populations but dependent on the release of certain soluble carbohydrates. The trapping devices may be grouped as below:

1. *Sticky branches*: Short lateral branches, often only a few cells long, are formed. These may anastomose to form loops which never form complex network.
2. *Sticky network*: The branches curl around and anastomose with similar branches or with parent hyphae. These loops produce complex network. The adhesive surface of the network helps to hold the prey.
3. *Sticky knobs*: Small spherical to subspherical lobes on one or two celled lateral hyphae are formed. Only the terminal cell is sticky. In the above sticky traps (*Arthrobotrys oligospora*, *Dactylella lobata*, *D. cinophaga*, *D. ellipsospora*) the wandering nematodes are caught in the loops and held thereby an adhesive fluid secreted by the fungal cells. The trapped nematode is immobilized within one or two hours by which time fungal hyphae have already penetrated the nematode cuticle. Inside the nematode body the fungus produces swollen bulb-like structures from which hyphae radiate throughout the lumen of the nematode body and absorb the body contents.
4. *Constricting rings*: The trap is formed when a short hyphal branch curls back on behalf and anastomoses, forming a three- or four-celled ring. The ring is borne on a slender one- or two- celled

stalk. When a nematode enters the ring and contacts the inner walls of the ring cells, the cells bulge inward filling the lumen of the ring and holding the prey fast. Such constricting rings are seen in *Dactylella bemicoides*.

5. *Non-constricting rings*: The non-constricting rings develop as the constricting rings but are neither constricting nor sticky. The nematodes get entrapped by the rings which have a diameter less than the diameter of the nematode body. In attempts to get away the namatode gets further entrapped by the snares. In constricting as well as non-constricting traps the method of feeding on the nematode body is the same as for the fungi producing sticky branches, rings and knobs.

Endoparasitic fungi

The endoparasitic (endozoic) fungi produce conidia as infection agents. They are obligate parasites and form vegetative mycelium only within the nematode body. There is little or no mycelial development outside the host. Conidia germinate and form an infection tube that penetrates the nematode cuticle and causes infection. The condia are generally sticky and adhere to the body of nematodes coming in contact with them. Hyphae ramify within the nematode body and absorb the body contents. Latter, spore bearing hyphae emerge from the body and help in dispersal of the fungus. The Phycomycetous endoparasites of nematodes are more common in soil and have studied in some detail. The examples are some species of *Catenaria*. These have been especially found attacking free-living and predatory nematodes (Dorylaimids and *Mononchus*).

The zoospores of these fungi stick to nematode body, germinate, and the germ tube penetrates the nematode cuticle causing infection. Within the body the ramifying hyphae produce encysted zoospores. Sporophores come out through the cuticle. The thallus of *Catenaria vermicola* develops within the nematode body. Fine strands of separate hyaline hyphae running along the length of the body develop. The rate of infection of plant parasitic nematode genera is slow. It has been observed that nematodes of genera in the Dorylaimoidae are more easily attacked by parasitic fungi with mobile zoospores than are the Tylenchida. This is especially evident among *Xiphinema* spp., which are members of the Dorylaimoidea. Certain phycomycetes attack eggs of nematodes in the soil and destroy them.

Some endozoic parasites such as *Nematoctonus heptocladus* and *N. concurrence* kill the nematodes on contact before penetrating the cuticle. *Curvularia pallescence* and *Fusarium* spp. have also been isolated from females of *M. javanica* extracted from root galls of tomato grown in soil amended with oil-cakes. The fungus destroys the body contents leaving only an empty sac or the infected females fail to lay eggs. There are many reports of parasitization of eggs and cysts of phytonematodes by fungi. The viability of eggs of the potato cyst nematode is known to decline over a time cysts are left in the soil for long.

The fungus *Verticillium chalmydosporum* and many other non-sporulating fungi are reported to the responsible for the decline of viability of eggs in cysts of *H. schachtii* and *H. avenae*. *Cylindrocapon destructans*, *Acremonium strictus* and *Fusarium oxysporum* have been isolated from cysts and are capable of destroying eggs of sugar-beet cyst nematode. The fungus *Phialophora heteroderae* enters the cysts of the potato cyst nematode, *G. rostochiensis*, through oral and vulval openings and penetrates the egg shell reducing the number of viable eggs in the cysts. Eggs of *M. incognita* are infected by *Paeciolmyces lilacinus*. The fungus invades both egg masses and females of *Meloidogyne* eggs. This fungus is capable of limited saprophytic growth and is a specialized hyphomycete parasite of nematode eggs. *Nematophthora gynophila* parasitises eggs of *H. avenae* and is the major factor in limiting the natural multiplication of this nematode.

Predatory Nematodes

Predatory nematodes can be considered as effective enemies of other nematode. Because of their size larger than the size of some of the destructive phytonematodes such as *Meloidogyne* which are not only small and slender but sluggish in movement the predatory nematodes can prove effective biocontrol agents if suitable soil conditions can be created for their multiplication. The predatory nematode *Seinura* feeds voraciously on nematodes. Some members of the superfamily Dorylaimoidea such as *Discolaimus*, *Dorylaimus*, and *Actinolaimus* pierce the nematode body and eggs and suck out the contents. These predatory nematodes can be conveniently grouped in categories based on their feeding habit:

1. The nematode prey is ingested *in toto* such as by *Tripyla* spp. and *Monohystera* spp. These nematodes have a very flexible buccal cavity and plain esophagus. The stoma is usually unarmed and often small.
2. Some nematodes feed by puncturing a hole in the nematode body with fixed or moveable stomal teeth. Example is *Mononchus papillatus*.

3. Predatory nematodes like *Seinura christiei* and *S. tenuicaudata* are equipped with stylet with which they feed on other nematodes in the same manner as phytonematodes feed on their host plants.

Prey preferences of the predaceous nematodes in water agar cultures have also been observed. The lesion nematodes, *Pratylenchus penetrans*, *P. vulnus*, *Paratylenchus curvitatus* and *Meloidogyne* spp. are susceptible to predation by several Dorylaimida whereas *Hoplolaimus tylenchiformis*, *Belonolaimus longicaudatus* and *Criconemoides* spp. are fairly resistant to predation. Despite the apparent success of predaceous nematodes they probably do not occur in soil in sufficiently large numbers in the crucial early phase of population build-up of the phytonematodes to be efficient natural regulators of nematodes.

Viruses and Bacteria

Unconfirmed evidence suggests that plant parasitic nematodes do suffer from virus infection. However, conclusive proof of pathogenicity and the presence of virus particles in the infected nematode are lacking. In *M. incognita incognita* it was reported that juveniles showing virus infection had a very sluggish motion and that the sickness was transferable to healthy juveniles when they came into contact. After eliminating the possibility of involvement of bacteria and fungi it was shown that the filtrates of a suspension of sick nematodes cold include the disease in healthy juveniles. In another case the phenomenon of swarming (aggregation of nematodes in masses) was also reported.

Electron microscopic studies of the cuticle of *Tylenchorhynchus martini* revealed abnormal and disruptive morphological changes in the cuticle layers that suggested a diseased condition. Virus-like particles were observed in the hypodermis and muscle layers, digestive and reproductive systems and on the surface of the cuticle. Normal healthy nematodes exhibited intact tissues without any suspected virus particle. However, in these studies neither the virus could be isolated nor the disease could be transmitted to healthy nematodes. Bacteria have been detected in the body of free-living nematodes but it is not clear whether they are saprophytes or parasites.

Bacillus penetrans is an obligate parasite of plant parasitic nematodes. It was first detected in *Pratylenchus pratensis* and described as a protozoan, *Duboscquia penetrans*. In addition to *P. pratensis* it also attacks *P. crenatus*, *P. penetrans*, *Rotylenchus robustus*, *Tylenchorhynchus dubius* and *Meloidogyne arenaria*. Spores of the bacterium adhere to second stage juveniles in soil and germinate after the nematodes enter the host roots and start feeding. Germ tubes penetrate the cuticle, colonies develop in the pseudocoelum and ultimately some of the vegetative cells differentiate into spores which fill the body cavity. Infected nematodes do not reproduce.

All stages in the process of infection of *M. javanica* by *B. penetrans* were favoured by temperature optimum for the nematode. At 30°C the parasite proliferates extensively in females before they reach maturity. Tomato roots containing large numbers of *Meloidogyne* females infected with *B. penetrans* were obtained by inoculating tomato plants with second stage larvae infected by the bacterium. Roots were then air-dried and powered. The preparation had the potential of microbial nematicide. However, this method cannot be a practical method of large scale production of the nematicide. The major problem in large-scale utilization of *B. penetrans* is, thus the inability to culture the bacterium *in vitro* on any of the standard bacterial culture media. Neither vegetative cells nor spores of the bacterium can be harvested in sufficient quality for extensive tests. Intracellular bacterium-like organisms have been reported also in *Heterodera goettingiana*, *H. glycines*, and *Globodera rostochiensis*.

Protozoa and Soil Microfauna

Protozoa

A large amoeboid organism, *Teratromyxa weberi* (Proteomyxa, Vampyrellidae), frequently observed ingesting nematodes, is considered of practical importance in the control of phytonematodes. It has seen engulfing larval stages of *G. rostochiensis*, *Meloidogyne* spp. and *Pratylenchus* spp. *Urostyla* spp. have also been found to feed on nematodes. Other soil protozoa probably have incidental predatory relationship with nematodes. In addition to predatory nematodes many other soil fauna are known as enemies of nematodes. These include tardigrades, collembola, turbellarians, mites and enchytraeids. Although occasionally seen destroying nematodes females, larvae and eggs in soil especially where organic matter is decomposing, there is not much information on their practical use.

Tardigrades

Tardigrades are known to feed on several nematodes and other soil organisms, *Hypsibius myrops*, a taridgrade, has been cultured on *Panagrellus redivivus* in a laboratory for several weeks and can reduce populations of *D. dipsaci*. Turbellarians such as *Adenoplea* spp. have also shown good promise as nematode controlling agents. They consume significantly high numbers of nematodes in certain soil types although under natural conditions they are likely to be present in sufficient numbers to regulate nematode populations.

Collembola

These organism are also active predators of nematodes feeding rapidly especially on eggs. *Onychiurus armatus*, *Isotoma viridis* and *Hypogastrura* spp. destroy cysts of *Heterodera cruciferae*. Mites such as *Pergalumma* spp. are reported as predators of *Tylenchorhynchus martini*. The enchytraeid worms *Fridericia* and *Enchytraeus* feed on sugar-beet cyst nematodes and in pot experiments were able to reduce the population of this nematode. The above account of the parasites and predators of nematodes leaves no doubt that these enemies of nematodes naturally exist in soil but are not effective in controlling the phytonematodes because of their low population. Their exploitation as biocontrol agents are reasonable cost can be achieved through manipulation of soil environments in their favour. This may increase chances of increase in their numbers to desirable level.

The modification of soil environment as an approach to achieving desirable populations of biocontrol agents in soil is designed in full accordance with the ecological principle that soil population at any time will be determined by habitat conditions and that the populations can, therefore, be changed in any direction by making appropriate changes in soil conditions. Major emphasis to bring about these changes in soil conditions at the microsite level has been given to the exploration of the effects of decomposition of crop residues and organic waste especially those commonly available at low cost in the area. This aspect of soil amendment has been discussed in the preceding section of this chapter.

Chemical Control of Plant Parasitic Nematodes

Soil disinfection for control of phytonematodes is used mainly when cultural and biological methods fail to reduce the nematode populations or when these methods are not economical for the growers. Chemicals are also used when there is very heavy infection of the soil and proper resistant varieties are not available to be grown satisfactorily. Under these situations chemical soil disinfestation is the most effective method to reduce nematode populations to a threshold level. These measures allow the farmer flexibility in choice of crops to be grown on nematode infested soil and the farmer can grow what he wants, where he wants and when he wants.

The main objectives of chemical soil treatment are to protect plants from nematode damage, prevent or restrict nematode multiplication, improve the growth of plants, and finally improve the quality of the produce. One should not aim at complete eradication of nematode populations since it is impossible to achieve this in a complex biological environments of the soil. The population should be brought below the injury or harmful level so that profitable crops can be grown in the infested land. Nematicides have been on a large-scale use since 1945 in Europe and North America and their use has been increasing every year. However, the nematicides are now generally used even in these countries only for cash crops which give high cash returns.

The chemicals are also used in nursery beds to obtain nematode-free plants. The earliest soil fumigants used against nematodes were formaldehyde, carbon disulphide and calcium cyanide. These were mostly used in glasshouses. At present these are not used on a large scale. Perhaps, the earliest report on the application of carbon disulphide as soil fumigant is that by Kuehn, who tried to control the sugar-beet cyst nematode with carbon disulphide in Germany. In 1919 Byars used hydrocyanic gas to control root knot. The nematicidal property of chloropicrin was discovered in the same year and this chemical was widely used in the controlling root knot nematodes in pineapple fields in Hawaii. Methyl bromide is widely used as an insecticide and nematicide for nursery beds and in glasshouse soil.

Hurst and Franklin and Smedly reported the nematicidal properties of cynamide and isothiocynate compounds, respectively. D-D was discovered in 1943 by Carter and was the first soil fumigant used on a large scale to control many species of plant parasitic nematodes. The discovery of ethylene dibromide (EDB) followed soon. Later, Mc Beth reported, 1, 2 dibromo-3 chloropropane (DBCP) as a potential nematicide. In India, the first serious attempt to demonstrate nematode control by soil fumigation with D-D were undertaken by Burmah Shell in 1953. Their efforts were primarily directed at nematode control in tea gardens and nurseries.

Types of Nematicides

Nematicides in commercial use basically belong to two groups, fumigants and non-fumigants. The fumigants consist of compounds belonging to halogenated hydrocarbon group and isothiocynate group. The non-fumigants consists of organo-phosphates and carbamates.

Fumigant Nematicides or Volatile Soil Fumigants

The fumigant nematicides possess a high vapour pressure are diffuse through soil pore spaces where nematodes live. They are dissolved and dispersed in soil moisture in concentrations high enough to kill the nematode larvae. These nematicides are not very specific. They can be used for the control of many kinds of nematodes and soil insects and, at high dosages, even fungal and bacterial pathogens of plants. There are two major types of fumigants in the market. Halogenated hydrocarbons such as methyl bromide, ethylene, dibromide, Telone, D-D mixture, Chloropicrin and methyl isothiocynate (MIT) and MIT releasing soil fumigants such as metham sodium or Vapam and Dazomet, Mylone or Basamid.

Contact and Systematic Nematicides

These nematicides have little fumigant action do not kill the nematodes directly. They include organophosphates such as Phenamiphos, Fensulphothion, Thionazin, Mocap, Diazinon, VC-13, Phorate etc., and Oxime N-methyl carbamates such as Aldicarb, Oxamyl and Carbofuran.

Soil Fumigants

D-D mixture ($CH_2Cl.CH\ CHCl + CH_2Cl\ CHCl\ CH_3$)

D-D is a mixture of 1, 3-dichloropropene and 1, 2-dichloropropane, in 2:1 ratio, with some other hydrocarbon impurities. Ever since its discovery in 1943 this fumigant has been widely used to control plant parasitic nematodes in different parts of the world. It kills nematodes, soil, insects and wireworms and also some fungi at high dosages. The nematicide is sold as undiluted or as an emulsifiable concentrate in drums. D-D is the trade name of the product of the Shell Chemical Co. The dosage rate is 400 to 500 lit per ha as a pre-plant soil fumigant. The best results of the treatment are obtained at soil temperatures between 10° and 27°. D-D has no permanent detrimental effect on beneficial soil organisms but there is a temporary reduction in the number of nitrifying bacteria and inhibition of nitrification results in high levels of ammonia. It is, therefore, desirable that nitrogen in the form of nitrate should be added to the soil, especially if it contains high levels of organic matter. Crops like sugarcane and pineapple can tolerate ammonia; hence, in their case there is no need to apply nitrates after fumigation with D-D. The nematicide is also sold under the names of Vidden-D, Nemafume and Nematox.

Telone (1, 3-dichloropropene. $Ch_2Cl..CH{=}CHCl$)

This nematicide is a mixture of *cis-* and *trans*-isomers of 1, 3-dichloropropene with small quantities of related hydrocarbons. The chemical was introduced in 1956 by Dow Chemical Company. The *cis*-isomer is more toxic than the *trans*-isomer. Dorolone is a mixture of EDB and Telone. The dosage rate of the nematicide is 200 lit/ha depending upon the soil type.

E D B (1, 2-dibromeothane or ethyl dibromide)

Christie had reported that this chemical had excellent nematicidal properties. It kills nematodes as well insects and has been used extensively in tobacco. Since EDB contains brominc it may cause phytotoxicity in bromine sensitive plants. The biological activity is similar to that D-D mixture. Rate of application is 18 to 120 lit/ha depending upon soil type and the crop. The nematicide is not effective when the soil temperatures are below 10°C. This nematicide also reduces nitrification.

D B C P (1, 2-dibromo 3-chloropropane, $CH_3Br.C\ Br.CH_2\ Cl$)

Under the name of Nemagen this chemical had been found very useful as pre-plant and post-plant treatment of established plantation crops. Due to its low phytotoxicity it was also used through irrigation water in citrus nursery soils. Rate of application was 20 to 50 lit/ha depending upon the crop and soil type. The use of Nemagen has now been restricted because of the residue problem.

Methyl bromide ($CH_3\ Br$)

This fumigant is very effective against nematodes, soil insects, fungi, rodents and weeds. The material is compressed into liquid and marketed in metal containers. Since methyl bromide is toxic to humans and is odourless a warning agent such as chloropicrin is usually added to it. For seen and nursery beds the treated surface is immediately covered with polythene sheets. In tobacco nursery beds methyl bromide is applied at the rate of 500 g per 5 to 20 sq m surface. The bed is covered with a polythene sheet to aid diffusion of the chemical. For large-scale field treatment methyl bromide is applied at 15 to 20 cm depth. This nematicide also destroys nitrifying bacteria and problems may occur in ammonia sensitive crops. It is marketed under the name of Dofume MC-2, Meth-O-gas and Terabol, etc.

Chloropicrin (Trichloronitromethane. CCl_3NO_2)

Chloropicrin was developed primarily as a soil fungicide to control fungal wilts of plants but was found extremely useful for controlling root knot nematodes. Sclerotia forming fungi are not affected. In addition to nematodes and fungi chloropicrin kills bacteria and weeds also. Rate of application is 500 lit/ha. Chloropicrin is also known as pic-fume and tear gas.

Dazomet (3, 5-dimethyl-tetrahydro 1, 3, 5.2 H Thiadiazine Thion)

On application Dazomet breaks down into methyl aminomethyl dithiocarbamate which in turn, yields isothiocynate. It is a good fungicide and nematicide. Rate of application is 20-35 g/sq m as a pre-plant soil fumigant. Other names of the nematicide are Mylone are Basamid.

Metham sodium or Vapam (sodium N-methyl dithio-carbamate. CH NH.CS.SNa.2HO)

Vapam releases methyl isothiocynate which is the active component. The chemical is used against fungi, nematodes, weeds and sometimes soil insects. Rate of application is 700 to 1000 lit/ha. The soil temperature should be reasonably warm (15° to 30°C) at the time of application.

Trapex (methyl isothiocynate. CH.NCS)

This nematicide is a 20 per cent (w/w) product of an organic solvent, generally xylol, with or without an emulsifier. In addition to nematodes, soil insects and weeds it has been found effective against cyst forming nematodes also.

Methods of Application of Soil Fumigants

There are several methods of application of fumigant nematicides to the soil for good results. Application of fumigants to the entire filed is known as area treatment. The field is prepared well in advance of the treatment into fine seed bed condition. Marker lines are drawn at 30 to 35 cm distance in either direction. The nematicide is injected into the soil to a depth of 22.5 to 30 cm at the intersecting points of lines. Hand-operated injector guns are used for the purpose. In case of large size areas (several acres) the operation is carried out with the help of tractor-mounted applicators and the nematicide is injected at a depth of 30 cm. Area treatment is undertaken where maximum control is desired especially in commercial crops like tobacco and vegetables. When the crops are planted in rows of 60 cm or more with, row treatment can be used. The rows are marked and the nematicide is applied under the planting row.

Row treatment is widely used in vegetable crops and tobacco to get reasonably good control of nematodes. Strip treatment is commonly used when perennial crops are planted in strips. The planting row of 2 to 3 m are treated by hand injector guns or a tractor applicator. Site treatment is done when individual trees are to be planted. The area around the trees can be treated with nematicide. This method has been used for large-scale treatment of citrus trees as pre-plant and post-plant treatment with Nemagon.

Criteria for Success of Soil Fumigation

The fumigant nematicides are usually injurious to seeds or plants if they are planted immediately after soil fumigation. In order to avoid the injurious effects, generally two to three weeks waiting period is required depending upon the type of nematicide used. A good fumigant must disperse and diffuse through the soil particles to kill nematodes. Newhall had discussed the theoretical considerations of soil fumigants. The initial stage is relatively short and characterized by radial diffusion from injection points. This is followed by transition stage and then by a uniform distribution of the fumigant through the soil.

The efficacy of fumigant nematicides in soil, therefore, varies with vapour pressure of the chemical, its molecular weight, soil conditions such as porosity, moisture content, temperature and organic matter content. For maximum efficiency the soil must be properly prepared for the treatment. It is also important to remove all weeds are undecomposed plant material including roots of the previous crop. Their presence interferes with normal application of the nematicide and may also protect the nematode stages present inside the plant tissues where the chemical cannot reach.

Mode of action of Fumigant Nematicides

The distinguishing feature of the action of fumigant nematicides form the non-fumigant nematicides is that when there is sufficient exposure of the nematodes to the fumigant nematicide they are killed whereas non-fumigant nematicides usually incapacitate them in one or more functions at field concentrations and the effects are reversible. Penetration of the nematode body by the chemical is a necessary prerequisite to physiological reactions associated with death. Chemical may enter the nematode body directly through the cuticle or through such orifices as mouth, anus, and vulva. EDB is known to enter the nematode body two and half times faster than water and produce a narcotic effect on the body. Eggs are less sensitive than juveniles and adults to the effects of nematicides.

The first reaction of the nematode to the entry of chemicals in the body is hyperactivity followed by gradual decrease in activity, leading to eventual paralysis but this narcotic effect may be temporary and the nematode may recover to complete its life cycle. Within the nematode body the chemical may act in many ways such as by precipitating proteins, blocking nerve may act in many ways such as by precipitating proteins, blocking nerve endings, destroying nerve sheaths, cellular membranes and amphids. The nematicidal activity has been suggested to involve alkylation of nucleophilic sites on proteins or oxidation of iron (Fe II) porphyrine and haemoproteins. There is no information on the mode of action of methyl isothiocynate or their generators or nematodes although it would be expected that isothiocynate molecules have an affinity to a variety of relativity nucleophilic sites within cells.

Beneficial Effects of Soil Fumigation

There is plenty of information in developed countries on the impact of soil fumigation on food production through control of nematodes. Improved plant growth in fumigated soil is the general effect observed in most crops. The increase in yield is reported to be as high as 87 per cent. Although it can be expected that nematode control is the major factor involved in this increase, there are many other associated factors that account for the yield increase. Fumigation reduces the losses from disease complex. Biologically bound nutrients are released after the effect of fumigation on soil microflora and fauna and are utilized by the plants. Plants in treated plots show distinctly improved and uniform growth with extensive root system and feeder roots. The increased root system enhances uptake of nutrients and water resulting in better yields. The healthy plans growing in treated plots are also less susceptible to other plant diseases and can tolerate adverse growing conditions better.

Side Effects of Fumigant Nematicides

All soil fumigant nematicides are general biocides. Halogenated compounds can temporarily upset normal soil nitrification process which results in increased ammonia accumulation with the result that plants may suffer from ammonia toxicity or nitrate starvation. Brominated compounds like methyl bromide or EDB have residue problem. Bromine residues have been found in a number of crops like citrus, carrots, potatoes, groundnut and tomato following fumigation with these nematicides. Increased uptake of bromine or chlorine by some plants has been reported. Increased halogen content in tobacco leaves reduces leaf quality and adversely affects the burning quality of tobacco grown in fumigated fields. Nematicides with fungicidal properties can adversely affect nematophagous fungi and other parasites and predators of nematodes. Normal infection and development of mycorrhizal fungi are also suppressed by nematicides having fungicidal property. Citrus seedlings grown in fumigated soil are commonly stunted and chlorotic.

Heavy fertilization partially overcomes stunting. For many years this problem was attributed to soil toxicity developed by fumigation. Kleinschmidt and Gerdemann later showed that stunting was caused by inadequate nutrition due to destruction of vesicular arbuscular mycorrhizal fungi. Inoculation of fumigated soil with VA-mycorrhizae partially solved stunting problem and reduced need for frequent fertilization. Little information is available on the long-term ecological effects of frequent and repeated use of soil fumigants in cultivated fields. Investigations at the University of California in the U.S.A. had indicated that EDB, DBCP and D-D mixture are not persistent in soil and have no lasting adverse effects on the physical and biological properties of the soil, on the environment and on the nutritional value of the crop grown in fumigated fields. The fumigants are quickly degraded by normal biological activities in the soil.

Contact and Systemic Nematicides

The introduction of granular contact and systemic nematicides in the seventies has provided very effective chemicals for nematode control. Some of these chemicals have replaced the fumigant nematicides in several countries. These non-volatile, non-fumigant nematicides are relatively less phytotoxic and can be applied before or at the time of planting or even on standing plants. They are effective at much lower dosages than the fumigants. Most systemic nematicides are taken up by roots from soil application and translocated up in the plant system giving the plant protection against nematodes and many insects including white flies, but not against soil-borne fungal plant pathogens. The applications of these nematicides does not require any special equipment. Commercial contact and systemic nematicides belong to the organophosphate and carbamate groups.

1. *Mocap* [0-ethyl S,S-dipropyl dithiophosphate]: Also known as Ethoprophos and Ethoprop, mocap is a phosphatic compound with a broad spectrum of activity as nematicide and soil insecticide. The chemical is non-fumigant, non-systemic and has efficient contact action with good soil movement and residual properties. It can be applied immediately before or at the time of planting or used for treatment of established plants. The chemical should be either applied with irrigation water or if applied direct it should be mixed with soil.
2. *Aldicarb* [2 methyl-2(methyl thio) propionaldehyde-0-(methyl carbomoyl) oxime]: Aldicarb is usually sold under the name Temik. It is a strong systemic insecticide-nematicide and acaricide. The

chemical was introduced in 1965 by Union Carbide Corporation. As a versatile insecticide-nematicide to control several nematode parasites of important crops it is used extensively.

3. *Diazinon* [0,0-diethyl-0 (2-isopropyl-4 methyl) pyriidinyl]: Diazinon (Basudin, Neocidol) is an organic phosphate insecticide, nematicide, and acaricide with contact action.
4. *Phenamiphos* [0-ethyl-0 (3-methyl thiophenyl) isopropyl amidophosphate]: Sold in the market under the name Nemacur, phenamiphos is a highly effective systemic nematicide against nearly all the economically important phytonematodes. The active ingredient simultaneously affords good protection against sucking insects for six to eight weeks. The chemical is readily absorbed by roots as well as by leaves and translocated both up and down in the plant system.
5. *Oxamyl* [methyl-N-N-methyl-N (methyl carbomoyl) oxy-1 thioxamidate]. Oxamyl is an oxime carbamate and sold as Thioxamyl, Vydate and Du Pont 140.
6. *VC-13* [0,0-diethyl-0 (2-4 dichlorophenyl) monothiophosphate]: This is also an organic phosphate contact nematicide-insecticide applied to soil.
7. *Carbofuran* (2, 3-dihydro-2,2-dimethyl 7-benzofuranyl N-methyl carbamate): Carbofuran is sold as Furadan or Curatter and is a highly effective soil nematicide-insecticide which has systemic action is plants.
8. *Fensulphothion* [0.0-diethyl-0(4-methyl sulphinyl) monothiophosphate]: Sold as Terracur P and Dasanit this non-systemic organo-phosphate can be used as an insecticide as well as nematicide. The active ingredient disperses in soil via the water system so that the chemical easily comes in contact with nematodes. The chemical easily comes in contact with nematodes. The chemical remains effective in soil for about three months at a temperature of 20°C.
9. *Phorate* [0, 0-diethyl-S (ethyl thiomethyl) dithio-phosphate]: Commonly sold in the market as Thimet, phorate is a systemic organic phosphate soil insecticide and nematicide. Absorbed by roots it is translocated in the plant system are persists for several weeks protecting it not only from nematodes but also sucking insects.
10. *Thionazin* [0,0-diethyl-0 (pyrazinyl) monothiophosphate]: This organophosphate contact nematicide and insecticide was introduced in 1961 as soil pesticide by American Cynamid Co. Its other names are Nemafos, Cynem and Zinophos.

Some of the organophosphate and carbamate nematicides are systemic but are only absorbed by roots and translocated up to the foliage. Few, such as oxamyl, phenamiphos and carbofuran have the capacity to move from foliage down to roots and from roots up to foliage under certain conditions in some plants and provide protection against nematodes. The movement of these nematicides in plant is, however, faster from roots to foliage than from foliage to roots.

Methods of Application of Systemic Nematicides and Response of Nematodes

The non-volatile nematicides can be applied to the soil at planting time and also around established plants without any phytotoxic effect. Application can be by spot treatment at the time of planting young seedlings in the main fields or by row treatment. Chemicals like oxamyl and phenamiphos can be used as foliar sprays also or through watering of the seedlings at the time of planting. This method saves the amount of chemicals and provides protection against initial infection. The rate of application may vary depending upon the method of treatment, nematode species to be controlled, depth of application necessary to reach the target nematode, soil conditions, environments and cultural practices.

Soil application

Broadcasting the granules or spraying the emulsifiable concentrates on the soil surface followed by through mixing with the soil is the normal method of application for achieving maximum protection of crops. Soil application places the nematicide in the top few inches of the soil and then the chemical moves downwards and reaches greater depths. In field trails conducted by Whitehead and his associates effective control of potato cyst nematodes was obtained when oxime carbamates (oxamyl, Aldicarb) were mixed with soil at the rate of 11 kg a.i./ha. Aldicarb affects the behaviour of the nematode by disrupting their orientation to potato roots. The chemical also causes a marked increase in the male/female ratio. Development of potatoes is reported to be much improved by soil application of fensulphothion at the rate of 10 kg a.i./ha. A high degree of control of various species of *Meloidogyne* (root knot) was reported under field conditions with less than 10 kg a.i./ha of phenamiphos.

According to Brodie and Good soil application of systemic nematicides has often proved better than fumigants in the control of root knot and in increasing tobacco yields. The method of soil application influences the performance of systemic nematicides. Reddy compared different methods are rates of application of granular nematicides for the control of root knot of tobacco. Spot application (at 1.0 kg a.i./ha) and row application (4.0 kg a.i./ha) of aldicarb 10 G, fensulphothion 5G or carbofuran 3 G to tobacco seedlings planted in soil infested with *M. incognita* and *M. javanica* increased yield of cured tobacco leaves and decreased root galling. Yields were 83 to 135 kg/ha higher in treated than in untreated plots.

Nematode control and yields from spot treatment were as high as those from row application and were higher for fensulphothion and aldicarb than for carbofuran. Spot application of aldicarb, carbofuran and phorate is reported to be superior to row treatment or broadcasting in the control of root knot of okra and tomato. Sitaramaiah and Vishwakarma obtained increased tomato yields by applying the nematicides to the soil surface and then mixing into the soil better than the row or spot treatment. Time factor is also important in application of granular nematicides. Application of aldicarb, phorate, diazinon and fensulphothion seven days before planting tomato seedlings gave better control of root knot nematodes than the soil treatment immediately before transplanting the seedlings.

Root treatment

Roots of young seedling or nursery planting material infested with endoparasitic nematodes can be disinfested by root treatment with systemic nematicides. Mostly bare root dips for specific durations have been tested. Depending upon the type of planting material the duration can be for a few minutes to hours. Nursery stock of perennials can be exposed to nematicide solution for as many as six hours while roots of tender seedlings require 10- 30-minute exposure. Although longer durations are more effective, there may be phytotoxicity.

Seed treatment

Some systemic nematicides have been tested as seed dressing material to reduce the cost of treatment without any deleterious effects on the plant growth. Among such nematicides carbofuran and aldicarb have been found effective against several plant parasitic nematodes. Seeds are uniformity coated with gum or starch (2 per cent) as stickers and the required quantity a wettable formulation of the nematicide is sprinkled over the seeds and mixed thoroughly to give a uniform smooth coating. The treated seeds and dried in shade before sowing. This method was used by Prasad and Mathur to treat sugar-beet seeds with carbofuran and aldicarb sulfone.

Sugar-beet seeds coated with carbofuran at the rate of 1 per cent a.i. and aldicarb sulphone 2 per cent a.i. reduced *M. incognita* populations in pot tests and improved the growth of plants. Sivakumar and associates tested carbofuran as okra seen treatment for the control of root knot and observed that such treatments do not give absolute protection but could be effectively employed to reduce the severity of infection. Potato tuber dip in phenamiphos at the rate of 4.75 a.i./lit water is reported to eliminate *M. hapla* from tubers. The duration of dip was 10 seconds only. There was no phytotoxicity.

Foliar treatment

Bunt had reported control of the lesion nematode, *Pratylenchus penetrans*, by foliage spray with oxamyl. Foliar application of promising nematicides could also be used in conjunction with soil application of granular nematicides which could provide initial seedling root protection until the foliage was sufficiently expanded to be sprayed with systemic nematicides. Alam and associates found the highest reduction of root galling by *M. incognita* when the seedlings were immersed in oxamyl for 20 to 30 minutes followed by five foliar sprays at weekly intervals. Spraying alone provide poor control. Prasad and Rao also fond foliar sprays with oxamyl at 100 to 2000 ppm effective in reducing populations of the stunt nematode. *Tylenchorhynchus* spp. Foliar treatments of plants for nematode control are not yet practical on a field scale and hence the traditional soil application in the common method.

Mode of Action of Contact and Systemic Nematicides

It is generally accepted that at field dosages the organo-phosphate and carbamate nematicides do not kill the nematodes directly. They act by impairing the nervous system of the nematodes, reducing the rate of hatching, motility, orientation to host roots, feeding and development. These nematicides primarily act by inhibiting the enzyme acetyl cholinesterase at cholinergic synapses in the nematode nervous system as is known in the case of mammals and insects. Soil application of thionazin reduces the fecundity of *M. incognita* females while aldicarb has no such effect. Larvae from eggs produced in thionazin and aldicarb treated plants do not show any adverse effect of the treatment.

Similarly, hatchability of *Meloidogyne* eggs is not influenced by these nematicides when they are subsequently transferred to water. For thionazin only the highest concentration of 1000 ppm tried has been found lethal to *M. incognita* but at the recommended field dosages a five-day exposure has no effect on the hatched larvae from egg masses. The gelatinous matrix of the egg sac acts as a physical barrier against the penetration of nematicides. The sedentary endoparasites such as root knot and cyst nematodes can be most effectively controlled only by treating the soil before the juveniles enter the host roots. Prevention of attack on roots and feeding on the host by the nematodes in treatments with systemic nematicides has been reported by many investigators.

Oxamyl and carbofuran are known to disrupt orientation and attack on roots by *M. incognita* and *Pratylenchus penetrans*, respectively. On reaching the roots the nematodes did not feed. Mc Leod and Khair used a precise test to determine the effect of nematicides on development of nematodes with tomato seedlings as indicators. Nematodes were allowed to invade the roots over a 24-hour period, the roots were then washed and the plants placed in tubes containing sand and nematicide solutions. The number of juveniles which had developed to mature saccate female stage was recorded after 10 days. Aldicarb and thionzin at 2 ppm completely prevented development, methomyl, ethoprophos, phenamiphos and phorate were progressively less effective.

Experimental Chemicals other than Commercial Nematicides

Plant growth regulators, phenolics, aromatic acids, amino acids and fatty acids have been tested either as foliage treatment, soil drench or as bare root dip treatment to reduce nematode infestation of plants. These chemicals, though effective, are still in experimental stage and no commercial formulations have been proposed. Foliar application of 2, 4-D and Cycocel reduced the egg mass production in *M. javanica* and *M. incognita*. The ability of grape fruit seedlings to grow well in the presence of the burrowing nematode, *Radopholus similis*, was increased by application of exogenous phenols. Cinnamic acid and p-caumeric acid sprayed on susceptible potato plants prevented susceptible responses to *Globodera rostochiensis*. Exposure of roots of susceptible tomato seedlings to phenols, aromatic acids and fatty acids imparted some resistance in invasion of juveniles of *M. javanica*.

Resistant Varieties for Nematode Control

Soil treatment with chemicals and employing practices such as crop rotation are common methods of nematode control. However, in a nematode control programme, or in any plant protection programme, these methods should not be the first choice to consider. A most effective and economic method is the use of resistant cultivars. Growing resistant varieties of crops in crop rotations is economical for the farmer for reducing nematode in populations gradually in the infested fields. According to Oostenbrink the cultivation of a resistant variety may suppress a nematode's population to 10 to 50 per cent of its harmful density. With reduced multiplication and population density through successive cultivation of resistant varieties for sometime, it is possible to grow even a susceptible cultivar at frequent intervals without extra cost on use of nematicides.

The use of nematicides can be directed to high value crops or where the farmer wants to grow a particular crop continuously in the same field year after year. The use of nematode resistant cultivars and chemical soil treatment should be considered complementary methods of control and not as comparative measures. Considerable progress has been made in the U.K., Europe and the U.S.A. in breeding crop varieties resistant to specific sedentary and migratory endoparasitic nematodes. Resistant varieties have been developed in cotton cowpea, tobacco, beans, soybean, tomato, grapes and peaches against the root knot nematodes in alfalfa for resistance to *Tylenchulus semipenetrans*. Nematode resistance has been discovered in both cultivated and wild plants.

In most of the economically important crops like potato, tomato, barley and oats, resistant genes to several endoparasitic nematodes have been identified. Resistance to the potato cyst nematode has been found in *Solanum verni*, *S. multidissectum*, *S. oplocense*, *S. spegazzinni*, *S. kurtzianum* and *S. tubersum* s.sp. *andigena*. Screening for resistance to potato cyst nematode was studied extensively in the U.K. and the Netherlands. Several genes for resistance (R genes) to many pathotypes of *Globodera restochiensis* have been found and used for developing commercial cultivars. Resistance to root knot nematode has been found in *Lycopersicon peruvianum*, *L. pimpinellifolium*, *Gossypium barbadense* and several species of *Nicotiana*.

Gilbert and Mc Guire studied the inheritance of resistance in tomato to root knot and found the presence of a single dominant resistance in tomato to root knot and found the presence of a single dominant resistance gene Mi which was effective against all species of *Meloidogyne* except *M. hapla*. Subsequently, two dominant and one recessive genes for resistance were designated as L Mi Rl, L Mi R2 (cultivars Nematex and Small Fry-1) and L Mi R3.

High resistance to the cereal cyst nematode, *H. avena*, was found in species of *Avenue* (*A. sterilis*, *A. byzantina*). These species possess resistance to several pathotypes of *H. avenae* and were effectively utilized to develop resistence in cultivated oats (*A. sativa*). Resistance in barely against this nematode has been detected in barely varieties Primus, Svanhals, Chevalier, etc. Two pathotypes (Race 1 and Race 2) were originally known in *Heterodera avenue*. Resistance to Race 1 is found in cultivar Drost. At present there are at least six resistance genes identified in various barley cultivars which have been designated differently in different countries.

High resistance to *Heterodera schachtii* has been found in *Beta petellaris*, *B. procumbens* and *B. webbiana*. These sources of resistance are useful for incorporation of resistance in the cultivated species. The resistance in plants to nematodes can be due to the presence or release of preformed antinematode compounds or it can be due host reaction to invasion wherein the host does not allow development of juveniles to adult eggs laying stages. Both these aspects have been discussed in some detail earlier under "antagonistic plants and under "Host-parasite Relationship-Resistance and Susceptibility."

Interesting information regarding the antinematode compounds produced by antagonistic plants is available but such compounds have not been identified in commercial crop varieties. The antinemic compound in *Asparagus officinalis* var. Mary Washington that kills the larvae or disrupts reproductive progress is a glycoside with a low molecular weight aglycone. It is toxic to *Trichodorus christiei*, *Paratylenchus projectus*, *Tylenchorhynchus claytoni*, *Helicotylenchus nannus*, and *Belonolaiamus gracilis*. The compound is reported to interfere with acetyl cholinesterase in the nervous system.

The nematicidal properties of marigold (*Tagetes erecta* and *T. patula*) are attributed to the presence of terthienyl and bithienyl compounds which are highly toxic to *Pratylencus* and many other nematodes including root knot. Pyrocatechol in *Eragrostis curvula* is toxic to *Meloidogyne* spp. Similarly, the alkaloids nimbidin and thionemone in *neem* tree (*Azadirachta indica*) are reported to have antinemic action. In most crop varieties possessing resistance to nematodes, there is hardly, well in resistant as well as susceptible plants. It is only after entry that the resistant plants exert their influence on nematode activity, mostly by not allowing the larvae to feed properly and develop into adult egg laying females. This could be, as discussed earlier, due to suppression of giant cell and nurse cell development which causes starvation of the nematodes.

There is no strong evidence to suggest the resistance in plants to attack and invasion of certain nematodes is due to an absence of a particular nutrient for the nematode in the plant. The foregoing account of possible approaches to solve the problem of nematode control discusses regulatory measure (quarantine, etc.), cultural practices including biological control, use of chemicals and growing of resistant varieties. How far these approaches can be easily adopted by the farmers in India is a difficult question to answer. The majority of farmers in India are interested in maximum return from their limited holdings with minimum investment. He is not voluntarily willing to obey certain strict regulatory laws that affect his annual income or modify his methods of crop culture, or invest in costly antinematode chemicals. It may be due to his need for short-term planning.

There is hardly any work in India to impress upon the farmer the need for long-term planning to avoid serious losses in future or losses already occurring in his field. There is no doubt that the farmer can be provided with resistant varieties of commonly grown crops, he will readily accept them because he has not to invest extra money. Therefore, availability of resistant varieties in the first and foremost suggestion that can be practical. Unfortunately, there is insufficient work in this direction. Because of intricacies involved in breeding for resistant cultivars of a crop some of such varieties introduced from abroad have failed in India.

Further, resistance breeding is a time consuming process. Indigenous and exotic varieties/lines resistant to specific nematodes are listed in the respective chapters. In the absence of resistant varieties, practical and chemical methods need emphasis. Flood fallowing, use of decoy or trap crops or growing of antagonistic plants are not practical methods under Indian conditions. Crop rotation is a time-tested successful method. However, its limitations have been pointed out. Choice of crops with the Indian farmer is very limited. For example, in north India, rice in *kharif* and wheat in *rabi* season are common crops except where a crop like sugar cane predominates. With vegetable growers the problem is more acute. Almost all the vegetables grown for the market are susceptible to root knot. The grower cannot switch over to cereal cultivation in place of vegetables. The old system of a legume following cereals is no more in practice.

In the absence of a wide range of crops for choice the best alternative could be the use of resistant cultivars of the main crop in the rotation but, as stated above, resistant cultivars, with a desire yielding quality, are not available. However, there is a good scope for using the crop-free period during summer as summer fallow with frequent ploughings to expose the nematodes to extremes of weather.

Using one *kharif* season occasionally for green manuring is also possible. The advantage offsets the loss of one crop.

There is a need for developing techniques for quick decomposition of crop residue in the field. This could encourage indirectly the biological control of nematodes. Finally, chemicals remain the only direct methods of nematode control. But the prohibitive of chemicals and lack of the knowledge about their proper use excludes a majority of the farmers who own limited areas of land. It, therefore, requires a good deal of work on the part of scientists of develop well-designed package of practices that could be integrated with the normal crop growing practices followed by the farmer.

INDEX